Biochemical, Immunological and Epidemiological Analysis of Parasitic Diseases

P. K. Bandyopadhyay • N. R. Das •
Amit Chattopadhyay

Biochemical, Immunological and Epidemiological Analysis of Parasitic Diseases

Springer

P. K. Bandyopadhyay (*deceased*)
Kalyani, West Bengal, India

N. R. Das
Department of Zoology
Serampore College
Hooghly, West Bengal, India

Amit Chattopadhyay
Department of Zoology
Serampore College
Hoogly, West Bengal, India

ISBN 978-981-16-4383-5 ISBN 978-981-16-4384-2 (eBook)
https://doi.org/10.1007/978-981-16-4384-2

This Springer imprint is published by the registered company Springer Nature Singapore Pte Ltd.
The registered company address is: 152 Beach Road, #21-01/04 Gateway East, Singapore 189721, Singapore

One of the Editors Dr. Das likes to dedicate his works to his great teacher ***Late Dr. H.N. Ray****, P.R.S., Ph.D. (Lond), F.Z.S., F. N.I., one of the internationally famed scientists of his time and to his father* ***Late Dr. M.R. Das****, M.B. (Cal) who was a legendary figure in the field of medical profession.*

Dr. Amit Chattopadhyay likes to dedicate his contribution in this book to his lovable son ***Master Ayush Chattopadhyay****.*

We, Dr. Nitisranjan Das and Dr. Amit Chattopadhyay, are most fortunate to work with the late Professor

Dr. P.K. Bandyopadhyay, a man of letters. He offered us to write with him the manuscript of this publication entitled Biochemical, Immunological and Epidemiological Analysis of Parasitic Diseases.

Dr. Nitisranjan Das is lucky to have his valuable and meaningful friendship. Dr. Amit Chattopadhyay is proud to have him as his Ph.D. supervisor. Both of us dedicate this volume to the memory of our great friend and mentor Late Professor

Dr. P.K. Bandyopadhyay who had a dream to write such a book.

Foreword

It is my utmost pleasure to dedicate the book on parasitology to the student and the scientific community. As a former undergraduate student of zoology at Calcutta University, I always looked out for a decent reference/textbook for each subject matter covered under the curriculum. I can confidently say that this book will definitely fill the gap in the subject matter of parasitology. Not only that, but it will also serve as a text/reference book for a more advanced postgraduate programme covering parasitology. Needless to add, this book is a comprehensive quencher for those who seek first-hand knowledge in parasitology.

The three authors of this book have vast experience in parasitology, and I would like to appreciate their hard work and thoughts to bring out this enlightenment in the field of parasitology. I wish them success in their great endeavour.

(I would like to declare that all the opinions expressed here are my own and has no bearing with my employer/sponsor.)

AstraZeneca
Gaithersburg, MD, USA
8 December 2020

Suddhasil Mookherjee

Preface

This textbook is written specially for students who are engaged in higher studies on parasitology. It comprises the history of discovery, geographical distribution, habitat, morphology, life cycle, energy production, pathogenesis and diagnosis of each and every parasite described. Host defence mechanisms, i.e. immune response by the host, are also included. It is seen that these topics are most important to the majority of students.

Parasites of the phylum Protozoa, Nemathelminthes, Platyhelminthes and Arthropoda are included in this edition. The edition integrates a precious wealth of information on epidemiology, biochemistry, molecular biology, physiology and nutrition, immunology, ecology and evolution of the parasites, so students will have the urge to study the fascinating world of parasites. Updated information on the prevalence of infections is also included in this book. The study of parasitology is very important from the point of human diseases as the parasites residing in our body cause disturbances to the normal physiology of us leading to pathological conditions. The parasites are dangerous relatives rather than villain of the human society.

The subject parasitology attracted attention during World War II and several years thereafter due to belated realizations of the importance of the subject as a factor in world health as well as in the welfare of the military personnel. The rapid advances in knowledge in the field of parasitology made it necessary to revise and take the help of different branches of life sciences to control the parasitic diseases. We have made a conscious effort to lay more stress on zoological and physiological aspects of parasitology. However systematic classification is not neglected but has been included at the end of appropriate chapters. The method of laboratory diagnosis of parasites is of paramount importance as only clinical symptoms of the diseases are not always sufficient for the detection of the causative agent of the disease. We have tried our best to give a broader perspective of the science of parasitology. Here Dr. Das records the help of Mrs Ruby Das, M.A., B.Ed., Diploma in Russian Language, in the composition of the manuscript of Dr. Das's contributed chapters. Dr. Das is really indebted to her for her service. Dr. Das and Dr. Chattopadhyay would like to express their sincere gratitude to Dr. Vansanglura, Principal, Serampore College, for his encouragement of this work.

About the Book

This volume pertaining to the disease causing organisms that changes the normal physiology of human beings leading to mortality which contributes the loss of labour force and man-days. The end result is the retardation of economic growth and lack of development of the country. All the advancement and unique discovery in the field of parasitology are presented here to understand the relationship and their detrimental effect not only on individuals but also on the economics of the country itself. Efforts have been made by the authors to cover all aspects of the age-old conflict between parasites and their hosts. The history and mystery of the life-threatening diseases of global concern have found places in this book. The different types of animal associations and host-parasite relationships along with the classification of the parasites are included in the book. Evolution of the parasitism and types of parasites and hosts are presented with the ecology of parasites; parasites from different phylum like Protozoa, Nemathelminthes, Platyhelminthes and Arthropoda are also discussed in the light of modern knowledge. The history of discovery, morphology, life history and pathogenicity of each and every parasite have been discussed with necessary details in a methodical way so that students and readers will not have to search for the required information in an ocean of unnecessary details. One of the most important aspects of parasitology is the vector and their role in transmission of a number of diseases which are still nightmares even in the era of modern science. So vector biology has also been incorporated in this book. Zoonosis, the most attractive and debatable subject of today, has been given its due honour in this book. Parasites of veterinary importance are also discussed for the benefit of veterinary students. We have also tried our best to discuss all the considerable aspects of parasitic nematodes of crop plants to have focused knowledge for agriculture students also. Last but not the least, host-parasite interactions, immune response and biochemical adaptations of parasites which are discovered by the effort of research workers provided the weapon to fight against disease causing organisms, which are highlighted here with all conceivable effort to have a relatively disease-free society and for the benefit of medical students. Modern knowledge and recent information are being incorporated in all the chapters mentioned in the contents. Now it is up to the reader to assess our shortcomings. As a whole this book offers a comprehensive survey of the knowledge on parasitology in lucid and understandable English with diagrams.

Contents

About the Authors

P. K. Bandyopadhyay was a professor in the Department of Zoology, University of Kalyani. He was a nominated member of the National Academy of Sciences, India, and was an executive and life member of the Indian Society for Parasitology. He was an adjudicatory person and was a board member of several universities and colleges in India and abroad. Professor Bandyopadhyay's latest research pertained to fish diseases and their plant-based control measures. He and his team isolated and characterized the bioactive compound, oleic acid, from Carica papaya seed and used the product as aqua-soluble drug in treating fish bacterial disease invivo. He had published more than 170 research articles on animal taxonomy.

N. R. Das is a guest professor in the postgraduate section of Serampore College, Serampore, West Bengal, India. He graduated and completed his Ph.D. from the University of Calcutta as a "Presidencian" under the guidance of Late Dr. H. N. Ray. He has worked as an assistant research officer with the Indian Council of Medical Research, New Delhi, from 1965 to 1970 at Presidency College, Calcutta, before he moved to Serampore College in 1970 as a lecturer. He became head of the department in 1995 as a reader and retired in 2001. Dr. Das was elected to the board of management of the Serampore College in 1998 and served the college as the bursar. He has been bestowed a number of special honors and has availed distinction from different organizations and institutions. Dr. Das is a fellow of the Zoological Society of India and the Royal Society of Tropical Medicine and Hygiene, London. He was also a resource person of the National Council of Educational Research and Training, New Delhi, India, for more than ten years. He has written a number of reference books on parasitology, wild life, embryology and treatment of tropical diseases published from Calcutta, West Germany, London and Poland.

Amit Chattopadhyay is an assistant professor in the Department of Zoology for UG and PG studies at Serampore College. He has completed M.Sc.in zoology and Ph.D. in the molecular basis of vector biology. Dr. Chattopadhyay joined the Serampore College in the year 2009 and served as head of the Department of zoology. Presently, he is the PG course coordinator for the Department of Zoology. He has been conferred with the Young Scientist of the Year 2016 award by the International Foundation for Environment and Ecology and Confederation of Indian Universities, New Delhi. He has written three books in regional language for undergraduate students of Zoology under University of Calcutta. Dr. Chattopadhyay is a life member of the Indian Society for Parasitology.

Introduction

1

Parasitology is a branch of Zoology as well as a part of the medical science also. The science of parasitology has a long history and has its foundations in zoology with its importance on identification, classification and the interpretation of lifecycles with their concern for the disease caused by parasites. Parasites are those organisms which cannot sustain without the help of others. They are dependent on others for food and shelter but the relation is heterospecific. The organisms which give them food and shelter are hosts. Once entered within the host, parasites take complete control on its host and establish an intimate relationship between them. Parasites establish combinations of physiological, biochemical and nutritional adaptations to set up themselves within the host. Parasites also create mechanism for evading the immune responses of the host so that they can reside permanently within the host until death without any resistance. Parasites are threat to the millions of human and animals throughout the world. They are the reason of human misery and immense income loss to the country. There are innumerable numbers of parasitic organisms in the globe. It is believed that the number and type of parasites are far more than non-parasitic organisms. Viruses and bacteria are not included though these are all parasites yet the number of parasites are far more than those of non-parasitic free living organisms.

The first text book on parasitology was entirely written on zoological aspect. Then a number of text books were written with the advancement of knowledge on parasites about the disease caused by them. Now different aspects of biochemistry, physiology, immunology and epidemiology have been incorporated which are the results of modern research on them.

The parasites are causative agents of different types of diseases of human beings and domestic animals. The evolution of parasitic way of life is so successful that they evolved independently in every phylum of animals and in many plant groups. The life and activity of a number of parasites of different groups are recorded here with their significance in relation to human beings.

P. K. Bandyopadhyay et al., *Biochemical, Immunological and Epidemiological Analysis of Parasitic Diseases*, https://doi.org/10.1007/978-981-16-4384-2_1

The study of the lifecycle of parasites is most important as the parasites for their maintenance of own species must have to transfer their offspring to a new host. The unique way of transfer, different in different species, is known as infection. All the parasites do not complete their lifecycle within a single host, a large number require two or three different hosts to complete their lifecycle. But each and every time they have to face environmental stress or hazards for safe transfer of their offspring to a new host.

The detailed study of the lifecycle of each and every parasite is necessary to understand the way of infection of them responsible for the particular disease they create or carry. Study of ecology and parasitic adaptations also has great importance from the point of human diseases they create or carry. Not only the lifecycle, parasitic adaptations and ecology, the laboratory diagnosis and recognition are of paramount importance because only symptoms of the diseases are not sufficient to detect the causative agent of the disease.

When the routine process in laboratory is failed to detect the organism, then antigen–antibody reaction is the last hope. Recognition of parasites by laboratory diagnosis, will give us the idea of preventives and treatment of the disease. The study of the host–parasite interaction of different parasites will also indicate the mischievous character of the parasites and will open the path for better recognition of the parasites and their hidden activity.

The study of the metabolism of the parasites will give us the idea of symptoms of the diseases and the detrimental effects on the hosts due to the presence of the different metabolic by-products produced by the parasites.

The antigen–antibody reaction helps us to identify the parasite and gives us the idea of the preparation of the vaccines.

The understanding of infective form of the parasite is very important to prevent the particular disease. The age old conflict between the host and parasites is now making the parasites cleverer and they try their best to establish themselves within the host hiding their identity. The antigen detection is now the most modern method to detect those resistant species.

If we cannot recognize them and prevent infection ultimately we have to suffer and face death. The death of the host does not interfere with the life and maintenance of the parasites as in the meantime the parasites will find newer way to establish themselves in another healthy host. In the eternal conflict between parasites and hosts the immune competent host ultimately becomes victorious and prevents the parasite from invasion and infection. But immune compromised host surrenders before the parasites and allows their way for invasion and infection to host.

The parasites during their transfer of offspring to new host face tremendous environmental stress and hazards but this is mandatory, so mother produces offspring in astronomical figure so that at least some will ultimately find their permanent address through thick and thin though most of them go to oblivion. This will save the species from its extinction.

Here in this book we first attempt to restrict ourselves on the classical aspects of parasitology like classification, systematic, lifecycles and diseases caused by them, then we will try to enter into the modern concept by elucidating aspects of biochemistry, molecular biology, physiology, immunology and epidemiology based on the knowledge of modern researches.

1.1 General Concept on Parasitism, Parasites and Host

The word parasitism is a type of interspecific relation in which one small individual of a species sustains at the expense of the other individual of another species comparatively large one for food and shelter to the parasite.

Parasites are those organisms which cannot live free without the help of other organism of different species. The small dependent individual is called parasite and the other large individual upon which parasites depend is called host. Due to parasitism the parasite is benefitted at the expense of the host and the presence of parasite creates detrimental effect on the host.

1.1.1 Different Interspecific Relation

The interspecific co-action is interaction between the organisms of two different species, a relation is formed by which both the species may be benefitted or one may benefit at the cost of the other or relation may be neutral where none is benefitted or causes any harmful effect on the other.

The interspecific coactions may be of different types like parasitism, symbiosis, commensalism, ammensalism predation, neutralism, etc. Due to this co-action the density of any one species may increase or decrease which maintains ecological balance of the nature.

1.1.2 Classification of Parasites

The association may be of different types and accordingly is called ectoparasite and endoparasite.

1. The parasites that remain on the body surface of the hosts are called ectoparasites. Example: ticks, mites, lice, etc.
2. The parasites that remain inside the body of the host in different organs or systems are called endoparasites. Example: *Ascaris*.

They may live in the body cavity, within the alimentary canal, in the blood or lymph vessels, within blood corpuscles or in various types of tissues.

Depending on the sites they reside, the parasites are called:

(a) *Haemoparasite:* The parasites living in the blood. Example: *Plasmodium*, *Trypanosoma*, etc.
(b) *Histoparasites:* The parasites that live in the tissues. Example: *Leishmania*.

(c) *Coeloparasites:* The parasites living in the body cavity or coelom of the host are called coeloparasite. Example: *Monocystis*.
(d) *Cellular parasites:* The parasites that reside within the cell as intracellular parasites are called cellular parasites. Example: Trophozoites of *Plasmodium* within the RBC of human beings.

1.1.2.1 Depending Upon the Time Span

Obligatory parasite: The parasites that cannot live independently in any time or throughout their lifecycle are called obligatory parasites. Example: *Taenia*, *Monocystis*, etc.

Facultative parasite: The organisms that are not dependent on others for their existence and can live independently, but sometimes in congenial circumstances act as parasite temporarily are called facultative parasites. Example: Bedbug, Leech, etc.

1.1.2.2 Others

Erratic parasite: Some parasites sometimes migrate to the unusual organs of the hosts and are called erratic parasites.

Sporadic parasite: Some parasites sometimes come closer to the host for collecting metabolic materials and they are called sporadic parasites.

Incidental or accidental parasite: The organisms which sometimes suddenly become parasitized to unnatural host are called incidental parasites.

Pathogenic parasite: The parasites that cause disease in the body of the hosts are called pathogenic parasites.

Hyperparasitism: When a parasite depends upon other parasite for their food and shelter they are called hyper parasites.

Example: Protozoan parasite *Nosema dollfusi* is parasitic on the eggs of Trematode parasite *Bucephalus cuculus*. On the other side *B. cuculus* is parasite on American snails.

1.1.3 Parasitoid

Some Dipteran and Hymenopteran flies lay eggs on the eggs and larva of another fly. The larvae after hatching from these eggs start eating their larva or eggs on which they are deposited. As a result if the hostfly dies the parasites remain alive. This phenomenon is known as parasitoidism and the benefitted species is called Parasitoid.

1.1.3.1 Brood Parasitism

The cuckoos cannot prepare their nest and at the same time cannot incubate their eggs to hatch. They enter into the nests of the crows stealthily and lay eggs there. The crow cannot recognize the eggs of cuckoo and so incubate to hatch them thinking that the eggs are their own. This is called Brood parasitism. Example: Cuckoo.

1.1.3.2 Social Parasitism

When a species exploit another species for their own benefit causing no harm to the other, the relation is called social parasitism.

The Cowbird of North America will not prepare their own nests and will not incubate their own eggs. But they incubate eggs and nurse the nestlings of surrogate father and mother.

General characters of Parasites:

1. The parasites and host must be of two different species.
2. Generally hosts are larger in size than the parasites.
3. Ideal parasites never cause that much harm to the host so that the host may die. Though the heavy infection by parasites and for secondary infection sometimes hosts may die.
4. The length of life of parasites is very much limited but the reproduction rate is very high.
5. There is always a balance of population between the parasites and hosts.
6. One parasite during his lifetime or lifecycle may remain within several hosts.
7. Parasitism may be partial or whole time.
8. The organs of parasites are highly specialized in relation to the parasitic adaptation within the hosts.
9. Most of the time parasites choose a particular place for their habitat.
10. Besides animal plants may also be parasitized.

1.1.4 Parasitic Adaptations

To cope with the microenvironment of host the parasites undergo changes in their structural, physiological characters and reproductional behaviour.

The features are:

Structural adaptations

1. Ectoparasites develop piercing and sucking mouth parts, while the endoparasites develop suckers, hooks, etc. for anchorage to the body of the host.
2. In the parasites sense organs, nervous system, alimentary system, etc. and organs of locomotion are ill developed or sometime absent.
3. To suck blood from the host or to suck cell sap from plants parasites develop sucking and boring apparatus.
4. Parasites living in the intestine of a host have an envelope of cuticle over them so that they will not be digested by the digestive enzymes of the host.

Physiological adaptations

1. Parasites those who suck blood from the body of the host have anticoagulant in their salivary secretion to prevent coagulation of blood during sucking.
2. Endoparasites exhibit anaerobic respiration.

3. Parasites are transferred from one host to other by ingestion of the eggs, cysts, etc. through food and drinks or by vectors. The route of infection is called faecal–oral route.

Reproductive adaptations

1. Parasites show high rate of reproduction, parthenogenesis, hermaphroditism polyembryony, intermediate hosts and complicated lifecycle.

1.1.5 Host

The hosts are those within whom the parasites live and grow.

1.1.6 Classification of Hosts

Definitive or final host: The hosts in which parasites undergo their sexual cycle are called final or definitive host.

Intermediate host: The hosts where parasites undergo asexual lifecycle are called intermediate hosts.

Paratenic host: To complete the lifecycle some parasites enter into some organisms, those are not usually their host but use the organisms to reach their final host, they are called paratenic hosts.

Reservoir host: Some animals store in their body a part of the lifecycle of some parasites and spread from there to another hosts are called reservoir hosts.

1.2 Need of the Study of Parasitology and Its Present Relevance

The study of parasitology is very much important because the parasites are disease causing organisms which change the normal physiology of human beings leading to morbidity and mortality that contributes diminished work capacity of human beings resulting into the loss of labour force and workers. The ultimate result is the retardation of economic growth and development of the country. Out of the all relations among non-sexual individuals parasitism is the worst and result of this queer relation is against the well being of the human society.

The parasitic infections occur throughout the world. But they are very common in third world, tropical and subtropical countries where the climate is congenial to the life and spread of parasitic infections. Besides, other reasons are there for high rate of parasitic infections. These are high population density, poor condition of sanitation, poor personal hygiene, extreme poverty, living in slums, ignorance and lack of community medical service.

Recently with all these factors some added reasons are deforestation, building of dams which help vector breeding and density of reservoir host. The life style of slum dwellers and poor municipal services greatly contributed to the spread of infections.

To alleviate the lead effects of infective diseases caused by parasites the study of parasites is very much important. All the advancement and unique discovery in the field of parasitology help us to understand the relationship and the detrimental effects of the parasitic infections not only on the individuals but economy of the country itself. Now the parasitologists are trying their best to know the interaction between host and parasites which are being traced up to the cellular and chemical level. The most important aspect of parasitology is the vector and their role in transmission of the diseases which is being pursued by the Parasitologists has helped us in finding the solution of this great problem. All these are opening a new path in the discovery of vaccine and medicine for the critical parasitic diseases.

These are the prime needs for study of the parasitology and the study will lead us one day to create a disease free life of human beings on our planet and hopefully we are marching on to that goal.

Protozoan Parasites

2

The protozoan parasites are known to parasitize both human beings and domesticated animals, thus causing immense loss to human society. These parasites cause different important diseases like malaria, Chagas disease, sleeping sickness, kala-azar, leishmaniasis, etc. those are major diseases of tropical countries bringing death to human beings and their domesticated animals.

In this chapter the flagellates like *Trichomonas vaginalis*, *Giardia intestinalis*, *Trichomonas hominis*, *Trypanosoma cruzi*, *Trypanosoma gambiense*, *Leishmania donovani* are discussed about their history of discovery, habit and habitat, morphology, modes of infection, life cycle, pathogenesis, energy production, diagnosis, epidemiology of the parasites and immune response of the host.

Parasitic amoebae like *Entamoeba histolytica*, *Acanthamoeba*, *Naegleria fowleri*, *Entamoeba coli*, *Dientamoeba fragilis*, *Iodamoeba butschlii* are described concerning their same factors as the parasitic flagellates.

Coccidia like *Cryptosporidium parvum*, four species of malarial parasites, *Toxoplasma gondii*, *Babesia* and an intestinal *ciliophoran*, *Balantidium coli* causing diseases like dysentery are also discussed under the same subheadings.

Systematics of all these protozoan parasites are written for the benefit of the students who are interested in the classification of these parasites to know their exact position in the animal kingdom.

Protozoa is the subkingdom under Kingdom Animalia. This scheme of classification is introduced by N.D. Levine in the year 1980. Before that Protozoa was regarded as a phylum in the scheme of classification by T.J. Parker and W.A. Haswell in 1940. In the year 1942 A.J. Marshall and W.D. Williams edited a book where Parker and Haswell's classification was accepted and described.

In the year 1675 Antony Van Leewenhoek found for the first time the microscopic, single celled organisms and described them. In the year 1818 Scientist Goldfuss for the first time coined the term Protozoa for this group of organisms. After the discovery of cell by Robert Hooke in the year 1839, Van Siebold in 1845

P. K. Bandyopadhyay et al., *Biochemical, Immunological and Epidemiological Analysis of Parasitic Diseases*, https://doi.org/10.1007/978-981-16-4384-2_2

understanding the characteristics of Protista and on the basis of this discovery he for the first time applied the name Protozoa for the microscopic unicellular organisms.

N.D. Levine in the year 1980 divided the subkingdom Protozoa into seven phyla. Out of these seven phyla the phylum Sarcomastigophora, phylum Apicomplexa, phylum Microspora, phylum Ascetospora, phylum Ciliophora contain parasitic organisms. Examples are in phylum Sarcomastigophora organisms like Giardia, Trypanosoma, Amoeba, etc. are parasites. In phylum Apicomplexa organisms like Gregarina, Monocystis, Plasmodium, Toxoplasma, etc. are the parasitic animals. In the phylum Microspora all the members of the phylum are endoparasites. The members of the phylum Ascetospora are parasites. In phylum ciliophora very few are parasitic organisms.

The Protozoa are unicellular organisms and the single cell performs all the vital functions of life. As protozoans are very very small they can be studied only under microscope. The main structure of any member of the parasitic Protozoa consists of protoplasm enclosed in a cell membrane, plasmalemma. The protoplasm is divided into cytoplasm and nucleus. Though parasites some have the power of movement. The movement is performed by either pseudopodia or cilia or flagella, etc.

Some of the protozoan parasites are:

Entamoeba histolytica	*Plasmodium falciparum*
Giardia intestinalis	*Toxoplasma gondii*
Trichomonas vaginalis	*Cryptosporidium parvum*
Balantidium coli	*Babesia bigemina*
Trypanosoma cruzi	
Trypanosoma brucei	
Leishmania donovani	
Plasmodium vivax	
All are infective to human beings	

The Phylum Protozoa though it is now called kingdom Protozoa as per the classification of N.D. Levine (1980) have more than 45,000 species of which 10,000 are parasitic in invertebrates and in all species of vertebrates. Human and their domesticated animals serve as the hosts of the protozoan parasites like malaria, Chagas disease, sleeping sickness, Leishmaniasis, etc. those are considered major diseases of the third world tropical countries. In poultry, dairy and domesticated animals theileriosis and coccidiosis present a continuous threat to the persons who are in these business and trying to produce food for human beings.

In the living world kingdom Protozoa have three Phyla under which there are a number of parasites those attack human beings and domesticated animals. They are:

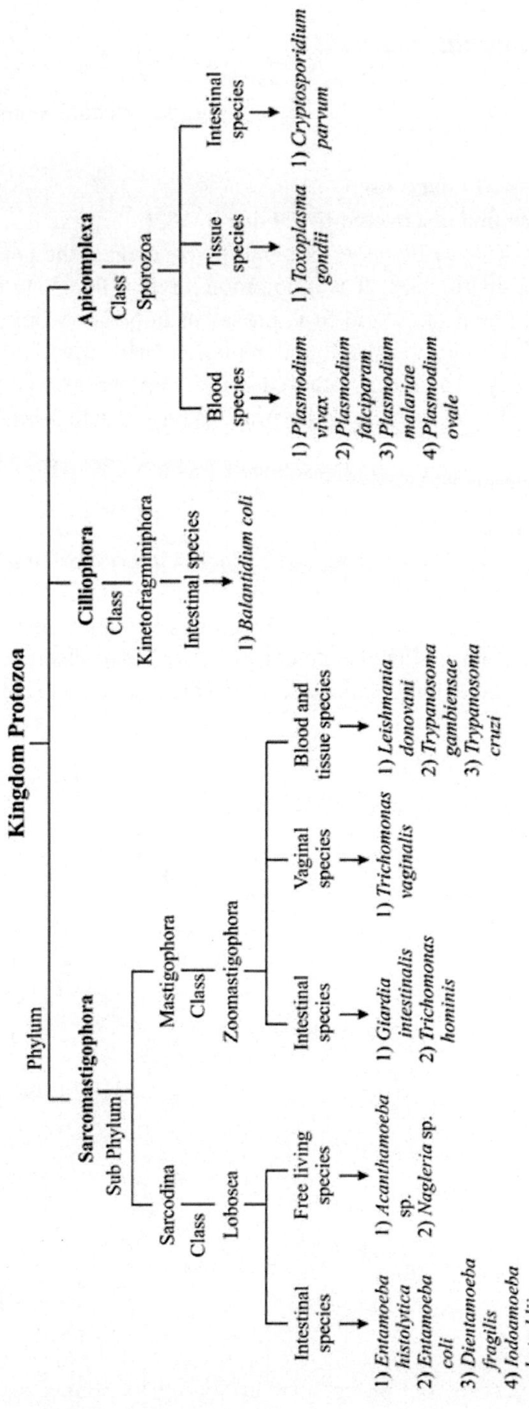
Kingdom Protozoa
Phylum
Sarcomastigophora
Cilliophora
Apicomplexa
Sub Phylum
Sarcodina
Mastigophora
Class
Class
Class
Class
Lobosea
Zoomastigophora
Kinetofragminiphora
Sporozoa
Intestinal species
Free living species
Intestinal species
Vaginal species
Blood and tissue species
Intestinal species
Blood species
Tissue species
Intestinal species
1) Entamoeba histolytica
2) Entamoeba coli
3) Dientamoeba fragilis
4) Iodoamoeba butschlii
1) Acanthamoeba sp.
2) Nagleria sp.
1) Giardia intestinalis
2) Trichomonas hominis
1) Trichomonas vaginalis
1) Leishmania donovani
2) Trypanosoma gambiensae
3) Trypanosoma cruzi
1) Balantidium coli
1) Plasmodium vivax
2) Plasmodium falciparam
3) Plasmodium malariae
4) Plasmodium ovale
1) Toxoplasma gondii
1) Cryptosporidium parvum

2.1 Flagellates

2.1.1 *Trichomonas vaginalis*

Trichomonas vaginalis is largest in size among the Trichomonads in human beings.

2.1.1.1 History of Discovery

This parasite was first discovered by Donne in 1836 in vaginal secretion. Next year he called this parasite *Trichomonas vaginalis*. He created the genus Trichomonas.

Geographical distribution: It is a common vaginal flagellate protozoan parasite distributed throughout the world. It is prevalent in Negro women twice as high as white women. *T. vaginalis* is also found in man occasionally though it is a parasite of women particularly. The disease caused by it is known as vaginitis. *Trichomonas vaginalis* is also found in man in the urine or in prostate secretion after prostatic massage.

2.1.1.2 Morphology

It is pear shaped, length varies from 10 to 30 μ but in average 20 μ long and 5–10 μ in breadth. This protozoan flagellate exists in trophozoite stage only, like other flagellates they do not form cysts. The trophozoites have five flagella (Fig. 2.1). Four of them are anterior flagella and one posterior. The posterior one remains along the margin of the undulating membrane which does not extend beyond the middle of

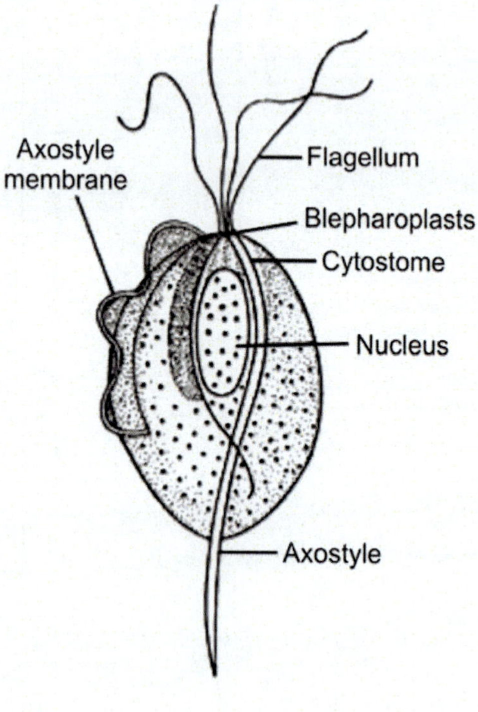

Fig. 2.1 *Trichomonas vaginalis*

the body. The body is supported by a stiff axostyle that originates from the anterior end and projects posterior beyond the body like a tail spine. The undulating membrane is attached to the body by a flexible rod known as costa. The projected axostyle is at posterior end and the organism seen to anchor themselves to the debris by this structure. The nucleus is oval in shape and is formed of scanty scattered chromatin granules. In the cytoplasm of the body some deep staining granules are seen and they are also found in rows along the costa known as metachromatic granules. The cytostome or cell mouth is not very distinct and the cell contains few food vacuoles. Beside the nucleus there is a structure called parabasal apparatus which is sausage shaped and light stained body. There is also a deep staining fibril that extends up to the middle of the body is the parabasal fibril.

The fresh active trophozoites exhibit jerky movements.

2.1.1.3 Habitat

T. vaginalis is primarily a parasite of vagina of women, it also invades the Skene's gland in urethra. In males they are found in urethra and prostate. *T. vaginalis* found in abundance in the upper part of the vagina up to the cervix of the uterus but do not enter into the uterus.

Scientists have found that *T. vaginalis* is associated with a low acidic creamy white frothy discharge.

2.1.1.4 Modes of Infection

T. vaginalis infection is accompanied by low acidity of the vagina, thin epithelium and less glycogen content in the cells. Normally the pH of the vaginal passage is highly acidic which prevents the infection. It is found that in children the vaginal passage is not highly acidic. The trophozoites of *T. vaginalis* are resistant to environmental changes. They may survive in urine and damp towel for a considerable period. These help in the spread of infection. Living trophozoites are seen on wet under cloths even after 24 h. *T. vaginalis* may infect the foetus during passing through the birth canal of the infected mother.

The trophozoites multiply by binary fission. The parasite is transmitted by sexual contact and also by sharing of towels or undergarments.

2.1.1.5 Life Cycle

The trophozoites are infective forms. It resides in vaginal passage, bartholin gland and the urethra of women and urethra, prostrate, seminal vesicles and epididymis in man. The multiplication of the trophozoites takes place by binary fission. The parasite is transmitted by sexual contact and also by sharing of towels or undergarments.

They cause vaginitis after an incubation period of about a month.

2.1.1.6 Pathogenicity

They cause vaginitis after an incubation period of about 1 month. The symptoms of vaginitis by *T. vaginalis* are a yellow frothy bad odorous discharge. The infection is also associated with leucorrhoeic condition. The vulva region becomes very much red due to congestion of capillaries, feeling of itching, painful micturition, tendency of frequent micturition and painful coitus also occur. In males the infection is asymptomatic but spreads the infection to female sexual partner during coitus.

2.1.1.7 Energy Production

Trichomonads are aerobic organisms. They acquire energy from incomplete degradation of simple sugars producing lactic acids and acetic acids. (see *Trichomonas hominis*).

In in vitro culture, it is found that these organisms grow most successfully in glucose and maltose.

ATP is formed in the cytoplasm by substrate level phosphorylation which is used by the organisms and energy is liberated from ATPs.

2.1.1.8 Diagnosis

In females *T. vaginalis* is found in sedimental urine, vaginal secretion swab but in males in the urine or in prostatic secretion after prostatic massage.

2.1.1.9 Prophylaxis

Male and female sexual partners are treated simultaneously. Its control and prevention needs time and patience.

2.2 Intestinal Flagellates

2.2.1 *Giardia intestinalis*

2.2.1.1 History of Discovery

The flagellate parasite was first seen by Leeuwenhoek in 1681 while examining his own stool. Author (Das) separated for the first time the *Giardia* in Indian Goats and Lions (Das and Jha 1967; Jha et al. 1968).

2.2.1.2 Geographical Distribution

These flagellate parasites are cosmopolitan in distribution.

2.2.1.3 Habit and Habitat

Giardia is a parasite of human beings and found in the duodenum and other parts of small intestine. Sometimes they are found in the colon and bile duct.

2.2.1.4 Structure

Giardia occurs in two different forms: Trophozoite and cyst.

Trophozoite

The body is bilaterally symmetrical and measures 10–18 μ in length. The shape of the body is like 'tear drop'. The dorsal surface is convex and ventral surface is concave. The concave ventral surface has two rigid bilobed concave adhesive discs.

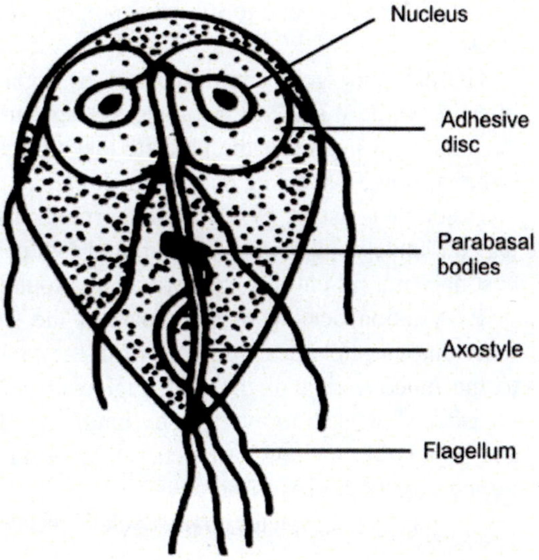

Fig. 2.2 Trophozoite of *Giardia intestinalis* (*Advanced Parasitology*, Das)

These two ventral discs occupy most of the anterior part on the ventral surface of the trophozoites. These adhesive discs are organelle of attachment which make contact and attach the parasite with intestinal wall of the host (Fig. 2.2). The attachment is also achieved by a hydrodynamic force generated by a pair of ventral flagella and a mechanical force developed by contractile protein like giardin present in the ventral disc (Karyakarta and Damle 2003).

Just below the adhesive discs there is a single sometimes double median bodies which stain black in iron-alum-haematoxylene. There are four pairs of flagella: anterior, posterior, ventral and caudal pair.

There are two nuclei placed at the broader end of the body containing haploid number of chromosomes. A single or double axostyle (Kinetosome) is seen in the mid ventral line of the body.

Giardia is unable to synthesize phospholipids and sterols which are necessary for its growth and metabolism. They use phospholipids and sterols found in the intestine of the host. It is proved that bile salts present in the intestine facilitate the entry of these organic chemicals within the trophozoites.

Cysts

The fully formed cyst is oval in shape and measures 12 μm × 7 μm in dimension. The cyst wall is thin and cytoplasm does not fill the entire cyst. There are four nuclei which remain clustered at one end or lie in pairs at opposite poles. The remains of disintegrated flagella are seen as streak in iodine preparation. The axostyle lies diagonally.

2.2.1.5 Life Cycle

Mature cysts are the infective forms, the normal dose is 10–100 cysts. The transmission route is faecal–oral. Excystation happens in small intestine of the host. The process is initiated due to the presence of pancreatic enzymes present in the small intestine.

Giardia undergoes multiplication by binary fission in the upper part of the intestine while it stays in trophozoite stage. In unfavourable condition in the small intestine encystment of the parasite takes place. Usually encystment takes place in large intestine. An acidic environment causes the *Giardia* to form cysts.

A tough resistant cyst wall is formed around the organism. Within the cyst parasite undergoes binary division and two cells are formed. The cytoplasm of the cyst does not fill entirely the cyst a gap around the inner side of cyst wall remains.

Excystation occurs in upper part of the small intestine of the new host. Two individual trophozoites come out from the cyst. The trophozoites attach themselves to the mucosa coat of the intestinal wall with the help of ventral sucker of the parasite. They used to multiply by longitudinal binary fission every 6–10 h interval and population increases tremendously within the intestine. The trophozoites feed through the processes of pinocytosis.

The parasite completes its life cycle in a single host. They are monogenic parasite.

2.2.1.6 Transmission

Trophozoites after encystation come out from its host along with the faeces of the host. They remain viable in soil and water for 2 weeks. Then the cysts transmit to the new healthy host with contaminated food and drink with cysts of *Giardia*.

2.2.1.7 Possible Pathway

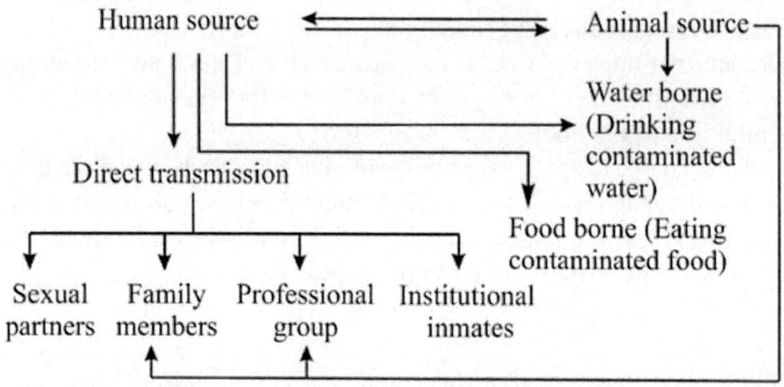

2.2.1.8 Pathogenicity

Most of the *Giardia* infection is symptomless. When the infection is symptomatic the incubation period is 1–3 weeks. Early symptoms included diarrhoea, abdominal pain, nausea, vomiting. Acute infection subsides within 2 weeks. In subacute infection to chronic one diarrhoea, abdominal irritation with excess of mucus production take place. Sometime inflammation of gall bladder occurs. If the population of parasite is large then malabsorption of fat may cause. Patient may complain of looseness of bowel and greasy stools with excess of fat.

2.2.1.9 Pathways Associated with Energy Production

Giardia is capable of incorporating certain monosaccharides into glycogen. Though the parasites have no mitochondria yet it can use oxygen if available. The organisms depend upon flavin-dependent substrate level phosphorylation for the requirement of energy. The end-products are ethanol, carbon dioxide and acetate.

2.2.1.10 Immunological Response

Giardia infection in human beings produces anti giardial antibodies and IgA-dependent host defence eradicate trophozoites. IgA binds to the surface of the trophozoites of *Giardia* which inhibits them to attach intestinal epithelium and expelled them from the body of the host.

2.2.1.11 Diagnosis

Diagnosis includes microscopical examination of freshly passed stool for trophozoites and cysts of Giardia.

Trophozoites of *Giardia* may be recovered from aspirated liquid from the duodenum or from the bile duct.

Control: Gastrointestinal parasites like *Giardia* may be controlled by strict method of assimilation and by encouraging hygienic habits in the community.

2.2.2 *Trichomonas hominis*

2.2.2.1 History

It was discovered by Davaine in the year 1860 and he called it *Cercomonas hominis.* As they possess five anterior flagella they are included in the genus Pentatrichomonas.

It is a non-pathogenic intestinal flagellate parasite. They are living as harmless commensals in the ileo-caecal region of human intestine. They may be seen in diarrohoeic stool. They exist only in trophozoite stage.

2.2.2.2 Morphology

The size ranges from 8 to 12 μm in length, the body is pear shaped. There is single ovoid nucleus with small central karyosome located at the round anterior end. They possess three to five anterior and one posterior flagellum. The undulating membrane is supported at the base by a rod like structure, the costa. The axostyle runs through the middle of the body and ends in a pointed extremity. The cytostome is conical.

The movement seen in freshly evacuated stool is jerky. They divide by binary fission. Infection is generally asymptomatic.

2.2.2.3 Habitat

T. hominis resides in the large intestine of human beings. Here they divide by binary fission and form a large colony. They consume bacteria and debris as food. *T. hominis* have been found from liver abscess. They have been found to withstand the acidic condition of the stomach.

2.2.2.4 Life Cycle

Transmission occurs through contamination. Flies also may act as mechanical vector. No cysts are formed and trophozoite is the infective form. The infection rate is high where sanitation is poor. It is observed that they cannot form colony in the mouth or urinogenital tract.

2.2.2.5 Diagnosis

The diagnosis is made from routine examination of diarrhoeic stool.

2.2.2.6 Pathogenesis

They are harmless commensal.

2.2.2.7 Pathways Associated with Energy Production

Trichomonads are anaerobic. They degrade carbohydrates to short chain organic acids, lactic and acetic acids and as usual carbon dioxide. Trychomonads in the absence of oxygen produce molecular hydrogen.

Trichomonads have organelles called hydrogenosomes because of their action, Hydrogenosomes are like mitochondria those are absent in the Trichomonads. In the hydrogenosomes DNA and cardiolipin are absent but these are present in the membranes of mitochondria. Two closely opposed membranes of 6 nm thick surround the hydrogenosomes (Schimidt and Roberts 2013).

By the process of glycolysis pyruvic acid is formed in the cytoplasm of Trichomonads and a portion of this pyruvic acid is reduced to lactic acid by the metabolic enzyme lactic dehydrogenase which is excreted but rest portion enters into hydrogenosomes. This is oxidatively decarboxylated. The electrons are accepted by ferredoxin. This is an iron sulphur protein that serves as electron acceptor in metabolic reactions. These electrons are changed to protons under anaerobic condition by the metabolic enzyme hydrogenase present in the hydrogenosomes to form molecular hydrogen. Oxidation of pyruvic acid to acetic acid generate ATP, i.e. generation of energy in the cell (Schimidt and Roberts 2013).

2.3 Parasitic Amoebae

2.3.1 *Entamoeba histolytica*

2.3.1.1 History

Entamoeba histolytica, the protozoan intestinal parasite was first discovered by Losch in 1873 in St. Petersburg, Russia. It was discovered from the faeces of a young peasant suffering from blood dysentery. Large number of parasites were found from his watery, bloody stool. Losch also observed that these parasitic amoebae contained RBCs in their food vacuoles.

Schaudin after 40 years in 1903 observed that intestinal amoebae can cause disease so they are pathogenic to man. This amoebic parasites are sometimes become tissue invader causing serious fatal disease. *Entamoeba histolytica* during its life time may become invasive form and secrete an enzyme which has lytic effect on tissues of the host and so they are called histolytica. But these parasitic organisms normally reside in the upper part of the large intestine as a commensal but sometimes invade the intestinal mucosa and cause amoebiasis.

2.3.1.2 Geographical Distribution

Entamoeba histolytica is cosmopolitan in distribution. But its prevalence is greater in tropical and sub-tropical countries and gangetic plain of India than temperate countries. The incidence is considerably higher in rural and densely populated urban areas than in the clean cities.

2.3.1.3 Habit and Habitat

E. histolytica is a unicellular, protozoa, microscopic endoparasite of human beings. They live as commensal but sometimes invade the mucosa and such invasion of the mucosa coat of the intestine of the hosts cause serious type of amoebiasis. In chronic cases, they enter into the blood circulation and migrate to the organs like spleen, liver, lungs and brain also, where they create ulcer and abscess causing death of the host. *E. histolytica* in its life cycle have trophozoite, cyst and pre-cystic stages. Cystic stage is the infective form and infection takes place by cyst from infected person to healthy persons. The harmful stage is the trophozoite.

2.3.1.4 Morphology

E. histolytica have two distinct forms: Trophozoite and cyst. But some are of opinion that they also have another form called pre-cystic forms but they remain in that form only for a very short time.

Trophozoite

Trophozoites of *E. histolytica* are very much active, motile and feeding forms. They sometime transfer into pathogenic forms to human beings.

The size of the trophozoite is 18–30 μ in diameter. The covering of the body, the plasma membrane is thin, elastic and semi-permeable in nature. The enclosed cytoplasm is differentiated into outer ectoplasm and inner endoplasm. The ectoplasm is clear and the endoplasm is granular. The body surface contains pseudopodia which are small, round or elongated projection formed from the ectoplasm.

The granular endoplasm contains a nucleus which is visible in a living specimen and can be studied from a fixed and stained one. Nucleus is enclosed by a thin and delicate membrane whose inner surface is beaded in appearance due to presence of chromatin granules arranged in a peripheral ring. In the nucleus there is a small dot like endosome surrounded by a clear space. The fluid filled space between the nuclear membrane and endosome is marked by spoke like striations of nucleoplasm giving the nucleus a cart wheel appearance. The size of the nucleus is 4–6 μ in diameter.

In the endoplasm there are food vacuoles with ingested RBCs, WBCs, fragments of epithelial cells and bacteria. But no contractile vacuoles are present, because osmotic concentration of protoplasm of *E. histolotica* is equal to that of the intestinal fluid of the host and so no water enters into the organism by osmosis. So there is no need for osmoregulation.

E. histolytica during movement produces broad, blunt pseudopodium. The **nutrition** is holozoic and it feeds by phagocytosis.

In this connection it may be mentioned that some scientists opined that there are two races: Magna and minuta. The magnas are 20–40 μ in diameter and are tissue invaders. The minutas are 13–15 μ in diameter and remain in the lumen of the intestine feeding on bacteria and organic materials.

Pre-cystic Phase

This is a phase between trophozoite and cyst. Before the formation of cyst trophozoites eliminate all food vacuoles, then round up and shrink by condensation of cytoplasm. It now measures 10–20 μ in diameter. The nuclear morphology remains same with that of trophozoite.

Cystic Phase

The cystic form is small, spherical, non-motile and non-feeding form. A delicate refractile cyst wall develops around the parasite. The cysts measure 12–15 μ in diameter. In case of minute forms cysts measure 6–9 μ in diameter.

The nuclear morphology is same with that of trophozoite. But appears large because of the condensation of the cytoplasm of trophozoites.

At the very onset, the cyst contains single nucleus, but soon it divides by binary fission and forms two nuclei and the cyst is called binucleate. Then again by another fission the nuclei become four in number. The cyst is called quadrinucleate. The cytoplasm of the cyst becomes clear and more or less transparent.

The measurement of nuclei is 2 μ in diameter each. In immature cyst there are broad rod shaped 1–4 bodies with blunt round ends called chromatoid bodies. These bodies stain deeply with iron-alum haematoxylene. These bodies are ribo-nucleo protein (RNP) in nature. These bodies disappear in the mature cyst. Disappearance is due to the dispersion of the RNP. The young cyst contains glycogen mass which also disappears gradually.

2.3.1.5 Life History

E. histolytica requires only one host to complete its life cycle and therefore *Entamoeba* is called monogenetic parasite.

Binary Fission

The trophozoites multiply by binary fission in the lumen of the large intestine. The scientists are of opinion that some trophozoites take mucus, blood corpuscles, bacteria, etc. from the wall of the intestine and forms ulcer there. Here in the mucosal folds they divide by binary fission. Some mature trophozoites due to peristaltic movement of the intestine dislodge from the mucus layer and become inactive instantly. These inactive parasites transform into pre-cystic stage. Some trophozoites when passed in faeces, die on exposure to air, gastric acid also destroy them. So trophozoites cannot transmit infection (Fig. 2.3).

During binary fission in the mucus wall of large intestine the nucleus of the trophozoite undergoes mitosis without the disappearance of the nuclear membrane. The mitosis of nucleus is followed by cytokinesis. Two daughter organisms are formed. They grow rapidly in size by taking food from the host and continue to form invasive forms.

For transmission of infection and continuation of species the pre-cystic forms again transform into cyst.

Encystation

In the process of encystation the trophozoite becomes non-motile, round and devoid of any food vacuole. This stage is called pre-cystic stage. The precysts are smaller than trophozoites and larger than cysts. They now secrete a highly refractile cyst wall around themselves and transform into cysts. Cysts are formed only in the lumen of the intestine and not in the tissues. The cyst at this stage is uninucleate and measures 8–22 μ in diameter and is spherical.

The immature cysts have a single nucleus of 5 μ in size, diffused glycogen and rod shaped bodies formed of RNP. These are called chromatoid bodies and stain black in iron-alum haematoxylene. The glycogen may reserve as food and chromatoid bodies contain ribonucleic protein.

The cysts mature after two mitotic divisions. The single nucleus of the cysts first become double and called binucleate cysts. Then after second mitotic division four nuclei are formed and the cysts are called quadrinucleate cysts. The nuclear size is successively reduced from 5 to 2 μ in diameter. The whole process is completed within a few hours. Simultaneously the glycogen mass and chromatoid bodies gradually disappear.

The tetranucleate cysts are infective forms. They are transmitted to new hosts. These tetranucleate cysts do not develop further and pass out of the body through faeces from the hosts. They are highly resistance to desiccation and survive about 10 days in moist outside environment. They withstand the gastric acid present in the stomach of the new healthy host. The strength of chlorination used in the municipal water supply cannot kill the cysts of *E. histolytica* only boiling and filtration can kill the cysts and prevent infection.

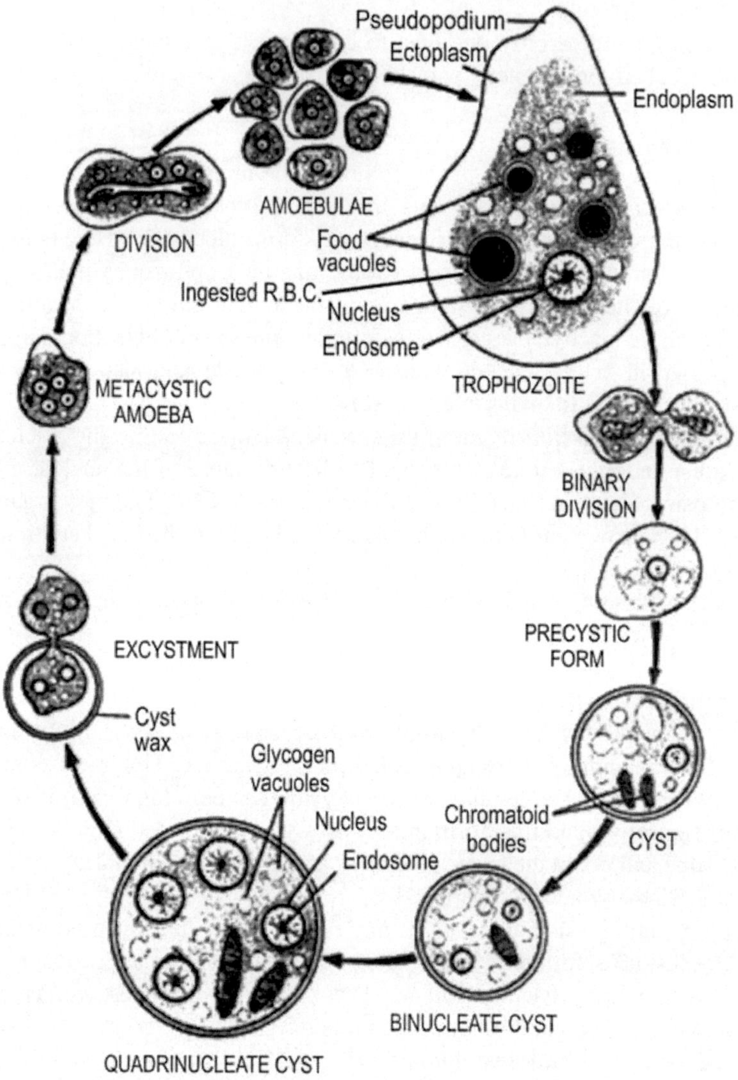

Fig. 2.3 Life cycle of *Entamoeba histolytica* (*Advanced Parasitology*, Das)

2.3.1.6 Pathways Associated with Energy Production

E. histolytica grows best in an oxygen free environment and for that reason it was once considered an obligate anaerobe. But it is proved that the organism can use oxygen in very low concentration. In vitro this parasitic amoebae cannot tolerate an oxygen concentration of 10% or higher and this is lethal to them. But carbon dioxide is required for their growth.

Glucose and galactose are the major carbohydrates, those are used by the *E. histolytica* from which they produce ethanol, acetate and carbon dioxide obviously the energy also. In the presence of sublethal concentrations of oxygen the end-products remain same but the proportions are different. If oxygen is present in the environment they store glycogen through E.M pathways also fructose phosphate is phosphorylated for energy.

2.3.1.7 Transmission of Infection

Infection takes place through ingestion of mature cysts in food and drink. The route of infection is called faecal–oral. The contamination takes place by flies, food handlers like cooks, food sellers, hawkers when they deal with unwashed hands. The unhygienic condition in the rural areas of some countries is a common source of infection. Cross-connection between sewage and waterlines may lead to water borne epidemics of amoebiasis.

After infection within the new host the cysts pass up to the ileo-caecal region, then alkaline environment and trypsin act on the cyst wall lead to the excystation in this region of the G.I. tract of the host.

A quadrinucleate metacyst emerges from each cyst. Each metacyst immediately divides by binary fission, its nuclei divide to produce eight small uninucleate metacystic trophozoites called amoebuli. These amoebuli pass into the large intestine and grow into mature trophozoites.

The mature trophozoites live in the lumen of the intestine without invading the intestinal mucosa in about 90% of infected persons. In such persons the trophozoites after sometime encyst themselves in certain condition like overpopulation of the parasite, pH change, changes in availability of food and oxygen. The cysts then pass outside along with the faeces of the host. These persons are called cyst passers they are actually the carrier and responsible for the spread of the disease.

In the remaining 10% of the infected persons the *Entamoeba histolytica* invade the host's tissue and cause tissue amoebiasis leading to liver abscess, brain abscess, etc. resulting in the death of the patient.

2.3.1.8 Advanced Idea About *E. histolytica*

They cause major calcium ion influx to the cells of the wall of large intestine resulting in cell death and ulcer formation.

2.3.1.9 Analysis of the Epithelial Damage Produced by *E. histolytica* Infection

Infection is initiated by interaction of the pathogen with intestinal epithelial cells. The interaction leads to disruption of intercellular structures such as tight junctions (TJ). TJ ensures sealing of the epithelial layer to separate host tissue from gut lumen. Recent studies provide evidence that disruption of TJ by the parasitic protein is prerequisite for *E. histolytica* invasion. Thus the analysis of molecular mechanisms involved in TJ disassembly during *E. histolytica* invasion is of paramount importance to improve our understanding of amoebic pathogenesis.

A comprehensive study has been made on the life style of *E. histolytica* on the whole genome level allowing identification of new virulent gene and signalling pathways and process relevant to amoebic biology.

The genome of *E. histolytica* is predicted to the 24 Mb with 14 chromosomes and is functionally tetraploid.

Much progress has been made in transcriptional profile of *E. histolytica* to study diverse aspects of amoebic biology including genes involved in amoebic pathogenesis assessing differences between virulent and nonvirulent amoebic strains.

In the last year multiple studies have reported a large prevalence of amoebic seroconversion of invasive amoebiasis and amoebic liver abscess among HIV positive individuals compared with HIV negative individuals.

Immune response by host: Acute and chronic *Entamoeba histolytica* infection in colon of human beings can initiate post-inflammatory response. In acute amoebiasis, it is seen an increase in the Th2 response and this is indicated by an increase production of IL-4 cytokine. In chronic condition patient exhibit little or no change in their CD4+/CD8+ ratio. It is also proved beyond doubt that Th1 and Th2 responses remain unchanged in these patients. But in the asymptomatic patients, it is observed that the level of IFN-y increases.

2.3.1.10 Pathogenesis

Trophozoites normally multiply within the large intestine in the crypts of mucus membrane. They feed on bacteria present in the gut, some available starches and mucus secretions. The invasive forms hydrolyse the mucus membrane. At this stage they do not need bacteria as their food. It is observed that oxidation–reduction potential initiate invasiveness of the parasites along with pH of the intestine.

The invasive forms produce ulcers on the intestinal wall and reach blood vessels. Now they swim in the blood circulation and reach lung, liver and brain also. The interesting phenomenon is these invasive forms reach their dead-end because they cannot come out of the host to infect others.

The cytoplasmic extension called filopodia are seen in electron micrographs attach to host cell and secrete cytotoxic substance for cytolysis of the host cells.

Intestinal ulcer is produced in the caecum, appendix, and throughout the length of the colon. Death may occur from perforation of colon. Sometime due to tissue response to chronic ulcer where active trophozoites may remain inflammated along with a granulomatus mass may form on the inner side of the intestinal wall called amoeboma which creates intestinal obstruction. This type of abnormality happens in central south America.

The chronic amoebic infection is so dangerous that secondary lesions are found in most of the organs of the body. Out of all organs liver is highly susceptible. Whatever may be the case the infection starts from intestinal abscess. The active trophozoites are carried by the bloodstream to the veinules of the mesentery and go to the liver via hepatic portal system. Here in the liver the parasites form abscess and the disease is now called hepatic amoebiasis.

The next vulnerable organ is the lung. Most of the time the infection comes from hepatic amoebiasis due to rupture of liver abscess. It is also observed that lung may affect independently. Other organs of the body may affect, but that is very rare, like brain, penis, and/or very rare site is the skin. The other very rare sites may be kidneys, glands like adrenal, spleen, male and female genital organs. But it is seen that all these abscesses in these organs remain bacteriologically free.

2.3.1.11 Diagnosis

The examination of stool smear preparation then fixed and stained preparation will help to observe the cysts. But repeated examination is essential as per the experience of laboratory technician. Cysts can be found in faeces during the convalescent stage of amoebic colitis. Cysts can be seen in saline preparation and in iodine preparation.

A special culture medium TYIS-33 is used to culture *E. histolytica*. DNA probes are used to distinguish between tissue invasive and non-invasive forms of *E. histolytica*.

Serological test is required to detect antiamoebic antibodies which will help to diagnose invasive forms of *E. histolytica*.

Another method of diagnosis in case of extra intestinal amoebiasis is the immunological test ELISA method application for detection of amoebic lectin antigens.

2.3.2 *Acanthamoeba*

This facultative parasite is under Phylum Sarcomastigophora, subphylum Sarcodina class Lobosa. Genus *Acanthamoeba* have five species. They are found in the tissues of human beings sometimes as facultative parasites.

These facultative parasites attack skin and/or central nervous system and cause infection in immunocompromised persons. The disease they cause is the corneal ulcer and keratitis, i.e. corneal inflammation and opacity of the cornea. It is also found that immunocompetent persons also sometimes suffer.

In 1990 it was estimated that so many cases of meningoencephalitis were detected which is due to *Acanthamoeba* infection.

The movement of the organism is very slow and has small spiky lobopodia. The five species have different invasive potential. The pathogenic organisms have ability to attach with host cells. Most of these organisms are found on beaches and multiply in saline water ranging from 0 to 3% salinity.

It was also discovered the persons with contact lenses are prone to *Acanthamoeba* infection due to washing of contact lenses with homemade saline water contaminated with *Acanthamoeba*. It is also discovered that public swimming pools are sometime source of infection. The parasites most of the time remain in fresh water and soil as free living organisms. Studies have demonstrated that if there is bacterial contamination in contact lens cleaning solution then these facultative parasites rapidly multiply increasing the rate of infection in the eyes.

Surprisingly it is also observed that some bacterial species like *Pseudomonas aeruginose* secretes toxin which is lethal to *Acanthamoeba*.

Acanthamoeba is a parasite of opportunistic infection in immune compromised persons with AIDS.

2.3.3 *Nagleria fowleri*

2.3.3.1 History

Nagleria fowleri is a free living amoeba belongs to excavate forms of Protista under the group Percolozoa or Heterolobosa. Excavates are a major assemblage of unicellular eukaryotes. Excavate contains a variety of free living and symbiotic forms of which some are important parasites of human beings.

N. fowleri though a free living amoeba but if found opportunity becomes pathogen causing the disease called primary amoebic meningo encephalitis or PAM.

The disease was first reported by Fowler and Carter in 1965.

Geographical distribution: The disease is reported from Australia, Czechoslovakia, the USA, Britain and New Zealand. Only two cases have been reported from India by Pan and Ghosh in the year 1971.

2.3.3.2 Morphology

N. fowleri is a soil amoeba. They are found in three forms: trophozoite, cyst and flagellate.

The trophozoites are elongated and measure 10–22 μ × 7 μ. They are actively motile with one broad and one pointed extremity. They move more than two body lengths per minute. The cytoplasm has a pulsating vacuole and a large nucleus. The nucleus is seen only in the fixed and stained specimen. The nucleus is single, large, 2 μm in diameter and has a large nucleolus and fine nuclear membrane. It has a large central karyosome and a perinuclear halo.

They convert from amoeboid trophozoites to biflagellate forms. The flagellated forms are 6–10 μm in diameter. They are more or less ovoid in shape having two flagella at the anterior side and a vacuole at the posterior side. The size and shape of the nucleus are same with that of the trophozoite.

2.3.3.3 Infection and Transmission

This parasite was first isolated by Carter in the year 1972. *Nagleria* is also known as brain eating amoeba. They are usually found in mud and warm fresh water from 25 to 35 °C in amoeboid or flagellated form. The amoeboid trophozoite converts from motile trophozoite to biflagellate form when it comes in contact with water. The non-feeding and non-dividing biflagellate helps the spread of *Nagleria* to fresh pools when it rains. The biflagellate forms do not multiply for multiplication they have to convert to trophozoite forms (Fig. 2.4).

The infective forms trophozoites enter through olfactory neuroepithellium of human beings during swimming in water heavily infested with these organisms. Sometimes infection also takes place by inhalation of particles of decaying animal manure which supports the growth of these amoebae.

They attack especially children and young healthy adults.

Besides trophozoite and biflagellate forms they may remain in the form of cyst. The cysts are uninucleate and possess double cyst wall. The structure of nucleus is same with that of the trophozoites.

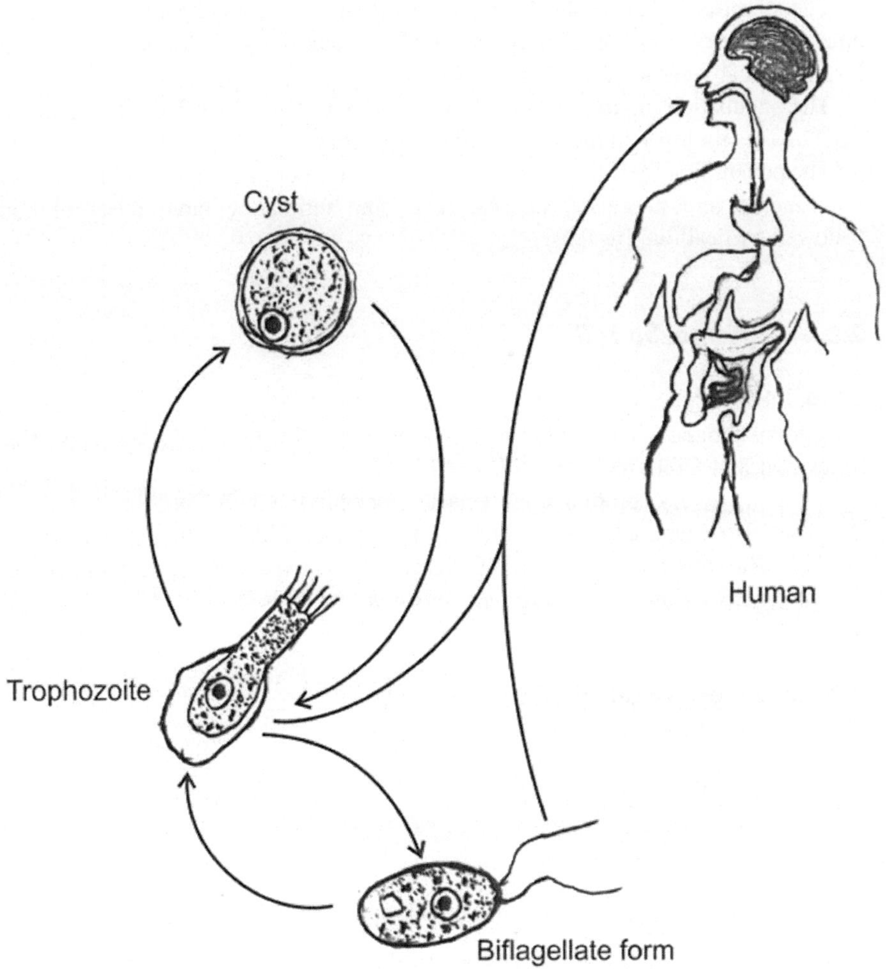

Fig. 2.4 Transmission of *N. fowleri* in human beings

2.3.3.4 Pathogenesis

The trophozoites and cysts are found in C5F and brain tissue, respectively, and flagellate form in CSF in the infected persons.

They invade and attack central nervous system via nose rather olfactory mucosa. As the amoebae penetrate the olfactory mucosa necrosis and haemorrhage start in the olfactory bulbs. Now amoebae climb along the olfactory nerve through floor of the cranium into the brain. The amoebae reaching the brain begin to consume the nervous tissue of the brain by means of sucker like structure extended from the cell surface. *Nagleria fowleri* in this condition is pathogenic to human beings causing primary amoebic meningitis or PAM disease.

The disease is known as PAM to distinguish it from the disease Amoebic encephalitis or Amoebic meningitis which is caused by *E. histolytica* in which lesions and abscess are formed in the brain.

The incubation period of PAM ranges from 2 to 15 days. The amoeboid trophozoites multiply in the grey matter of the brain.

The patient first feels the change in the perception of taste and smell followed by vomiting, severe headache, nausea, fever and ultimately rapid onset of coma followed by death in 2 weeks.

2.3.4 *Entamoeba coli*

2.3.4.1 History

This nonpathogenic amoeba was first discovered by Grassi in the year 1879. Mackinon and Dible found *E. coli* in Chimpanzee and other monkeys in the year 1938. This is a very common commensal amoeba living in the intestine of man, but they are also found in other primates. The author (Das) with his mentor Late Dr. H.N. Ray observed *E. coli* in the faeces of *Macaca speciosa*, *Papio porcarius, Cynocephalus mormon* and *Cercopithecus mona* (Thesis submitted for Ph.D in the year, 1970).

2.3.4.2 Geographical Distribution

It is distributed worldwide.

2.3.4.3 Morphology

The *Entamoeba coli* are found in two stages, Trophozoite and cyst with a transitional stage known as precyst are also found.

2.3.4.4 Trophozoite

It is said to be one of the largest amoeba residing in the colon. The more or less round or oval trophozoite varies in size from 15 to 50 µm in diameter. The movement of trophozoite is sluggish. The cytoplasm is not so clear. The endoplasm is granular and opaque. Within endoplasm there are a number of food vacuoles containing bacteria but no RBCs have ever found.

The nucleus of the trophozoite sometimes may be seen in unstained preparations. The karyosome is eccentric, the coarse chromatin granules remain arranged on the thick nuclear membrane.

The trophozoites of *E. coli* observed by the author (Das) in the faeces of stump tailed monkey were more or less similar to those recovered from man (Markel and Voge 1958; Noble and Noble 1961). Dobell demonstrated that *E. coli* of man may be experimentally transmitted to macaques.

2.3.4.5 Cyst

The round cyst varies from 15 to 20 µm in diameter. There are 1–8 karyosome placed eccentrically. Glycogen mass is visible in binucleate stage. The chromatoid bodies are square or with pointed ends. In the mature cysts glycogen mass or chromatoid bodies are not found. The size of the cysts varies from 10 to 33 μm in diameter. Author (Das) worked with these amoebae of primates. The investigation recorded the size ranging from 16.8 to 36.0 μm in diameter of the cysts of *E. coli* instead 10–33 μm as recorded in different literature (Levine 1961).

The chromatoid bars with broken, pointed or square ends varied in size and shape were very much similar with those of *E. coli* of man.

Ectocystic envelop quite similar to *E. histolytica* was found over the cysts of *E. coli* recovered from Stump tailed monkey and Mandrill. In nature it is strongly positive and also resisted saliva digestion (Das and Ray 1967). The function might be protective to prevent destruction when it comes in contact with soil. A PAS positive cystic wall and innumerable PAS positive granules are seen scattered within the cytoplasm.

The endosome is dot shaped and surrounded by loosely arranged fine granules of chromatin often the presence of a thin membrane around the endosome was also noticed. The peripheral chromatin was usually arranged in a beaded manner. Periendosomal granules appeared as delicate fibrils or sometime completely absent. The absence of periendosomal granules was mostly found in the cysts recovered from Stump tailed monkey. The nuclear structure of *E. coli* cysts of man was similar to those found in the cysts recovered from monkeys.

The trophozoites of *E. coli* seen in the faeces of Stump tailed monkey were more or less similar to those recovered from man (Markel and Voge 1958; Noble and Noble 1961).

Dobell demonstrated that *E. coli* of man can be experimentally transmitted to macaques.

Regarding pathogenesis, *E. coli* infection is nonpathogenic and shows no symptoms of infection in monkeys.

2.3.5 *Dienmtaoeba fragilis*

2.3.5.1 History

This harmless commensal was first discovered by Jepps and Dobell in the year 1918. About 60 years ago Dobell was of opinion that *D. fragilis* was very closely linked to the genus *Histomonas*. But Camp and his co-workers on the basis of immunological evidence positioned *Dientamoeba* within a sub family of Monocercomonadidae. In the genus the species *Dientamoeba fragilis* infects about 4% of human beings.

2.3.5.2 Geographical Distribution

It is cosmopolitan in distribution.

2.3.5.3 Morphology

Trophozoites are only observed and cysts are still not known of this type of amoebae. It is also a harmless commensal of human beings and a number of primates.

The trophozoites are sensitive and decompose very quickly in faeces or water. The size of the trophozoite is 6–12 μm in diameter and the rarest intestinal amoebae of human beings. The cytoplasm is differentiated into ecto and endoplasm. Single pseudopodium is present. The cytoplasm contains food vacuoles containing bacteria, cellular debris, starch granules, etc., but never RBCs.

The nuclear character shows the large chromatin granules are usually 6 in number which forms a star shaped cluster.

The salient feature is two nuclei connected by a filament which may be seen under light microscope and so the name is given *Dientamoeba*. The endosome is present eccentrically.

2.3.5.4 Habitat

D. fragilis resides in the caecum of human beings. As they are harmless commensals they devour cellular debris and bacteria. Some study carried in some parts of Canada showed some people remain infected with *D. fragilis*.

2.3.5.5 Symptoms

The symptoms of the disease found in the study of Canada were diarrhoea, abdominal pain, liquid stool, etc. scientists are of opinion that *D. fragilis* infection often occurs with other parasitic amoebae, so it cannot be ascertained that all the symptoms are caused by *D. fragilis*.

The mode of infection cannot be described as there is no cysts present and trophozoites can survive in the upper digestive tract.

The binucleate trophozoites of *D. fragilis* were encountered in the faeces of *Papio hamadryas, Papio porcariaus, Cynocephalus mormon* and *Cercopithecus nictitans* in our study.

The study done by one of the authors (N. Das) recorded primates as new hosts because no such hosts of *D. fragilis* are recorded by any earlier investigators. But presence of *D. fragilis* in the sheep of California was recorded by Noble and Noble.

The trophozoites found in our study were unusually oblong or ovoid measured 7.2–14.4 μm in length and 4.8–10.8 μm in breadth. *D. fragilis* from human sources are usually 9.0–18.0 μm in diameter. The two nuclei were small and connected by a delicate filament. This structure degenerates quickly with the death of the amoeba.

The endosome varied a great deal in morphology. It may be compact or loosely arranged, composed of six to eight chromatin granules placed centrally or it may be represented by chromatin plaques near the margin of the nuclei. Sometimes the endosome was very large, filling up almost the entire nuclear space. Peripheral chromatin was never observed. Periendosomal granules were present in a fibrillar form connecting the endosome with the nuclear membrane.

2.3.5.6 Pathogenesis

It was thought to be nonpathogenic. There were reports of mucus diarrhoea, sometimes virulent with these amoebae only.

D. fragilis is generally regarded as a rare species, but we found in 5 out of 12 hosts examined.

2.3.6 *Iodoamoeba butschlii*

2.3.6.1 History

The amoebae were first discovered by Prowazek in 1912 and it was first described by Dobell in 1919.

2.3.6.2 Habitat

They reside as harmless commensals in the colon of man and other primates like *Macaca speciosa, Papio hamadryas, Papio porcarius, Semnopithecus entellus, Cercopithecus nictitans* and *Pan satyrus.*

The only species of *Iodoamoeba* known as *butschlii* is capable of infecting human beings and other primates and mammal like pigs. The main or original host of this species of amoeba is the pig. The propensity of this parasite in human beings is 4–8%.

2.3.6.3 Geographical Distribution

It is cosmopolitan in distribution.

2.3.6.4 Morphology

The parasite exists in two forms trophozoite and cyst.

2.3.6.5 Trophozoite

It measures 9–14 μm in length and 4–10 μm in breadth. They have blunt pseudopodia and crawl slowly with the help of pseudopodia. The trophozoites have clear cytoplasm and granular endoplasm. The nucleus is vesicular and has a large endosome. The midway between nucleus and nuclear membrane there are lightly stained granules between the nuclear membrane and the endosome, there are some strands which are achromatic.

The cytoplasm of trophozoites has food vacuoles which are full of bacteria and yeasts.

The *Iodoamoeba butschlii* have pre-cystic stage. The shape is ovoid that contains no undigested food. The precyst secretes a hard cyst wall and the shape becomes ovoid. They measure 6–15 μm in length and 4–10 μm in breadth.

2.3.6.6 Cyst

The nucleus of the cyst is 9–12 μm having a glycogen vacuole which stains with iodine. So the name is iodoamoeba. In the matured cyst the nucleus is pushed towards the nuclear membrane. The karyosome is placed more or less near the nuclear membrane. This parasite is nonpathogenic to human beings.

2.3.6.7 Biology

I. butschlii resides in the large intestine especially near the caecum. As the route of infection is faecal–oral so infection spreads through food and drink contaminated with cysts. It is observed that infection may be due to faeces of infected human beings and pig.

During our study (Das and Ray) this common intestinal nonpathogenic parasite was observed in the *Macaca speciosa, Papio hamadryas, Papio porcarius, Semnopithecus entellus, Cercopithecus nictitans* and *Pan satyrus.*

The trophozoites recovered from different primates in our (Das and Ray) investigation ranged from 8.4 to 21.6 μm in length and 7.2–18 μm in breadth. But earlier the measurement recorded was 9.00–14 μm (Wenrich 1937) or from 4.00 to 20.00 μm (Levine 1961). Simitch et al. (1959) isolated trophozoites of *I. butschlii* from 156 pigs were morphologically similar with that of *I. butschlii* of man. But between two hosts there were differences in character and so they named them *I. suis.*

The nucleus measured 4.8–8.2 μm in diameter containing a single round endosome The earlier record was 2–6 μm in diameter (Wenrich 1937). The peripheral chromatin was present in trophozoites in the form of a thick ring. But its presence in the trophic form was not mentioned earlier (Wenrich 1937; Levine 1961). The periendosomal granules was not observed always but when present, it was in the form of fibrils extending from the endosome to the nuclear membrane (Das and Ray 1970).

The size of the ovoid cysts of *I. butschlii* was 8.4–26.6 μm long and 8.4–26.6 μm long and 8.4–15.6 μm broad. The measurement recorded by Hoare (1959), Markel and Voge (1958), Wenrich (1937) and others did not agree with our observation.

The cyst wall of *I. butschlii* from *Pan satyrus* was alcian blue positive suggesting acid mucopolysaccharide in nature.

Granular or alveolar cytoplasm of the cyst of *I. butschlii* contained PAS positive glycogen vacuole which sometime seemed to have pushed the nucleus to one side of the cyst.

The chromatoid bars, very slender, small and curved were seen in the cysts recovered from white nosed monkey only. Levine (1961) was of opinion that there were no chromatoid bars in the cysts but contained deeply stained volutin like granules.

Single large round nucleus measuring 3.0–7.2 μm in diameter showed a delicate, alcian blue positive nuclear membrane. This feature was not noticed by earlier workers.

The peripheral chromatin of the nuclear membrane of *I. butschlii* was seen to occur in the cysts from Langur, white nosed monkey and Chimpanzee. The presence of the chromatin granules in the *I. butschlii* cysts was not reported earlier.

Regarding pathogenicity, no pathogenic symptoms have been seen in the primates harbouring *I. butschlii*. There was an unique instance of fatal infection by *I. butschlii* in a Japanese soldier captured in New guinea.

2.4 Haemoflagellates

2.4.1 *Trypanosoma cruzi*

Trypanosoma cruzi is a haemoflagellate parasitic protozoa, the causative agent of the disease called Chagas disease or American trypanosomiasis in central and South America.

Trypanosoma cruzi is transmitted by the insect vector *Triatoma infestans, Panstrongylus megistus* and *Rhodinus prolixyus*, the blood sucking bugs of family Reduviidae.

The disease may also spread by blood transfusion, organ transplantation, ingestion of food contaminated with parasite and from mother to her foetus.

2.4.1.1 History

The disease was named after the Brazilian Physician, Carlos Chagas, who first reported the disease in 1909. But the disease was not recognized as a major disease in human beings until 1960. Chagas discovered that in the intestine of Triatoma, a reduviid bug, a flagellate protozoan parasite of genus *Trypanosoma* resides. He experimentally proved that the disease can be transmitted to marmoset monkeys taking the help of infected reduviid bugs. Chagas for the first time named the species *Trypanosoma cruzi* after the name of famous Brazilian Physician Oswald Cruz. Oswald Cruz successfully eradicated the epidemic of yellow fever, small pox and bubonic plague in Rio de Janeréo and he is famous for his such work.

2.4.1.2 Symptoms

The symptoms of Chagas disease vary in course of infection local swelling at the site of bite of the bug is seen. With the progress of the disease chronic symptoms of heart disease, malformation and malfunction of the intestine of the host appear. If untreated the disease becomes fatal.

In human beings the disease manifests in two stages: an acute stage appears just after initial infection and chronic stage that develops over a period of many years.

The acute stage continues for the first few weeks or months after infection. The symptoms manifested are fever, fatigue, body aches, headache, rash on the skin, loss of appetite, diarrhoea, nausea and vomiting. On physical and clinical examination mild enlargement of liver and spleen, swollen glands and local swelling at the site of infection are found. Most important symptom is swelling of the eyelid on the side of the face near the bite of the bug. If untreated patients die from severe inflammation of the heart muscles or brain. Some patients develop neurological disorders.

2.4.1.3 Morphology of the Parasite

The causative agent of the disease is a parasitic haemoflagellate, *Trypanosoma*. This parasitic haemoflagellate lives in the blood of man and other animals causing disease. The *T. cruzi* is a curved, stumpy trypanosome measures about 20 μ in length with a pointed posterior end and an elongated nucleus at the centre of the body. A large egg shaped kinetoplast is present at the posterior end. A narrow slightly wavy

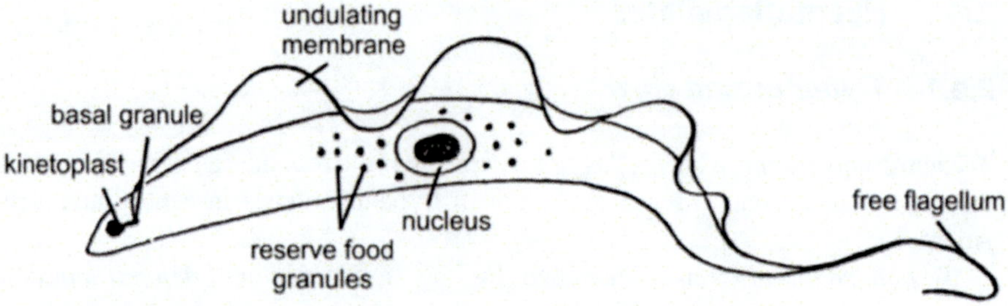

Fig. 2.5 Morphology of *Trypanosoma cruzi* (*Advanced Parasitology*, Das)

undulating membrane and a long free flagellum are the salient features of the said *Trypanosoma* (Fig. 2.5).

The morphology changes during their development. They remain as trypomastigote form in the peripheral blood of vertebrate host in acute stage of the disease and do not multiply within the blood of human beings.

When they enter into the bloodstream of the vertebrate host, trypomastigotes are phagocytosed by different cells of the immunity system including macrophages and quickly transform into amastigote form. The change is caused by the low pH of the lysosomal contents during phagocytosis. The resulting amastigote avoids the lysosomal action by entering into the cytoplasm of the infected cell. Amastigote forms live in muscles of heart, skeletal system, nerve cells and cells of R.E. system which is formed from metacyclic trypomastigote form. Amastigote forms are spherical and measure 2–4 μm in diameter and are the multiplying forms.

They complete their life cycle in two hosts: man and Reduviid bugs. Within the insect host they change from amastigote to promastigote, promastigote to epimastigote, epimastigote to metacyclic form of trypomastigote. Multiplication can take place in any of these developmental stages.

Horizontal transmission takes place from one vertebrate host to another by blood sucking bug *Triatoma*. After an infective feeding development takes place within the invertebrate host, reduviid bug. So certain time is required before it can infect another healthy individual. Infectiveness of the bugs develops only after the development of metacyclic form of trypomastigote within the bugs which requires some time.

Ingested trypomastigote in the sucked blood develops in the intestine of the insect vector from where they go backwards to the hind gut of the infected insect. Transmission happens mostly by rubbing the faecal matter of the insect contaminated with metacyclic trypomastigotes into the wound caused by the insect bite.

2.4.1.4 Developmental Stages

Amastigote form: Represented by round forms without any external flagellum and have a nucleus and a kinetoplast.

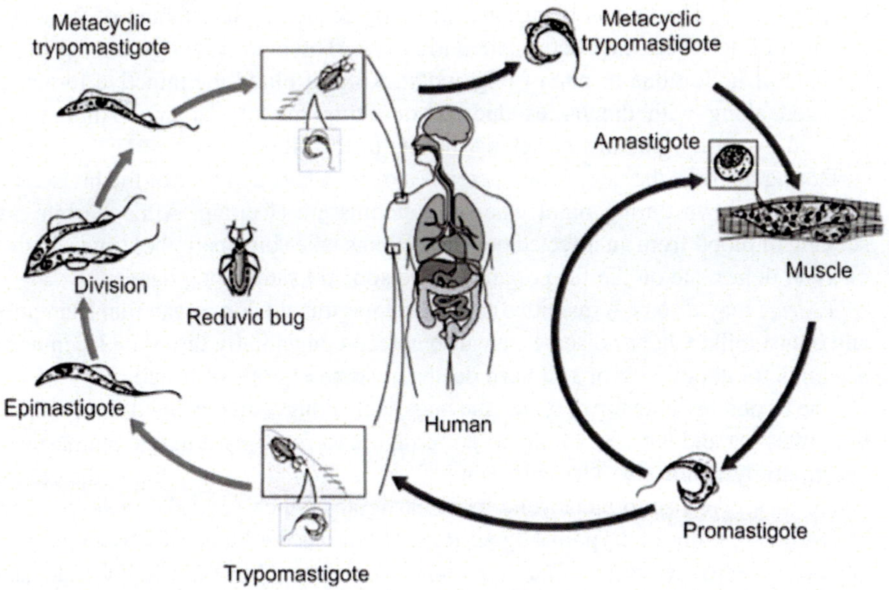

Fig. 2.6 Life cycle of *T. cruzi*

Promastigote form: Represented by forms with the kinotoplast present anterior to the nucleus, a single long slender flagellum which arises from a basal granule very near to the kinetoplast.

Epimastigote form: Represented by forms with kinetoplast located anterior and close to the nucleus the flagellum arises from the basal granule very near to the kinetoplast and is connected with the body up to the anterior end by an undulating membrane.

Trypomastigote form: Represented by forms with post nucleus kinetoplast located at the posterior end of the body and flagellum is attached to the body for most of its length with an undulating membrane.

2.4.1.5 Life Cycle

Trypanosoma cruzi requires two hosts, definitive and intermediate, to complete its life cycle. Human beings are the definitive hosts and reduviid bugs like *Triatoma infestans*, *Panstrongylus megistus* and *Rhodinis prolixus* are the intermediate hosts.

T. cruzi does not multiply in the blood of vertebrate host though it is a haemoparasite but divides rapidly in the tissue of human beings and the animals. The parasites multiply in the tissues of the heart and voluntary muscles, central nervous system and in different glands of the body.

The infective form of the *T. cruzi* is called metacyclic trypomastigote. These forms are liberated by the infected intermediate host in their excreta (Fig. 2.6). The intermediate hosts, the reduviid bugs of genus Triatoma and others bite human beings and suck blood from the face specially the eyelids. The bugs are known as kissing bugs for their special choice of the face of the vertebrate host for bitting and sucking blood. The human beings are infected because of the rubbing of the eye at the site of bite where they defaeciate during bite. The human beings rub the site as they feel irritation due to bite of bug and sucking of blood the infective forms are liberated along with the faeces due to squeezing out of slapping. Animals are infected by eating the bugs or licking the site of the bite.

During the day, the bugs hide in crevices of the walls and roof of the house. The bugs come down during night when the victims are sleeping. After the bite and sucking of blood from an infected person the bugs take some period of time and then bite and defaeciate on the face of another person and same thing happens.

T. cruzi may also be transmitted through blood transfusion, organ transplantation and breast milk. Chagas disease may also spread congenitally through placenta and accounts for about 13% of still born deaths in some regions of Brazil.

The trypomastigote forms which the bugs suck along with the blood meal reach in their intestine and within 24 h the trypomastigote forms transform into epimastigote and multiply profusely. The epimastigote forms after division pass to the rectum where they give rise to metacyclic trypomastigotes which are same in shapes, size and structure with the trypomastigote forms found in the blood of vertebrate hosts and are the infective forms. They are found in the larval bugs in about 6th day and 10th to 15th day in adults after the sucking of blood. As many as 3500 of metacyclic trypomastigote forms are found in 1 cubic millimetre of the excreta of the insect.

Within the human beings the metacyclic trypomastigotes enter into different tissues of definitive host where they divide quickly. The organs most vulnerable are the spleen, liver, lymph glands and all types of muscle cells. Besides nervous and reproductive systems, intestine and bone marrow are sometimes invaded. During the stage of rapid intracellular division the parasites transform into amastigote form then change into promastigote, epimastigote and trypomastigote form. The tissues or cells loaded with the parasites rupture liberating the parasites which either infect another tissue or disperse to other parts of the body through blood or lymph. In the early acute stage the parasites can live in the blood but in chronic stage when antibody reaction started in the vertebrate host the blood forms are not seen but can be found in large number in the tissues.

The ultimate trypomastigote forms in the blood are metacyclic trypomastigotes.

2.4.1.6 Pathways Associated with Energy Production

American *Trypanosoma cruzi* differs from African *Trypanosoma gambiensae* in their physiological function. The parasite is a 'partial aerobic fermenter'. The parasites exhibit Krebs cycle and classical cytochrome system in their bloodstream forms. In all the stages of life cycle of *T. cruzi* the well-developed mitochondrial cristae suggest that oxygen metabolism does not differ at all. It is proved that oxygen consumption is same in the intracellular stages and bloodstream trypomastigote and the stages that are found in insect also.

The mitochondria of Trypanosomes change in function and structure in each stage of their life cycle. Within the insect vector the parasites enter into the salivary glands. Here the mitochondria develop branches with large numbers of cristae and fully functional TCA cycle. It is able to oxidise proline and other amino acids present in the haemolymph of the insect vector.

A complete glycolytic pathway remains in all the stages with glucose as the major carbohydrate. Some portions of the consumed glucose are degenerated entirely into carbon dioxide while the rest portion is incompletely degraded to organic acids like succinic acid and acetic acid.

2.4.1.7 Pathogenicity

Chagas disease affects 16–18 million people in the year 2008 and 100 million people are at risk. It is estimated that the disease kills about 20,000 people annually.

The disease is present in 18 countries in the American continent ranging from Southern USA to Southern Argentina.

In the South America the wild reservoir of *T. cruzi* are opossums, raccoons, armadillos, squirrels and mice. Opossums are interestingly important as reservoir host because the *Trypanosoma* can complete its life cycle in the anal glands of the animal without the help of insect vector.

The first entry of the metacyclic trypomastigotes within the human body induce a local inflammation and swelling at the site of bite of the vector.

Acute causes of the disease are found in infants or young children. The first sign is oedematous swelling of the eyelid and sometime other parts of the face usually only on one side of the face where the bug bites for blood sucking. The first sign of swelling due to inflammation is called 'Chagoma' because the parasites after entry into blood go to the local lymph nodes where they are colonizing in tissues mainly in subcutaneous fat and begin to multiply in the phagocytic cells of the immune system. In later stages other chagoma develops in different parts of the body. In early infection patients complain severe headache with continuous fever and extreme exhaustion.

After sometime the acute stage disappears and chronic stage starts which according to many physicians persists for life. There is a long latent period with a few or no symptoms. Actually at that time the disease is progressing. In the chronic stage the symptoms are oedema, inflamed lymph glands and enlargement of liver and spleen in the patients. The patients who are suffering for a long period with the infection of *T. cruzi* anaemia develops sometimes nervous disturbance also as the parasite colonize in the muscle cells and nerve cells. In severe cases death may occur within 2 or 3 weeks.

Diseased condition of the heart is the common symptom in the patient because the parasites choose heart muscle and attack them. Heart block and other abnormalities are seen in electrocardiogram.

When acute phase progresses, i.e. when eyelid and other parts of eye are parasitized then oedema and swelling take place. Then the symptom is known as 'Roman's sign'.

2.4.1.8 Host Defence Mechanism

For all the stages of life cycle of *Trypanosoma cruzi* causing Chagas disease, cell mediated immunity is very much important. The investigation shows that membrane glycoproteins of parasite activate host cytokine production as a result increases macrophage killing capacity linked to NO production.

The response of the infected host is found in the preparation of specific antibodies and the activation of phagocytes stimulated by IFN-γ. In the early stages of acute infection increase the production of IL-2 and NK cell activity. During acute phase of infection intracellular amastigote secretes glutathione *S*-transferase in the bloodstream of host. T cell response to the infection decreases with decreased secretion of IL-2 and IFN-γ by which the parasites can avoid close observation of immune system.

Postacute or chronic infection is influenced by humoral response. It is seen that circulating IgG acts as protective immunoglobulin. Investigation on this has proved that nonprotective antibodies are also produced and they remain in bloodstream even after the patient is cured.

2.4.1.9 Diagnosis

Diagnosis is direct by examining the thick blood film under microscope. Besides blood film the parasites may be found from cerebrospinal fluid, fixed tissues and lymph. During fever of the patients in peripheral blood a large number of trypomastigotes may be found.

It is very difficult to find the parasites in the patients suffering in the chronic stage of the disease.

There is another method of diagnosis, the method is called xenodiagnosis. In this method the bug, *Triatoma* which are reared in laboratory are subjected to feed on the suspected patient and then after 10–15 days the intestine of bugs are taken out and a smear is prepared of the gut contents, if trypanosomes are present they may be seen under microscope.

In chronic cases complement fixation test or other immunologistic tests are performed to be sure of *Trypanosoma cruzi* infection.

2.4.1.10 Prophylaxis

Parasites should be controlled by treating the disease with specific drugs.

Destruction of vectors by spraying insecticides like DDT, or Dieldrin or BMC is very effective.

Personal guard is very important by using mosquito net.

2.4.2 *Tryponosoma gambiense*

2.4.2.1 History

Joseph Everett Dutton identified the parasite and in 1902 he proposed the name *Trypanosoma gambiense. The Trypanosoma* causing sleeping sickness in man was first discovered in Gambia by Dutta and Forde in 1902.

2.4.2.2 Geographical Distribution

Trypanosoma gambiense is found in Western Africa from Senegal to London. They are prevalent along the river side of Cairo and Nigeria and extended to Lake Tanganyika and south Sudan and *T. brucei* is prevalent in East Africa. *T. brucei* is found in many African wild animals and domestic animals like horses, camels and is infective for almost all types of mammals except baboons and man.

T. gambiense and *T. brucei* both are similar in all respects except their geographical distribution and some clinical manifestations.

2.4.2.3 Habitat

T. gambiense is the causative agent of the disease African sleeping sickness in man. The parasites reside in the bloodstream and in the lymph glands of human beings and are called haemoflagellates. It is also seen that in late stage they may invade the CSF.

They are mainly found in the lymph glands like spleen and cause enlargement of spleen. After 3 months they are found in CSF even in the brain tissue and spinal cord.

2.4.2.4 Morphology

The African polymorphic Tryponosomes vary in their measurement from 15 to 30 μ with exceptionally long or short forms.

The haemoflagellate *Trypanosoma* have polymorphic forms, all these are developmental stages. The different forms are:

1. Trypanosoma or Trypomastigote form: This form measures 12–30 μm in length and 2–4 μm in breadth. Nucleus is placed at the centre and kinetoplast is located at the posterior end of the cell. Undulating membrane nearly spread almost the whole length of the cell with a free flagellum.
2. Epimastogote or crithidial form: Cell body elongates with kinetoplast lying anterior and close to the nucleus and the flagellum arises from the basal granules very near to kinetoplast and is connected with the cell body up to the anterior end by an undulating membrane.
3. Promastigote or Leptomonad form: Represented by forms with kinetoplast anterior to the nucleus, a single slender long flagellum arises from a basal granule very near to kinetoplast.
4. Amastigote or Leishmanial form: Represented by round forms without any external flagellum and contains a nucleus and a kinetoplast.

The *Trypanosoma gambiense* when resides in vertebrate host like man it remains in trypomastigote form in the bloodstream and lymph glands. It is slender, elongated having a blunt posterior end and a finely pointed anterior end. It is 10–30 μm in length and 2.5–10 μm in breadth. The end from where the flagellum arises is the anterior end which is pointed (Fig. 2.7).

The large, oval, conspicuous nucleus located centrally. The kinetoplast which is situated at the posterior end is very small. The flagellum arises from a basal granule which is very near to the kinotoplast. The undulating membrane runs along the

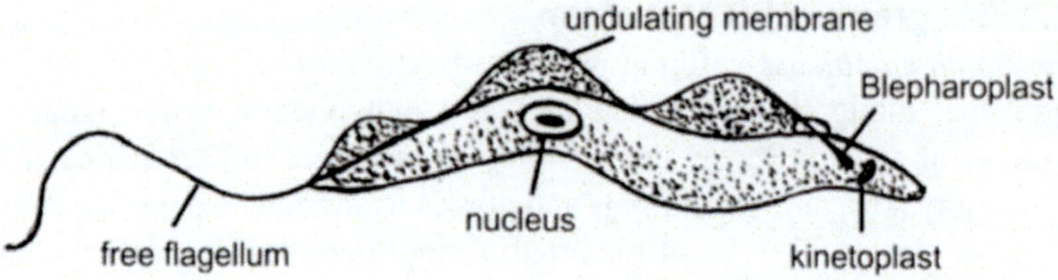

Fig. 2.7 Morphology of *Trypanosoma gambiense* (*Advanced Parasitology*, Das)

almost entire length of the body and extends beyond the anterior end as a free flagellum.

The *Trypanosoma gambiense* exists in various forms during their development, i.e. they are polymorphic.

2.4.2.5 Life Cycle

Trypanosoma gambiense requires two different hosts vertebrate and invertebrate, man and insect to complete its life cycle. The definitive host is the human beings or other mammals. The intermediate host is the insect vector tsetse fly, *Glossina palpalis*. When the infected insect vector bites healthy human beings the infective forms of trypanosome the metacyclic trypomastigotes are introduced into the blood of human beings which are stored in the salivary gland of the tsetse fly. The action is just like female anopheles mosquitoes perform in case of malarial infection.

Trypanosoma is an extracellular parasite resides in the bloodstream. The metacyclic forms when enter into the blood of human beings remain without free flagellum. They develop into flagellated long slender form, the trypomastigote form. Within the bloodstream they divide longitudinally by longitudinal binary fission. The division starts from kinetoplast which divides into two first. The nucleus divides by the process of mitotic karyokinesis. The cytokinesis starts from anterior end. Ultimately it is divided into two. After a number of divisions the long, slender forms become short stumpy forms with flagellum.

These stumpy forms enter into the gut of the vector during sucking of blood. These forms when reach the mid-gut become transformed into long, slender trypomastigote forms. These long, slender trypomastigotes multiply in the blood meal of mid-gut and ultimately reach salivary glands of the vector where they transform into epimastigote forms. Epimastigotes soon multiply and create metacyclic trypomastigotes (Fig. 2.8).

The metacyclic trypomastigotes waiting in the salivary gland of the vector reach human beings with the bite of the insect vector *Glossina palpalis*. The cycle is completed.

According to the investigators the trypanosomes stop multiplying when the glycolysis process is disturbed. The long slender forms become short stumpy forms due to loss of energy and want of food. The stumpy forms ultimately die if they are not sucked by the tsetse fly.

When the insect vector sucks the blood from the infected person the stumpy forms reach mid-gut along with the sucked blood and development within the insect vector

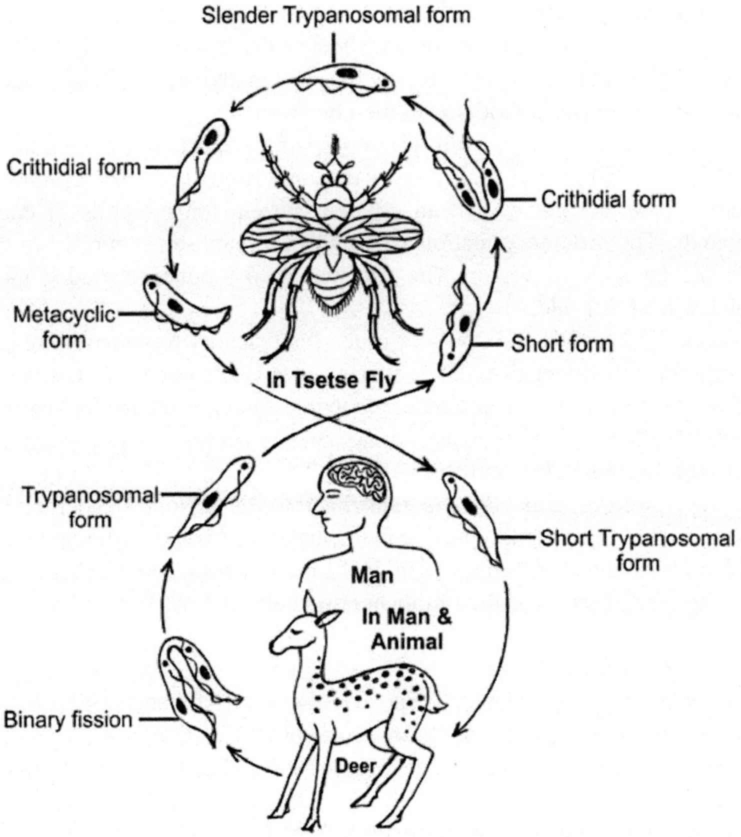

Fig. 2.8 Life cycle of *T. gambiense*

starts. It is found that in case of trypanosome the stumpy forms develop within the cavity of the peritrophic membrane of the insect. The stumpy forms transform into long, slender forms.

These long, slender forms go to the extra peritrophic cavity, i.e. the space between the peritrophic membrane and the body wall here they multiply for some days. By the 15th day they liberate from this space and enter into the proventriculus.

These long, slender forms from the proventriculus go straight to the salivary glands. Here in the salivary glands they multiply again and transform into first epimastogotes and then into metacyclic trypomastigotes which are the infective forms. The time taken by the parasite for development within the vector is about 20 days.

The vector now becomes fully ready to infect another healthy person to start new another healthy person to start new infection.

2.4.2.6 Pathways Associated with Energy Production

The long slender trypomastigote obtains energy rich compounds by substrate level phosphorylation through glycolysis which occurs in glycosomes. The reduced NAD produced in glycolysis is oxidised in the glycosomes.

The epimastigote with its well-developed mitochondria utilizes oxidative phosphorylation for ATP synthesis. Energy is procured from glycolysis. Epimastigotes in the lumen of insect gut evolve an efficient system for synthesis of energy rich compounds. The different situation arises when trypanosomes remain in the blood clot in the intestine of vector. The parasites then produce degraded glucose by glycolysis in TCA cycle.

A single mitochondrion is present in the long slender trypomastigote extending anteriorly from the kinetoplst. Their cristae are few in number, short and tubular in size. But in the parasites when remain in the intestine of vector mitochondria is well developed and extends both posterior and anterior from kinetoplast also cristae are innumerable and plate like in form and size. The infective forms for the tsetse flies are the short stumpy forms. But the transitional forms are long slender, noninfective forms. Whereas mitochondria present in the metacyclic trypomastigote form is more or less similar to those of forms present in the bloodstream. The stages in the salivary gland with tubular cristae in their mitochondria indicate less electron transfer system.

2.4.2.7 Pathogenicity

The *Trypanosoma gambiense* is the causative agent of African sleeping sickness, the human disease. It is the agent of chronic form of sleeping sickness. These flagellate parasitic protozoa enter the central nervous system and produce the chronic sleeping sickness.

A red inflamed area called trypanosome chencre develops after 1 week of the bite of tsetse fly. This chancre or wound heals but in the meantime the parasites (metacyclic trypanosomes) invade blood and lymph channels produce parasitemia. The large number of parasite invades all the organs of the body of hosts. The infected lymph nodes are congested and become swollen. Especially the nodes of neck, groin and legs are swollen. These swollen nodes are termed Winterbothom's sign. When the Trypanosomes in the circulating blood increases in number is in the early stage of infection fever starts. Then the symptoms are generalized pain, persistent headache, weakness and cramps in the body.

Parasitemia is seen until the specific antibodies are produced. Most of the trypanosomes degenerate due to specific antibodies produced in the host.

When the parasites enter central nervous system the symptoms are tremor of the tongue and trunk of the patient and then paralysis follows. This condition arises when the patient feels sleepiness and found falling asleep even while standing or eating. Ultimately dies due to either malnutrition or pneumonia or myocarditis with severe congestive heart failure.

2.4.2.8 Immune Response by the Host

It is a very peculiar phenomenon the parasites can change their outer variant surface glycoprotein (VSG) coat. Each VSG is immunogenic but different from the

Table 2.1 The differences of features between African and American *Trypanosoma*

Features	African	American
Distribution	Western and Central Africa	Central and South America
Disease causing organism	*Trypanosoma gambiense*	*Trypanosoma cruzi*
Vector	*Glossina palpalis* (Tsetse fly)	*Triatoma infestans* (Reduviid bug)
Reservoir	Human	Cats and dogs
Multiplication	Divides in blood	Does not divide in blood
Infective forms	Metacyclic trypomastigote	Metacyclic trypomastigote
Route of infection	Biting of insect vector	Biting of insect vector
Polymorphic form	Amastigote forms are not seen	All forms are seen
Depiction	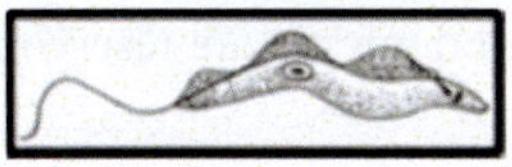	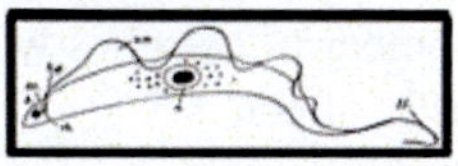

preceding VSG. They change their VSG every 8–10 days to hide themselves from immune response of the host. Each time the new VSG variant produces repeated parasitemia. This goes on for months. Then the parasites invade central nervous system.

As the number of parasites increases the VSG of the parasites creates IgM response by the host. The complement produces lysis effect on the parasites which results decrease in the number of parasitic population.

It is found that less than 1% of the surviving parasites, those who possess a variant VSG, again proliferate until they produce host response. In this way a cycle of remission alternating with parasitaemia occurs until the death of the host.

2.4.2.9 Comparative Account of African and American Sleeping Sickness

African sleeping sickness is somewhat different from American sleeping sickness which is called 'Chagas disease'. Though, both are caused by haemoflagellate of genus *Trypanosoma* (Table 2.1).

2.4.3 *Leishmania donovani*

2.4.3.1 History of Discovery

Leishmania donovani was discovered by Leishman and Donovan, both of them reported the presence of this organism simultaneously, Leishman from London in May1903 and Donovan from Madras in July1903.

Leishmania donovani is generally of two types: visceral Leishmaniasis known as Kala-Azar and cutaneous leishmaniasis produce lesion or local sores on the skin.

2.4.3.2 Geographical Distribution

It is found in many places in India, China, Africa, Southern Europe, South America and Russia.

In India it was once very common infection in Assam and Bengal along the coasts of the Ganges and Brahmaputra. It was also endemic in Bihar, Jharkhand, Orissa, Madras and the eastern parts of Uttar Pradesh.

2.4.3.3 Habitat

It is a parasite of blood and tissues of human beings and is intracellular living within WBC or hepatocytes, spleen, bone marrow, lymph glands, etc. They are found both inside and outside the cells and some are present in bloodstream and inside monocytes. The parasite is the causative agent of the disease Kala-azar.

A number of canaines wild and domestic are reservoir hosts and young children are mostly affected.

The vector of the disease is *Phlebotomus argentipes.*

2.4.3.4 Different Strains of *Leishmania*

The amastigote forms of the parasite infect adults and adolescent of human beings and there is no reservoir host of the strain found in India. But a more virulent, clinically similar strain is transmitted by *Phlebotomus* species in east Africa and wild rodents serve as their reservoir hosts.

A third strain is prevalent in Central and South America whose vector is *Phlebotomus longiceps* and their reservoir hosts are wild and pet dogs.

Leishmania donovani causing the visceral leishmaniasis showing symptoms of fever and enlargement of liver and spleen, cutaneous leishmaniasis caused by *Leishmania tropica* showing sores on the skin, lesions on mucus membrane of the nose, mouth, etc.

Leishmania is an important pathogenic zooflagellate. Actually there are two species *Leishmania donovani* and *L. tropica. L. donovani* is responsible for various types of visceral leishmaniasis.

In the year 1953 Biagi described all types of visceral Leishmaniasis which are known as Indian kala-azar, Chinese kala-azar, Mediterranean kala-azar, African kala-azar, Russian kala-azar and American kala-azar. The Indian kala-azar is transmitted by sandfly, *Phlebotomus argentipes.*

2.4.3.5 Disease

The symptoms of the disease are very misleading and same with that of typhoid, malaria or dysentery but no symptoms appear until the body resistance is lowered. The incubation period is generally 8 or 9 months.

At the initial stage there is an irregular fever along with hepatomegaly and spleenomegaly, anaemia, rheumatic fever and weakness. Cirrhosis of liver and hypertension may also result. The number of WBC population decreases. If the patient is not treated properly usually die within months. Sometimes after treatment and recovery post kala-azar dermal leishmanial nodules in the skin appear.

In Sudan a number of cases of extensive mouth lesions have been seen in which parasites were found in abundance though not in liver or spleen.

2.4.3.6 Morphology

The parasite exists in two forms: amastigote and promastigote.

Amastigote forms are found in human bodies as L.D. bodies and promastogote forms are found in insect vector sandfly.

The amastigote of *L. donovani* are small, ovoid and 2–4 μ in diameter. The L.D. bodies divide by binary fission and are mainly found in macrophages. The infected macrophages are found in quite considerable number in liver and spleen.

The promastigotes are found in the insect vector *Phlebotomus argentipes*. Promastigotes are small, oval and 15–20 μm × 1–2 μm in size.

Amastigote form: At this stage parasites are intracellular and reside within R.E system of man and other vertebrates.

It is round or oval bodies. Nucleus is located more or less at the centre or periphery. Absence of flagellum, kinetoplast present at right angle to the nucleus. There is a delicate filament present from kinetoplast to the margin of the cell called axonema. Sometimes there is a clear vacuole located along side the axonema. Axonema is the root of the flagellum.

Promastigote form: This stage is found only in the stomach tissue of sandfly, *Phlebotomus argentipes*.

This stage is represented by long, slender, spindle shaped body. A mature promastigote is 15–20 μm in length and 1–2 μm in breadth.

Nucleus is large, round and present either in the centre or at periphery. Kinotoplast is distinct and present posterior to the basal granule from where the flagellum arises. Flagellum is long either measures the same length of the body or more and present at the anterior end.

2.4.3.7 Life Cycle

Leishmania is a zooflagellate which requires two different hosts to complete its life cycle.

The main or definitive host is man or other vertebrates in which the parasite reproduces asexually (Fig. 2.9).

In the life cycle of *Leishmania*, when they reside within the mammalian host the amastigote infects macrophages, the cells of immune response, it is quite paradoxical. The parasites after entering the macrophages establishes itself in an intracytoplasmic vacuole called parasitophorous vacuole. Lysosomes fuse with this vacuole. The amastigotes defend themselves from the enzymatic actions of lysosomes and reside and reproduce within the parasitophorous vacuole. It is an interesting observation.

A number of mammals act as natural reservoir hosts of *Leishmania*. The most common of them are wild and domestic dogs and rodents. Leishmaniasis is therefore a zoonosis for human beings.

The intermediate host an insect vector known as sandfly, *Phlebotomus argentipes* is a blood sucking insect.

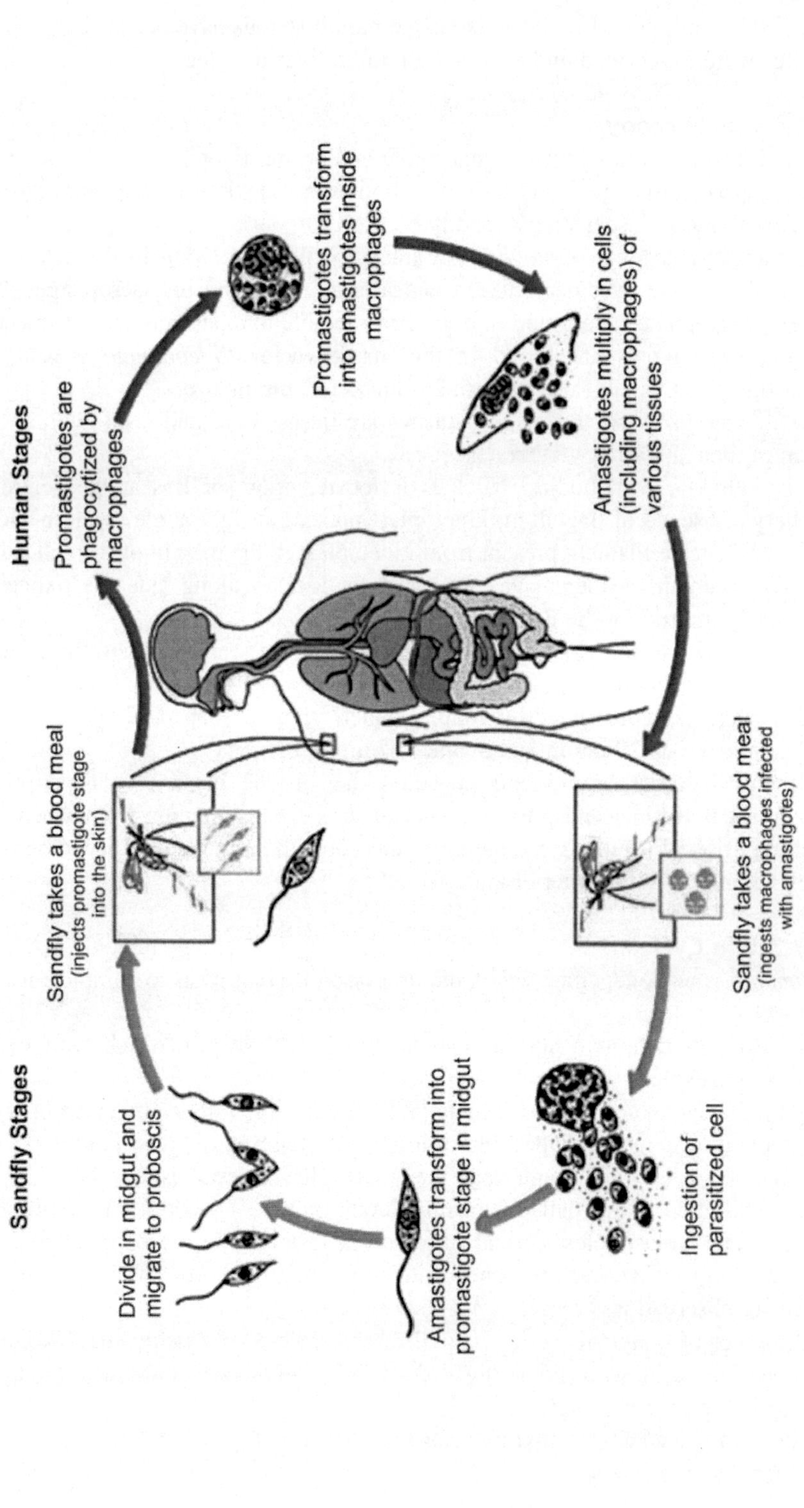

Fig. 2.9 Life cycle of *Leishmania donovani* (*Advanced Parasitology*, Das)

The reservoir hosts are dogs, squirrels, jackals, etc. in which the parasites do not undergo any change in morphology but help in spreading the disease from them to human beings. In the Mediterranean countries transmission from dog to man is more frequent than from man to man.

The infective forms of man are the promastigotes which are developed in the stomach wall of the insect vector. During bite of the sandfly the saliva causes vasodilation of the blood vessels where they bite human beings. The parasites are injected into the human beings. The parasites straight enter into the macrophages. Within the macrophages the promastigotes lose their flagella and develop into amastigote form. These amastigote forms undergo asexual binary fission to increase their population.

The macrophages have receptor on their surface, the parasites become attached to the macrophages and macrophages by their process of phagocytosis take the parasites within them and the amastigotes reproduce in large number.

The enormous population is formed within the macrophages make the cell to rupture. To increase their population the librated amastigotes again enter into fresh macrophages and in this way the cycle goes on. Some amastigotes swim free in the bloodstream, some are present in the neutrophils and monocytes.

At this stage a sandfly bites the infected person and sucks blood for her blood meal which is required by the female insect for the maturation of her ova. During sucking of blood the amastigote forms present in the bloodstream and tissue fluid of human beings are sucked in into the stomach of the sandfly. The amastigotes reach the mid-gut of the fly and transform here into promastigote forms.

The promastigote forms in the mid-gut tissue reproduce by longitudinal binary fission. The promastigote form with free flagellum migrates to the anterior part of the fly in the pharynx and buccal cavity. The pharynx becomes loaded with the promastigotes between 6 and 9 days after the sucking of blood meal from the infected person. In the case of sandfly salivary gland is not affected. Now when the sandfly again bites and sucks blood from a healthy person horizontal transmission takes place by introducing promastigotes into the blood of the vertebrates including man. The cycle is completed.

2.4.3.8 Pathogenicity

The incubation period of kala-azar of Indian origin is 3–6 months. But many times it remains asymptomatic. The fever like rheumatic fever sometimes is the early symptom. Then the hepatomegaly and spleenomegaly occur. Anaemia starts due to destruction of blood corpuscles in enormous number. The skin of the patient becomes pigmented, dry, rough and harsh. In case of some patients alopecia starts.

If not treated the infected person dies within 2 years.

Leishmaniasis affects mostly children in between 1 and 4 years in Mediterranean, south-west Asia, China. In East Africa and in India the children suffer from Leishmaniasis at the age of 5–9 years. The usual symptoms are fever, weight loss, diarrhoea, anorexia along with spleenomegaly, anaemia and pigmentation in the skin.

Post kala-azar symptoms sometimes involve lesions in the mouth cavity, nasal cavity and around mouth. Face may be mutilated so badly that people may hate themselves and outcast.

2.4.3.9 Enzyme Activities and Pathways Associated with Energy Production

In the genus *Leishmania* carbohydrate metabolism takes place in kinetoplast, mitochondria and glycosomes of the amastigote and promastigote forms.

In the amastigote forms carbohydrate metabolism takes place partially by anaerobic metabolism as there is lack of cytochrome system. The process occurs in glycosomes and cytoplasm produce acetic acid and lactic acids as end-products. ATP is formed by substrate level phosphorylation. When the amastigote forms are ingested by the insect vector they (amastigote forms) change into promastigote forms. With the change, the mitochondria grow, increase and become functionally and morphologically well developed with an active cytochrome system and functional TCA cycle. In this atmosphere, the cell utilizes aerobic metabolism producing ATP by oxidative phosphorylation.

The mitochondrial proliferation is directly linked to the kinetoplastic DNA. The chemical components of the pellicle of amastigote forms protect the cell from the hydrolytic action of the macrophage lysosomal enzymes.

2.4.3.10 Immune Response by the Host

Leishmaniasis according to WHO is an AIDS related opportunistic disease. *Leishmania* infection is controlled by the infected macrophages which induce cell mediated immunity by stimulating Th1 cells to produce IFN-y. Leishmaniasis and HIV simultaneous infection change the Th1/Th2 cell balance. For example, this type of double infection depresses Th1 cell activity and increases Th2 cell population. Inhibition of Th1 cell activity decreases the possibility of cure, while increased activity of the Th2 cell (IL-10) tends to depress macrophage activation. So here in this case the *Leishmania* infection becomes more virulent. On the other hand stimulates HIV infection by stimulating the secretion of IFN-y factor which stimulates HIV replication.

HIV infection suppresses cell mediated immunity due to T cells inhibition and promotes *Leishmania* multiplication. So this co-action resulting out of HIV and *Leishmania* infection simultaneously becomes fatal for the patients.

2.4.3.11 Diagnosis

Direct evidence: By microscopical examination of the stained blood film taken from the patient. It is often successful because of the presence of amastigote forms of the parasite in the peripheral blood.

But because of the less number of parasites present in the blood this may not produce positive result.

For positive result preparation of thick blood film and/or by centrifuging citrated blood or by making a blood culture taking blood from the patients is done. But it takes long time to prepare blood culture.

2.4.3.12 Indirect Evidence

By blood count, i.e. TC, DC, serological test and complement fixation test,

2.4.3.13 Prophylaxis

Eradication of vector Sandfly, the infective agent, control of reservoir hosts and the use of mosquito net during sleeping are the best procedure.

2.5 Coccidia

2.5.1 *Cryptosporidium parvum*

2.5.1.1 History

Tyzzar first observed *Cryptosporidium* in the year 1907 in the crypts of the gastric mucus layer. It was discovered in the laboratory mice. It was reported from human beings in the year 1976. In 1980 it is found in HIV patient.

Cryptosporidium is a genus of subkingdom Protozoa. It is an apicomplexa and contains 24 species still known today. Out of these 24 species only *C. parvum, C. homini, C. canis, C. felis, C. meleagridis* and *C. muris* have the ability to infect human beings. The disease caused by the protozoan parasites known as cryptosporidiosis which is a type of gastrointestinal illness with diarrhoea.

It is also found in HIV positive patients having symptoms of copious diarrhoea.

2.5.1.2 Geographical Distribution

It is cosmopolitan in distribution.

It is not only a parasite of human beings it is found in birds, cats, turkeys, cattle, sheep and goats, etc. The infection can spread from one host species to some other animals

2.5.1.3 Characteristics and Life Cycle

Though *P. vivax*, an apicomplexan needs mosquito as vector for the spread of the diseases. *Cryptosporidia* do not need an insect vector for completing its life cycle. It completes its life cycle within a single host so they are called monogenetic parasite.

Cryptosporidium infection happens by ingesting the oocysts through contaminated food and/or drinks. Sometimes through food sources like chicken salad.

The infective form is the oocyst. The oocyst is 2–5 μm in diameter. There are two types of oocysts: one thick walled and other thin walled oocysts. The oocysts are refractile and have some prominent granules within them and remain clustered near the margin of the cell.

The thin walled oocysts reinfect the same host but thick walled oocysts come out from the host through faecal matters and possibly respiratory secretions in the outside environment (Fig. 2.10).

Cryptosporidiosis is typically an acute short-term disease.

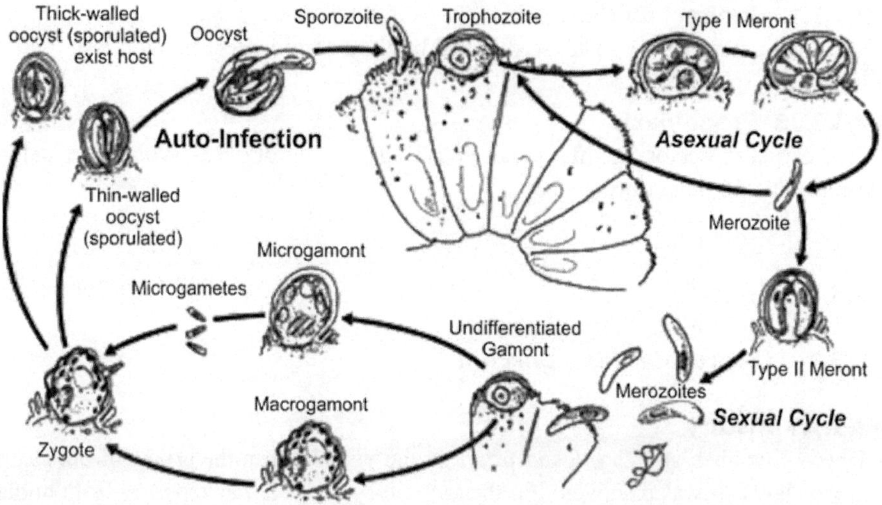

Fig. 2.10 Life cycle of *Cryptosporidium parvum* (*Advanced Parasitology*, Das)

The disease sometimes becomes severe in children and immunocompromised persons like AIDS patients. It is reported that there were many out breaks of cryptosporidiosis in water parks, community swimming pools and day care centres. Zoonotic and anthroponotic transmission of *C. parvum* happens through intimate contact with infected pet animals and vector contaminated with the faeces of contaminated animals.

The parasites remain in lower intestine and may remain there for 5 weeks.

The transmission of the disease takes place through thick walled hard oocysts.

Both the thin walled and thick walled oocysts contain within them four sporozoites.

The four sporozoites are liberated in the intestine of the host. The sporozoites enter through the outer membrane of the gastrointestinal brush bordered epithelial cells and remain there.

Here they develop into trophozoites. These trophozoites are 3–6 μm in size. These trophozoites here undergo asexual multiplication either schizogony or merogony. The result of the schizogony produces merozoites those infect new cell of the intestinal wall repeatedly, i.e. autoinfection takes place. The merozoites are 7 μm in size and form 8 banana shaped meronts.

The trophozoites multiply by schizogony and like *Plasmodium* also produce micro- and macrogametocytes. From microgametocytes microgamets are formed and fertilize macrogamets. Microgamets form non flagellated rod shaped 16 microgamets those are 1.5–2 μm in size. The fertilization produces oocysts both thick and thin walled. Oocysts are resistant to most disinfectants.

The thick walled oocysts escape from the host along with the faeces and other routes of the host after 5 days of infection and seek opportunity to infect new healthy individuals.

2.5.1.4 Pathogenicity

In developing countries where sanitation is poor *Cryptosporidia* infection is common in infants.

In developed countries the infection may take place in children. Cryptosporidiosis is typically an acute, short-term infection, but may become severe in immunocompromised persons.

C. parvum produces voluminous watery diarrhoea and the patient suffers for several months. The incubation period is 2 weeks. The severity of infection depends upon the immunity status of the patient.

In an immunocompromised person the symptoms that accompany diarrhoea are abdominal pain, weight loss, flatulence, nausea and vomiting.

The risks of infections in certain group of people by cryptosporiodosis are:

1. The people who swim regularly in different pools where sanitation is not so perfect. The faeces with oocysts deposited on the soil may mix with the water of the pool during rainy season.
2. Parents of infected children may be infected with the disease due to intimate contact and handling of the children.
3. Nurses or people who take care of the people infected with cryptosporidiosis.
4. People handling infected cattle.

Host defence mechanism: It appears that at the initial stage of infection, IL-15 secretion occurs for initial clearance of the parasite. Human memory response takes place by CD4+ cells and IFN-γ. Very little is known about the activity of B cells and CD8+ cells. In autoinflammatory and healing process TGF-β cells and IL-10 secretion are involved. Reinfection is prevented by the production of IgA.

2.5.1.5 Epidemiology

Reservoirs of *Cryptosporidium* infection are a number of animals. The route of infection is faecal–oral. Current and his co-workers in the year 2000 infected puppies, kittens, goats, etc. experimentally with oocysts taking from an immunodeficient person. They discovered that 12 immunocompetent persons were infected with *C. parvum* who were handling calves. So cryptosporidiosis is a zoonotic disease.

The zoonotic capability of *C. parvum* is discovered by surveys on cattle. It was found that 10,000 cattle, 65% of the diaries have infected animals. Cryptosporidiosis is an opportunistic infection in persons with AIDS. Parasites are distributed in the immune competent and asymptomatic population. It may be considered a usual causative agent of chronic diarrhoea in human population.

2.5.1.6 Diagnosis

Oocysts can be seen in faecal smears after staining with Jenner-Giemsa stain. Oocysts are seen as blue spherical structure under microscope.

Instead of Giemsa stain fluorescent staining with auramine phenol is also applied.

Indirect immunofluorescence using specific antibody may be used as a special technique.

In acute diarrhoea, the presence of oocysts is large in number but if their number is less then concentration technique may be applied with formal ether solution.

ELISA technique may be applied to detect antibodies.

ELISA technique in stool using monoclonal antibody is highly specific.

Diagnosis is done by observing oocysts in faeces. The oocysts may be induced to sporulate in 5% potassium dichromate solution. Concentration of oocysts with the help of centrifugation technique is also useful.

2.6 Malarial Parasites

2.6.1 *Plasmodium vivax*

Plasmodium is a protozoan parasite, a causative agent of the notorious disease malaria. The malaria recognized as the disease which played an important role in the progress and development of a country. The Japanese soldiers retreat in World War II due to wide spread infection of their soldiers with malaria and the defeat of Japan is due to the sufferings of large number of soldiers with malaria. Malaria caused more than five times casualties as did combat. A country cannot prosper without sound public health. But malaria caused disaster in the several Asian and African countries. Malaria is responsible for unemployment, communism and war opposed to peace, prosperity and contentment (Chandler and Read 1961).

2.6.1.1 History

Malaria means mala aria that is bad air. People at that time believed that bad air coming at evening through window is the cause of the disease until Laveran discovered the parasites in the blood of a patient in the year 1880. In the year 1891 Romanowsky for the first time stained the parasite and in 1898 Sir Ronald Ross discovered the mosquito cycle of the bird malaria working at P.G. Hospital in Calcutta and for this work he was awarded with the noble prize in 1902. But according to the Encyclopaedia Britannica Sir Ronald Ross discovered that the anopheles mosquito as the malaria vector on 20th August, 1897 in Secunderabad. Ross made this landmark discovery while dissecting the stomach tissue of female anopheles mosquito blood fed 4 days previously on a malarial patient.

2.6.1.2 Geographical Distribution

It is cosmopolitan in distribution especially in tropical countries.

2.6.1.3 Habitat

The malarial parasites reside in liver and red blood corpuscles of human beings. It is an intracellular parasite and mainly found within the RBCs where they develop through different stages and destroy the R.B.Cs in large number during their development. The parasites consume haemoglobin and produce malarial pigment due to metabolism of the parasite.

The malaria parasite completes their life cycle in two hosts: vertebrate and invertebrate hosts. Sexual reproduction (sporogony) of the parasites takes place within the stomach of female mosquito but asexual reproduction (schizogony) takes place in the blood of human beings exactly RBC and liver parenchyma.

2.6.1.4 Different Species of *Plasmodium*

The disease malaria in man is caused by the protozoan parasite of the genus *Plasmodium* and there are four different species of genus *Plasmodium*.

Plasmodium vivax: The disease is called Benign tertian malaria. Here the paroxysm of chill and fever is felt at the interval of 48 h.

Plasmodium falciparum: Malignant tertian malaria is caused by the parasite *Plasmodium falciparum*. Here also the paroxysm of chill and fever is felt every 48 h. But the disease is dangerous with high mortality rate.

Plasmodium malariae: Quartan malaria is caused by *P. malariae*. The feeling of chill and fever occur every 72 h interval, i.e. on every 4th day.

Plasmodium ovale: Mild tertian malaria is caused by *P. ovale*. Here fever occurs every 48 h interval.

2.6.1.5 Life Cycle

To complete the life cycle the malaria parasite requires two different hosts vertebrate and invertebrate, man and female Anopheles mosquito.

The parasite reproduces asexually in the liver cells and RBCs of man and is called asexual generation. The process of asexual reproduction is known as schizogony.

Most of the parasites infect human beings or vertebrate animals, the sexual reproduction takes place in the vertebrate host and they are called definitive host. But in case of malaria parasite the sexual reproduction or gametogony takes place in the invertebrate vector, female anopheles mosquito.

The life history of Plasmodium is divided into four stages:

1. Pre-erythrocytic stage: The development and reproduction within the liver cells before actual infection of RBC.
2. Erythrocytic stage: The growth and reproduction within the red blood corpuscles, RBCs.
3. Exoerythrocytic stage: The development and reproduction within the liver cell after the infection in RBCs the effect of which is relapse.
4. Sexual cycle: The fertilization of anisogamates and consequent growth and reproduction of the parasites in the stomach of the female Anopheles mosquito.

1. Pre-erythrocytic stage: Human beings are infected when the infected female anopheles mosquito bites a healthy person and sucks blood by the help of sucking mouth parts, proboscis. During sucking of blood to prevent coagulation in the proboscis mosquitoes inject saliva in which there is an anticoagulant. With

the saliva the infective forms of malarial parasites enter into the human blood as the infective form, sporozoites remain stored in the salivary glands of mosquitoes. The sporozoites after entering into the bloodstream swim in the blood for 30 min or so and hide themselves in the parenchyma cells of liver of the vertebrate host to protect from immune reaction of the host.

Sporozoites stay here for nearly 8 days growth and reproduction of the parasites take place here.

The infective forms sporozoites measure 12–14 μ in length and 1 μ in breadth. They are very slender and curved structure. The nucleus is situated at the middle. There is a complex structure called apicomplex present at the anterior terminal end or at apex of the sporozoites. The apical complex helps the parasites to enter within the liver parenchyma cells.

Then the sporozoites transform into schizont within liver cells. The schizonts are round or oval. The nucleus remains at one side called growing schizont. The growing schizont then matures and becomes 42 μ in diameter in average. Each mature schizont asexually divides, the process is known as schizogony into 12,000 small merozoites.

According to Huff (1946) there occur two successive generations of merozoites in the liver cells of man. The first generation is called cryptozoites. Cryptozoites then again enter into other liver cells and second generations of merozoites are formed called metacryptozoites. The metacryptozoites then invade the bloodstream and enter within RBCs.

In some species the pre-erythrocytic metacryptozoites continue to multiply indefinitely within the liver cells and they are the source of exoerythrocytic cycle.

The pre-erythrocytic merozoites or metacryptozoites are 1.5 μm in length and 0.5 μm in breadth. They are oval in shape and nucleus is present at the centre. The time taken by the parasites to develop and reproduce within the liver cells is called incubation period and no symptoms are witnessed at that time. It is found experimentally that if quinine is administered the parasites are not destroyed at this time.

2. Erythrocytic stages: The metacryptozoites now enter into the bloodstream in innumerable number like shower. They invade RBCs by making an invagination of the plasma membrane of the RBCs. It is found experimentally that the invasion of parasites into the RBCs takes place in three consecutive stages and takes about 10–20 s.
   Now after entering within RBCs the metacryptozoites start consuming haemoglobin and develop into trophozoites.

   Trophozoites measure 2.5–3 μm in diameter. At the initial stage it looks like a ring so they are called signet ring stage. It appears like a little signet ring as the centre of the parasite is occupied by a vacuole surrounded by a delicate ring of cytoplasm which stains blue and small nucleus at one side which stains red with usual Giemsa's stain. The ring of *P. vivax* is 1/3rd of the diameter of RBCs. Haematin pigment appears in the cytoplasm at this time. Haematin or haemozoin pigment is produced when haemoglobin is digested by the parasites and it is the insoluble polymer of haeme. The trophozoites now grow and become double in size. The RBCs now become pale in colour and some granules appear at the margin of the RBCs called Schuffner's dots stain red. After 48 h the trophozoites mature and transform into schizonts. Each and every erythrocytic schizont by the process of schizogony creates 12–14 merozoites.

The erythrocytic merozoites formed by the division of nucleus of the trophozoites into two, then four and ultimately 12–14. The nuclei take up peripheral position in the schizont and small portion of cytoplasm concentrates around each fragment of chromatin material. The matured merozoites then rupture the RBC and come out of the corpuscles. These erythrocytic merozoites then attack new young RBCs and the erythrocytic cycle is repeated again and again.

From the ruptured RBCs toxin, the pigment and other waste products left behind by the parasites also come out of the RBCs and mix with the bloodstream. Just at the time of rupture of thousands of RBCs, the paroxisms of chill and fever is felt by the patients. The pigments and waste products that left behind when the merozoites are liberated and deposited in the spleen or under the skin cause pale yellow colour characteristic of malaria patient.

Some of the erythrocytic merozoites instead of forming schizonts again and again develop into single nucleated organisms, gamonts. There are two types of gamonts: microgametocytes or male gametocytes and macrogametocytes or female gametocytes.

The female gametocyte is 10–12 μ in diameter. Presence of pigments in the cytoplasm and the nucleus remains at the margin of the gametocyte.

The male gametocyte is 9–10 μ in diameter. Granule in the cytoplasm is less in number and the nucleus is present at the centre.

3. Exocrythrocytic stage: The pre-erythrocytic merozoites and erythrocytic merozoites sometimes again attack the liver cells to continue the liver cycle. The product of the exocrythrocytic merozoites again attacks RBCs if the erythrocytic cycle is destroyed by using medicine and this is called relapse. The exocrythrocytic products are known as hypnozoites or phenerozoites.
4. Sexual cycle: Both the gametocytes do not develop further in the human body. They wait for the mosquitoes to suck themselves up in their (vectors') stomach. The difference in temperature initiates formation of active gamets from the gametocytes. When a female anopheles mosquito for their maturation of ovum sucks warm blood from the infected human beings, the gametocytes come out from the RBCs and transform into gamets. When removed from the warm blood gametocytes only develop further. But other asexual forms perish very soon in the stomach of the mosquito.

   At first the male gametocytes through a complex process transform into eight flagellated male gametes after coming out from the RBCs. From the gametocytes eight flagella like structure develop, the nucleus divides thrice to form eight nuclear fragments, around each nuclear fragment cytoplasm concentrates and ultimately break free. The process of formation of flagellated male gamets from male gametocyte is called exflagellation.

The female gametocytes in the meantime just come out from the RBCs. Now with the help of meiosis and creation of polar body the gamets mature and form female gamets.

According to Garnham the male and female gamets present in the blood meal of mosquito fuse and fertilization occurs resulting in the formation of zygote. After sometime the zygotes become active and motile. The motile zygote is called ookinete. These motile ookinete then touch the wall of the stomach of the mosquito and form oocyst by invading the stomach wall and forming a cyst around it at the outerside of the stomach wall.

But Howard (1960) was of opinion that round inactive zygotes are formed on the periphery of blood meal lodged in the fold of the stomach wall formed due to digestion of the blood meal and contraction of the stomach wall which was distended due to sucking of blood. But zygotes which are formed in the centre of the blood meal do not have the chance to come in touch with the stomach wall and perish. They are unfortunate zygotes. They become extremely flattened and are called ookinetes. But ookinetes, the unfortunate zygotes are voided with the faeces of the mosquito.

But according to Garnham et al. the ookinetes actively move and ultimately come in touch with the mosquito's stomach wall and further development takes place.

The oocysts now develop on the outside of the stomach wall of the mosquito, they grow in size from 6 to 60 μ in diameter. There may be as many as 50 oocysts in a single mosquito. Within the oocysts cell division takes place the first is meiosis followed by mitosis. Within the oocysts gradually develop sporoblasts. Within sporoblasts sporozoites are formed by the process of sporogony. In each oocyst there may be 10,000 sporozoites, very slender, curved and crescent shaped. When the sporozoites mature they rupture the oocysts and are liberated in the body cavity of the mosquito. Then the thousands of sporozoites migrate towards salivary gland and stay there. During the bite of the mosquito the sporozoites are injected into a healthy person along with the saliva. The cycle is completed (Fig. 2.11).

2.6.1.6 Enzyme Activities and Pathways Associated with Energy Production

Glucose is the main carbohydrate from which the parasites acquire energy through glycolysis. In the erythrocytic stages when the parasites remain within the RBCs, they are considered as facultative anaerobes.

The end-products of carbohydrate metabolism are lactic acid, formic acid and acetic acid. The parasites fix carbon dioxide at this stage and the enzymes that help the process are vulnerable to quinine and chloroquinine.

2.6.1.7 Pathogenicity

As the malaria parasites draw nutrition from the haemoglobin and RBC ruptures every 48 h the large number of RBCs are destroyed leading to anaemia of the patient. Haemolytic jaundice also results in the malaria disease.

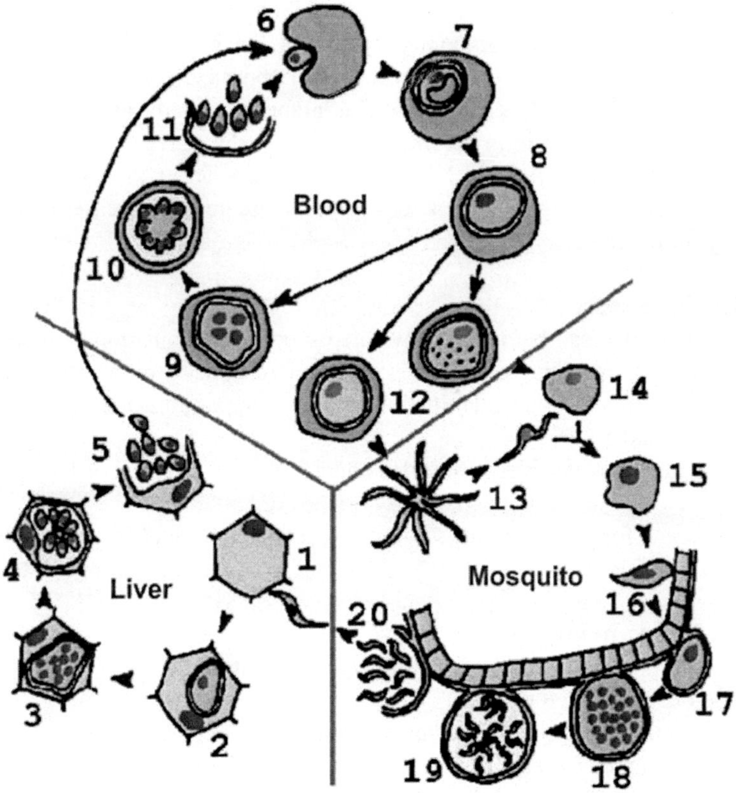

Fig. 2.11 Asexual and sexual cycle of *Plasmodium vivax*. 1. Sporozoite entering hepatocyte, 2. Trophozoite within hepatocyte, 3. Growing schizont in hepatocyte, 4. Merozoites in hepatocytes, 5. Merozoites liberating from hepatocyte, 6. Merozoite entering RBC, 7. Ring stage within RBC, 8. Schizont in RBC, 9. Merozoite in RBC, 10. Rosette formation, 11. Merozoites liberating from RBC, 12. Micro and Macro-gametocyte, 13. Exflagellation, 14. Microgametocyte, 15. Zygote, 16. Ookinete, 17. Oocyst, 18. Sporoblast, 19. Sporozoites within sporoblast, 20. Sporozoites liberating

The malaria patient may feel malaise, loss of appetite, headache and slight fever before the first paroxysm. Then patient suffers from shivering, nausea and vomiting.

Due to destruction of large number of RBCs and liberation of toxin the enlargement of liver and spleen of the patient is observed. The degenerated and ruptured RBCs may clump and create obstruction in the normal flow of blood. Due to excessive haemolysis bilious vomiting may occur. In chronic cases hepatomegaly and spleenomegaly are observed in the patients.

2.6.1.8 Host Defence Mechanism

The immune response by the host is quite different in two stages: Pre-erythrocytic stage within liver and erythrocytic stage within RBCs.

It is believed that CD8+ T cells are the main components that play a part in pre-erythrocytic immunity. On the other hand CD4+ T cells control the immunity in erythrocytic stages. Protection from malaria infection is that the number of parasites are held in check by the immunity system of the host. Therefore host remains asymptomatic. It is found that when a person completely cured then again he or she may be susceptible. It is also found that infants in endemic areas are protected by the antibodies formed within the mother.

In case of severe infection balance between the Th1 (IFN-γ) and Th2 (IL-4) mediated immune response affects seriously. IL-4 secretion increases along with the severity of infection.

2.6.1.9 Diagnosis

Malarial parasites can be detected by microscopical examination of blood film. In most of the cases parasites can be seen under microscope by preparing thick and thin blood film. But before the examination of blood film antimalarial drugs should not be administered.

In this film parasites can be detected and at the same time can be identified which of the species is infected.

Other tests for detection are serological test, blood count, complement fixation test, etc.

2.6.1.10 Prophylaxis

(a) Use of mosquito net at night to prevent mosquito bite.
(b) Use of antimalarial drugs.
(c) Destruction of adult mosquitoes by spraying DDT or gammaxine.
(d) Destruction of larval stages by using larvicides.
(e) Elimination of breeding places of mosquitoes.

2.6.2 *Plasmodium falciparum*

2.6.2.1 History

The name of the disease malaria was given in the year 1753. At that time people believed that the disease is related to the bad air. In the year 1880 Laveran for the first time discovered the parasite in the blood film preparation. In 1886 Golgi demonstrated the erythrocytic schizogony of benign tertian malaria and differentiated between benign and quartan malaria. In 1949 Short and others demonstrated the pre-erythrocytic schizogony of *P. falciparum*. In 1897 Willium H. Welch gave the name *Plasmodium falciparum* which ICZN formally adopted the name in 1954. But it is proved that a German physician Johann Freidrich Meckel was the first to observe *P. falciparum*, at that time he did not know what it is.

2.6.2.2 Geographical Distribution

Malarial parasites are found in all the countries of the world extending from 40° south to 60° north. The endemic areas are the tropical zone.

The *Plasmodium falciparum* is the causative agent of malignant malaria or cerebral malaria causes clinical illness, often very severe, affects 100 million and over a million people die from malaria. Malaria threatens 2200 million persons, about 40% of the world population undermining health and welfare of the families, endangering the survival of children, debilitating the active population and straining both countries and people's scare resources. The vector in our country is *Anopheles stephensi.*

Like *Plasmodium vivax, P. falciparum* has pre-erythrocytic, erythrocytic stage and sexual cycle but exoerythrocytic stage is absent in case of *P. falciparum.*

2.6.2.3 Pre-erythrocytic Stage

Malaria sporozoites are released into the blood with the bite of an infected female Anopheles mosquito and within minutes attach to and invade liver cells by binding to the hepatocyte receptor located on the basolateral surface of liver cells. Within the liver cells malaria parasites multiply rapidly, so as many as 40,000 merozoites are released when the hepatocytes rupture. They attack new and old RBCs. It is found that a type of glycoprotein, namely glycophorin A present on the membrane of the RBCs acts as receptor of *P. falciparum.* Within the RBCs the parasites multiply in a membrane bound digestive vacuole hydrolysing haemoglobin by secreting enzymes that include haem polymerase.

As the parasites mature within RBCs they change morphology from ring to schizont form. The ring stage is small in shape 1.2 μ in diameter. It consists of a fine and thread like uniform cytoplasmic ring with a distinct nucleus often projecting beyond the ring or lying outside the ring. The parasites after attaching itself to the margins or the edge of the host cell the nucleus and a small part of the cytoplasm remaining almost outside. There may be two nuclei in a ring. There may be more than one ring stage in a single RBC. The infected RBCs remain unchanged in shape and size. The colour of the cell is reddish violet in Leishman stain. Schuffner's dots are not seen but instead 6–12 Maurer's dots staining brick red are seen. The ring stage and mature trophozoites disappear within 24 h and are sequestrated in the internal organs like brain, heart, spleen, bone marrow, etc. because the outer surface membrane of RBCs produce proteins and form knob like structure. For this protein the infected RBCs bind together with the post capillary venular epithelium. This binding causes sequestration of the infected RBCs in the venules. But surprisingly the RBCs those are with gametocytes have no knob like structure on their surface and so they do not attach with the venular endothelium and pass freely within the venules. So only early ring stages and gametocytes are seen in the peripheral blood of patients infected with *P. falciparum* 8–32 merozoites usually 16 merozoites are formed within RBCs.

In falciparum infection only the ring stages are seen in the slide preparation because all other stages are sequestrated in the internal organs of the patient. Mature schizont occupied 2/3rd of the RBC. Within the schizont 8–32 (generally 8–18) merozoites are formed. The merozoites come out by bursting the RBCs and attack new RBCs. If the rate of infection is high there may be 500,000 parasites in 1 mL of blood.

Some of the merozoites as usual form gametocytes. The gametocytes are seen in the peripheral blood though they develop within internal organs. They are elongated and crescent shaped. So the species is called *falciparum*. With the growth of the gametocytes, the Red Blood Cells are gradually used up only their membrane remains in the form of a covering enclosing the gametocytes.

The size of the mature gametocytes are about one and half times large then the RBCs harbouring them, so the RBCs are stretched beyond recognition and can only be detected by the concave side of the parasites projecting outward in the form of arched rim.

The first appearance of the gametocytes takes place after 21 days of inoculation of sporozoites by the mosquito. The female gametocyte is generally more slender and elongated than the male counterpart. The cytoplasm of the female gametocyte stains deep blue, small nucleus is placed at the centre. It measures 10–12 μ by 2–3 pigments are distributed around nucleus.

The male gametocyte measuring 8–10 μ by 2–3 μ is sausage shaped, cytoplasm stains faint blue, nucleus is deep pink and pigments are scatteredly distributed.

Sometimes the number of gametocytes may be as high as 50,000–150,000 per μL of blood. The gametocytes remain viable for 30–60 days sometimes 120 days in peripheral blood. The mosquito cycle can start only if the infection rate is moderate, i.e. in the slide preparation there is 1 parasite in every 20 WBC count.

In case of *P. falciparum* relapse does not occur as there is no exoerythrocytic cycle. But recrudescence may occur due to late bursting of merozoites. In the laboratory it is found that during production of parasites para-aminobenzoic acid PABA is necessary. The research workers are now trying to develop medicine which will disturb the PABA formation or restrict use of PABA by the parasites.

2.6.2.4 Sexual Cycle

The sexual cycle of *P. falciparum* is same like that of *P. vivax*.

2.6.2.5 Pathogenicity

As the malarial parasites draw nutrition from the haemoglobin so RBC ruptures every 48 h. The large numbers of RBCs are destroyed leading to anaemia. Haemolytic jaundice also occurs in the *falciparum* malaria.

In the *falciparum* malaria sometimes severe complications may arise. Death occurs due to severe anaemia and blockage of capillaries and small blood vessels in the brain. A breakdown product of haemoglobin called bilirubin cause jaundice in the patient. In the falciparum malaria a common symptom is hypoglycaemia which may leads to coma. In recent years it is discovered that release of TNF and IFN-γ cause serious changes in metabolic process. It is found that the meninges, the covering of the brain, becomes red due to excess load of parasites in the capillary network. The circulation is choked resulting in decrease in the oxygen supply to the brain tissues. The lack of oxygen starts breakdown of glycogen into lactic acid and necrosis develop.

The temperature of the patient increases because of the presence of malarial toxin in large quantity. The WBCs become excited and liberate pyrogen. The pyrogen dissolves in plasma and reacts on the hypothalamus of brain. Constriction of vasomotor starts and pyroxysm of chill and fever is felt by the patients. The

Table 2.2 Comparison of different species of malarial parasites (Erythrocytic forms)

Features	*P. falciparum*	*P. malariae*	*P. ovale*
Ring	More than one in one corpuscle. Presence of two nuclei frequent, 1/6th to 1/5th of the diameter of corpuscle	Same as *P. vivax*	Same as *P. vivax*, Schuffner's dots present even at this stage
Corpuscles infected with schizonts	Not found in peripheral circulation, size normal, presence of Maurer's dots instead of Schuffner's dots	Size and colour normal, absence of Schuffner's dots	Oval in shape, often fabricated, corpuscle not much enlarged, presence of Schuffner's dots
Growing schizont	Usually compact, round, pigment coarse and black, not seen in peripheral circulation	More compact and round, pigment coarse and black	Usually round, pigment brownish, coarse and scattered
Merozoites	Occupy 2/3rd to 3/4th of corpuscle, number of merozoites from 8 to 32, not seen in peripheral circulation	Nearly fill normal size corpuscle, 6–12 merozoites arranged in petals of rose	Occupy 3/4th of oval corpuscle, 8–10 merozoites arranged like a bunch of grapes
Microgametocytes	Microgametocyte small with pale blue cytoplasm, nucleus pink in colour, pigment granules scattered. Crescent shaped	Same as *P. vivax*	Same as *P. vivax*
Macrogametocytes	Macrogametocytes longer and more slender, deep blue cytoplasm, small red nucleus, pigments more concentrated near centre. Crescent shaped	Same as *P. vivax*	Same as *P. vivax*
Interval between cycle	48 h	72 h	48 h

symptoms of cerebral malaria are also hyperpyrexia and coma. Acute *P. falciparum* infection produces high temperature, severe anaemia, cerebral symptoms, renal failure, pulmonary oedema and death (Table 2.2).

2.6.2.6 Resistance and Resurgence

Very recently it is seen that several parasitic diseases which had been considered quite under control or eradicated have been found to infect with stronger pathogenicity. This upsurge in malaria is due to the emergence of new strain of malarial parasites which are now resistant to available drugs as well as insecticide resistant vectors. The crowded condition of human inhabitation coming from resettlement of refuges due to political disturbance, famine, etc. hasten the progress of transmission.

Recrudescence and relapse are for lowering of antibody titre. Large family of genes encode variant antigens in *P. falciparum*. Tolerance of the parasites is due to loss of reactivity to TNF.

2.6.2.7 Host Immune Response

Plasmodium has two cycles in human beings—one pre-erythrocytic phase which starts when sporozoites infect hepatocytes and second cycle is an erythrocytic cycle when merozoites infect RBCs.

Suppression of CD4+ T helper cells function, decrease CD8+ cytotoxic cell differentiation and release of IFN-γ. The above mentioned two processes are important for immunity against cerebral malaria for extra erythrocytic stages.

It is found that the antigens from asexual erythrocytic stages of the parasites stimulate the immune system of the patient to produce specific antibodies which are present in IgG and IgM of the serum gamma globulin. The cell mediated and humoral defence mechanisms of the patient are effective only against the asexual erythrocytic stages of the parasites. They are not effective on the gametocytes. Decrease activity of cytotoxic cell functions prevents degeneration of sporozoites infected hepatocytes. While suppression of IFN-γ prevents stimulation of macrophages resulting release of uncontrolled number of merozoites during extra erythrocytic phase. One of the consequences of these changes is that AIDS patient exhibits increase incidence of cerebral malaria.

2.6.2.8 Energy Metabolism

Though the oxygen is available parasites acquire energy from the degradation of glucose to lactic acid. It is found that *Plasmodium* species of birds have mitochondria with cristae. But asexual stages of mammalian *Plasmodium* parasites have no cristae in their mitochondria but asexual stages within the stomach wall of mosquitoes have mitochondria with prominent cristae.

Erythrocytic stages of the parasites are facultative anaerobes. They consume oxygen when it is available. They use oxygen for synthesis of nucleic acids. But a classical cytochrome system is not found in the malarial parasites. The parasites cannot synthesize coenzyme A and it is obtained from the host. The Plasmodial parasites of mammals fix carbon dioxide into phosphoenol pyruvate. The antimalarial drugs actually inhibit the enzymes, carboxykinase or phosphoenol pyruvate carboxylase.

It is quite interesting that 25 different proteins have been found in different species of *Plasmodium*. These enzymes are very important for maturation and release of merozoites from infected RBCs and haemoglobin digestion within RBCs. It is found that ferriprotoporphyrin (FP) is formed from the digestion of haemoglobin by the parasites this FP is believed to inhibit secretion of plasmodial proteins.

A membraneous structure noticed under EM in Apicomplexa. They are called Apicoplasts which have four membranes. Parasitologists have suggested that apicoplasts have a number of functions one of which is survival and transmission of *P. falciparum*.

2.6.2.9 Diagnosis

Microscopical examination of blood film taken during pyrexial interval of falciparum malaria will show the presence of parasites. In acute falciparum malaria blood picture is normal or shows less number of leucocytes.

Passive haemagglutination test: If the infection is recent it shows positive result. Here infected erythrocytes are agglutinised. This test gives result in *P. falciparum* infection because a number of infected erythrocytes are present in this case.

2.6.3 *Plasmodium malariae*

2.6.3.1 History

Like other *Plasmodium*, *P. malariae* also cause the disease malaria in human beings. The disease caused by this parasite is called Quartan malaria. Golgi in the year 1885 gave an accurate description of *P. malariae*.

2.6.3.2 Geographical Distribution

It is cosmopolitan in distribution. It is mainly found in India, tropical Africa, Sri Lanka, Malaya, Java, New guinea and some parts of Europe.

It is the only human malaria species that is regularly found in wild animals also.

2.6.3.3 Life Cycle

Like all other species of malaria they have also pre-erythrocytic, erythrocytic, exoerythrocytic and sexual cycle.

Pre-erythrocytic Cycle

The pre-erythrocytic schizogony has not always seen in man but Parasitologists believe that this tissue cycle is present in man.

Probably the duration of tissue phase of this parasite is nearly 15 days. Garnham noticed that the tissue phase of this parasite is like *P. inui* (a parasite of monkeys). Bray in 1960 by an experiment showed that pre-erythrocytic schizont of *P. malariae* in the liver of Chimpanzee by inoculating sporozoites taking from salivary glands of infected mosquito.

The pre-erythrocytic schizont in early stage is 5.5 μm in diameter with 5 nuclei while the mature schizont is about 22 μm and merozoites are formed which are over 2000 in number.

Erythrocytic schizogony takes 13–16 days to complete within RBCs of man. The young ring form, i.e. trophozoite has the same appearance like the signet ring stage of *P. vivax*. It is less amoeboid and cytoplasm is somewhat thicker than that of *P. vivax*. The trophozoite of *P. malariae* often stretches across the RBCs and takes a band like appearance. Now they begin to collect pigment when they are 6–8 h old. In the cytoplasm of the parasites dark brown or black colour or course pigment granules appear. The infected RBCs are not enlarged in size.

Schizont is round in shape and measures 6.5–7 μm in diameter. After 45–54 h the nucleus divides resulting in 6–12 merozoites. They arrange themselves around the pigment at centre the parasites look like daisy head. Merozoites measure 2–2.5 μm in diameter.

Gametocytes are round and measure 7–7.5 μm in diameter. The female gametocytes are somewhat large than male gametocytes. Gametocytes mature within 6 days. The gametocytes are not found in peripheral blood they develop within internal organs.

It is found that the gametocytes when become matured come out in the peripheral blood and remain there for a long time.

Exoerythrocytic Schizogony

It is found that after the disappearance of erythrocytic stages of *P. malariae* relapse may occur even up to 32 or 55 years (Garnham 1966).

2.6.4 *Plasmodium ovale*

It is very rare and because of its oval structure they are called *Plasmodium ovale.* The parasites are the causative agent of the ovale tertian malaria.

2.6.4.1 Geographical Distribution

They are distributed in tropical countries but are found also in Europe and the USA. They are mostly found in the west coast of Africa.

2.6.4.2 Life Cycle

Pre-erythrocytic Schizogony

Their morphology is more or less same with *P. vivax* so the diagnosis is very difficult. The time taken to complete pre-erythrocytic cycle is more or less 9 days. Some schizonts have been found on 5th and 9th day in the tissue of liver.

A mature schizont measures about 70–80 μm in length and 40–50 μm in breadth. Merozoites are about 15,000 within a schizont. They are spherical about 1.8 μm in length and with a nucleus on one side.

Erythrocytic Schizogony

The schizogony completes in about 48 h. The ring stage in its initial phase has a round nucleus and the ring stage measures 2–2.5 μm in diameter. The pigments at this stage are coarse and dark brown in colour. In these parasites some granules appear and called James's dots.

Mature schizonts are round or oval and measure 6.2 μm in diameter. Four to 16 merozoites are formed but usually eight nuclei are produced due to asexual division.

The merozoites measure 2–2.5 μm in diameter. The shape of the nucleus is crescent.

2.6.4.3 Gametocyte

Gametocytes develop from the erythrocytic merozoites. The gametocytes of *P. ovale* are oval in shape and in the infected RBCs James's dots are seen. The infected RBCs are irregular in outline. Gametocytes appear in the peripheral blood after a long period. They appear in large number after 3 weeks of infection and at that time mosquitoes get infected from the infected persons.

2.6.4.4 Sexual Cycle

Same as *P. vivax*.

2.6.4.5 Exoerythrocytic Cycle

The exoerthrocytic forms of *P. ovale* have been demonstrated in the liver tissue of Chimpanzee after 18–40 days of sporozoites infection (Bray 1957).

2.7 Piroplasm

2.7.1 *Babesia* sp.

Babesia is a genus of protozoan haemoparasite (blood parasite) residing within the RBCs of human and nonhuman mammals. It is the causative agent of the disease babesiosis. The disease was called earlier as Piroplasmosis. More than 100 species of *Babesia* are known still today but only a few are responsible for the disease in human beings. Human infection is accidental and rare.

The distribution is very much restricted, in the USA, *Babesia microti* and in the Europe *Babesia divergens* are found. The most well-known species is *Babesia bigemina* which can cause mild to severe sometimes fatal disease and have worldwide distribution.

The disease babesiosis is also called Red water fever as the patients excrete red urine due to presence of haemoglobin in the urine known as haemoglobinourea.

At the onset of the disease, the babesiosis in human beings is shown symptoms like malaria. So most of the time the disease is misdiagnosed.

The symptoms of Babesiosis are irregular fever, chill, headache, general lethargy, pain and bodily discomfort. In severe cases cause haemolytic anaemia, jaundice, laboured breathing and haemoglobinourea. Haemoglobinourea occurs due to lytic effect of parasites on RBCs.

Resistant persons with healthy spleen most of the time recover without any treatment. But in splenectomised persons the infection is severe and becomes fatal within 5–10 days of infection.

Splenectomised persons suffer from severe haemolytic anaemia with hepatomegaly and spleenomegaly.

In *B. microti* infection which takes place mainly in the USA, sometimes complications arise like acute respiratory failure, congestive heart failure and renal failure. Among the hospitalized patients with babesiosis 5–10% becomes fatal. There is always a higher risk of death among the elderly persons infected with severe babesiosis. Babesiosis is characterized by destruction of RBCs and excretion of haemoglobin with urine. Babesiosis is also called Red water fever.

In *B. divergens* infection restricted in Europe mortality rate is about 42% and always have severe symptoms. Here also infected persons suffer from haemoglobinourea, jaundice, persistent high fever, chill and sweat. In untreated individuals develop pulmonary oedema and ultimately renal failure occurs.

2.7.1.1 History

The species *Babesia* was first discovered by Romanian Scientist Victor Babes in 1888. He discovered severe haemolytic anaemia in cattle and sheep due to babesiosis. But he identified the causative agent as bacteria and named it *Haematococcus bovis*.

In 1893 Theobald Smith and Fred Kilborne identified the parasite as the cause of Texas Cattle fever which is later known as babesiosis.

Smith and Kilborne are the first scientists to discover that the disease is transmitted by a tick of the *Boophelus* group called *Rhipicephalus*. This discovery for the first time recognized the role of arthropod as vector of the disease in the field of medical science.

In 1957 human infection by *Babesia* was discovered. The very first patient was a splenectomised person. In nonsplenectomised person the infection was first observed in 1969 and it proved that the protozoan parasite *Babesia* is the cause of the babesiosis in human beings.

2.7.1.2 Morphology

Babesia represents itself in three forms: Sporozoites, trophozoites and merozoites. Sporozoites are the infective stage of the parasite. *Babesia* enters into the RBCs at sporozoite stage. Within the RBCs the sporozoites transform into trophozoites. Sporozoite becomes round and forms a ring. The trophozoites undergo asexual division schizogony within the RBC and convert into merozoites. The RBCs when become filled up with merozoites rupture and large number of infected RBCs burst resulting haemolytic anaemia.

In the meantime the parasite assumes tetrad structure. The tetrad morphology of the parasite present in the RBCs makes the diagnosis very easy and differentiate *Babesia* from *Plasmodium*.

Trophozoites appear as ring forms. They are 2–4 μm in diameter and have a small chromatin dot with scanty cytoplasm. Trophozoites start feeding haemoglobulin but no pigment is produced. The tetrad morphology is best seen in Giemsa's stain in thin blood smear. They multiply by budding process and form merozoites. The merozoites coming out from the infected RBCs again attack new and uninfected RBCs to repeat the cycle.

B. bigemina is the causative agent of the disease red water fever or Texas fever in cattle throughout the world.

Infection of *Babesia* in human beings is very rare but infection takes place when the tick bites for blood meal. It is seen that if the biting tick is not removed from the

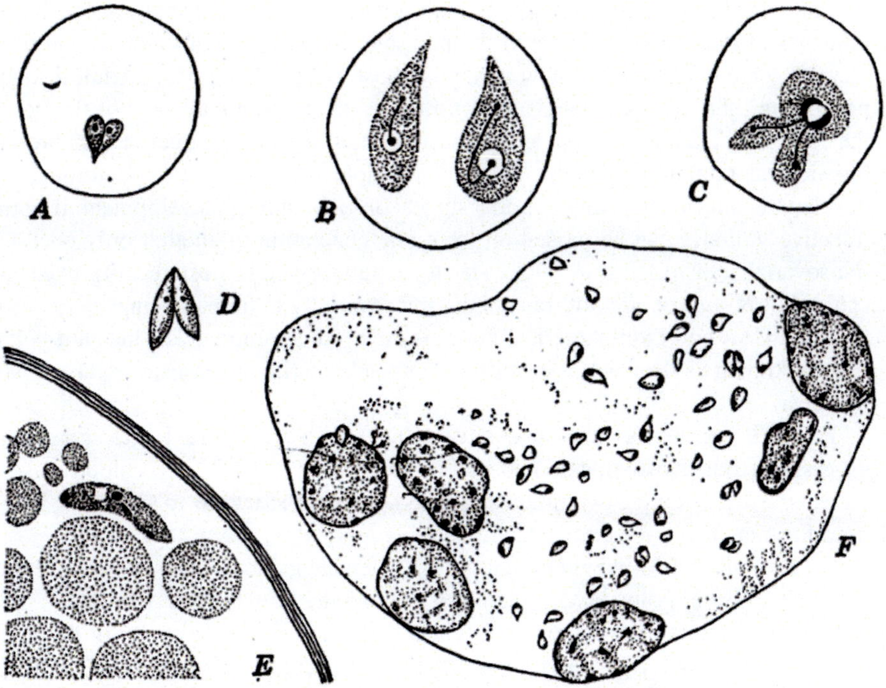

Fig. 2.12 Life history of *Babesia bigemina.* A. Trophozoite in RBC during binary fission, B. Pyriform in RBC, C. Division in RBC, D. Associated isogametes in the gut of tick, E. Ookinete in section of ovum of tick, F. Sporozoites in salivary gland of tick

skin remain there for 3–10 days. The longer period of biting is associated with higher probability of acquiring the parasite from the tick.

2.7.1.3 Life Cycle

Life cycle of *B. microti* requires a biological vector like rodent or tick. The transmission of infection occurs by arthropodan host tick of genus *Boophilus*. The ticks of genus *Boophilus* or Rhipicephilus are parasite on rat or rodents, when they suck blood meal from the rat the infected tick introduces sporozoite into the blood of the rodent. The sporozoites within a short time directly enter into the RBCs and undergo development via trophozoites and merozoites formed within the infected RBCs. Some of the merozoites like *Plasmodium* develop into gametocytes (Fig. 2.12).

The tick, the definitive host, as the sexual reproduction of *Babesia* takes place within ticks, during their blood meal from infected rat or rodent all the stages of *Babesia* like sporozoites, merozoites, gametocytes are sucked into the alimentary canal of the ticks. But except the gametocytes all other stages degenerate. Only the gametocytes develop into gamets and fertilization takes place in the hind gut of the tick and then through gametogony, the sexual reproduction, the parasites invade the reproductive organ of tick. They multiply extensively and migrate to all the tissue of the developing tick embryo through ovary. So the transmission is called transovarial transmission. Some of the parasites enter the salivary glands and from there they are transmitted to new vertebrate host by the young ticks. Adult ticks do not transmit the disease (Dennis 1932). Besides at the nymphal stage of the tick if they bite infected human beings the parasites enter into the g.i. tract of tick through blood meal remain within the tick and multiply within the tick all through its developmental stages. Now all the stages of tick during development become infective. *Babesia* can be passed on from one generation to another only once via transovarial transmission as Ticks strictly maintain single host feeding cycle on cattle. But *B. microti* are not transmitted transovarially in human beings.

In the USA and Latin America Ixodes is the most common tick. This hard tick is usually known as deer tick transmits various tick bearing diseases of wild animals like deer.

In 2003 Centre for disease control (CDC), U.S.A. reported that more than 40 cases of Babesiosis transmitted due to transfusion of packed red blood cells in patients. After that the Government of the USA made a strict law to screen the blood used for transfusion for *Babesia* antibodies.

On the wall of the hind gut of tick fertilized female gametes form oocytes. Within the oocytes sporogony takes place and sporozoites are formed ultimately. The sporozoites by rupturing the oocytes released into the coelomic cavity of the body and migrate to the reproductive organ and to the eggs. In the embryo the parasites increase in almost all tissues including salivary gland.

In some cases the infected seed tick bite human beings to suck blood. In this process the sporozoites of *Babesia* are introduced into the blood of the human beings along with the salivary secretion of the ticks. Human beings act as intermediate host as the asexual reproduction of the parasite takes place in the RBCs of human beings. The adult tick is not responsible for the transmission of the infection.

2.7.1.4 Pathogenicity

The symptoms are irregular fever, chill, sweating, muscle pain, fatigue, mild enlargement of spleen and mild haemolytic anaemia. The illness continues for weeks and months. The disease is severe in adult cattle than in calves. The incubation period of the disease is 8–15 days after infection.

In severe cases 75% of the RBCs may be degenerated, no severe anaemia happens. Due to overloading of breakdown products of haemoglobin jaundice occurs and due to clearance of haemoglobin through kidney the colour of the urine becomes red.

The disease may be controlled through eradication of ticks.

2.7.1.5 Host Defence Mechanism

The infection is sometime fatal for cattle and human beings. The parasites multiply within RBCs causing destruction of RBCs. But the disease is long lasting. The recovered animals become resistant to reinfection for acquired immunity. It is found that immune response of sporozoites is antibody response. There is some proof of cell mediated killing involving macrophages. As the antibodies have been found so immunization against Babesiosis in cattle and dogs is successful.

2.8 Ciliophora

2.8.1 *Balantidium coli*

2.8.1.1 History

In the year 1857 Malmsten was first to discover the organism in two human beings with dysentery. He identified it as a species of *Paramoecium* and called it *Paramoecium coli*. Leuckart described morphologically similar species from the intestine of pig in 1861.

A parasitic Ciliophoran intestinal protozoa is the *Balantidium coli*. It is the largest protozoan ciliate parasitic of human beings. The parasite causes dysentery in man. The disease is called balantidiasis.

It resides in large intestine of human beings, monkeys and pigs.

2.8.1.2 Geographical Distribution

It is cosmopolitan in distribution.

2.8.1.3 Morphology

The parasite has two forms: trophozoite and cyst. The trophozoite of the species is large in size measuring about 28–152, 7.22–12.3 μm in dimension. The shape is almost round or ovoid. At the anterior end there is a depression called peristome. The entire body is covered with a delicate pellicle having longitudinal rows of fine cilia. Cilia are the organs of locomotion.

At the bottom of the peristome there is a cytostome. The cytostome is the cell mouth from here the food is taken within the organism. The food vacuoles circulate within the endoplasm. There are two nuclei present in the cell body: macro and micro nuclei. The macronucleus is large slightly curved and concave at both sides. It is about 2/5th of the length of the body. The micro nucleus is very small and situated at the concavity of the macronucleus. At the anterior end from the peristome large cilia come out they are called adoral cilia which help during ingestion (Fig. 2.13).

The cysts those are commonly found in faeces are ovoid and about 45–55 μm in diameter. The cyst wall is tough and made of one or two layers. The nuclear shape and structure is same with that of trophozoites. The cysts have a very small contractile vacuole and one or two refractile body. No multiplication occurs within the cyst. The cysts are infective forms and are found in the faeces of human beings. Transmission of infection takes place through cysts. Surviving trophozoites and cysts are greenish or yellow in colour seen under microscope.

2.8.1.4 Life Cycle

Balantidium coli resides in the lumen of the large intestine specially caecum and colon of man and other animals transform into cysts and they escape through the

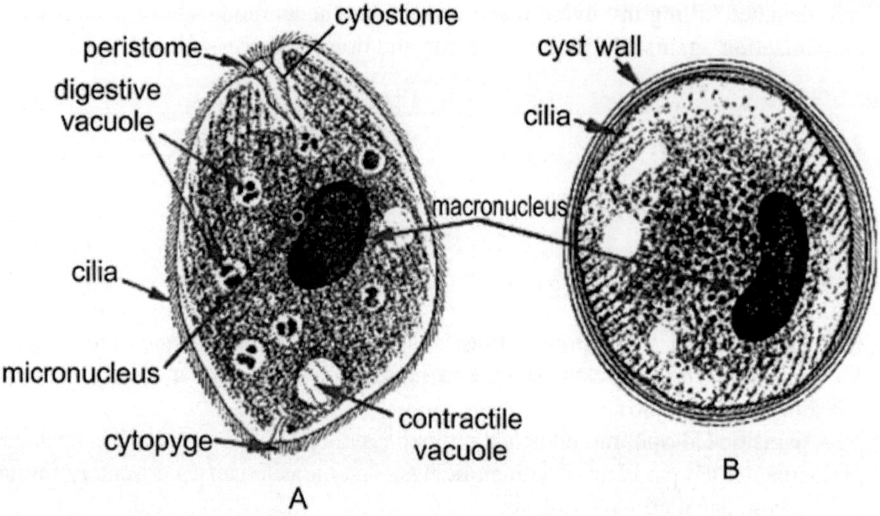

Fig. 2.13 (**a**) Trophozoite of *Balantidium coli*, (**b**) cyst

faeces of pig and are ingested by children. Because of the question of adaptation in new environment of intestinal symbiotic flora in a new host it cannot be transmitted easily. But when they become adapted to the new intestinal environment of human beings it becomes a pathogen which can create havoc. *B. coli* is seen that it cannot produce lesions within the intestine of animals except Primates. But they may infect secondarily in the lesions produced otherwise.

Trophozoites divide by transverse binary fission within the lumen of the intestine. During binary fission the peristome and cytostome remain present at the anterior end and they are formed at the anterior end of the newly formed posterior daughter. Some reported that conjugation is also seen sometimes especially in culture media.

The transmission route is faecal–oral. The *Balantidium* infection in man only occurs when cysts from animal sources are ingested through contaminated food and drink in large numbers.

Infection takes place when a host ingests cysts during the consumption of food and drink contaminated with cysts. Once the cysts are ingested they pass through the G.I. tract of the hosts. The cysts receive some protection from degradation of acidic environment of the stomach. The tough hard cyst wall gives protection from the acid of the stomach. It is proved that the cysts remain healthy at a pH greater than 5 but at less than 5 the cyst wall is damaged and destroyed. Infection occurs in malnourished individuals who have less acidic environment in their stomach.

Once the cysts reach the small intestine excystation occur and trophozoites are produced within the intestine. The trophozoites colonized in the large intestine where they live in the lumen and feed on the intestinal flora.

Some trophozoites undergo encystation after a period of growth and multiplication. Some trophozoites invade the wall of the colon of the host using proteolytic enzymes and reproduce there and some of them return in the lumen again. In the lumen of the intestine of the host the trophozoites undergo encystation. Encystation takes place due to dehydration of the intestinal contents and occurs in the distal part of the large intestine and the process may also occur in the outside environment in faeces.

2.8.1.5 Pathogenicity

Balatidium coli lives in the caecum and colon of the hosts like human beings, pigs, rats and other mammals. The trophozoites devour like other ciliates through cytostome. It is found that sometimes the organism, i.e. trophozoites of *B. coli* produce proteolytic enzymes which act on cracked or damaged intestinal epithelium of the host. At the same time hyaluronidase is also secreted and this enzyme helps in the enlargement of the ulcer in the intestinal wall. The ulcers are like the amoebic ulcers usually flask shaped with a narrow neck proceeding into a sac like depression in the submucosa layer of the intestine. Here the parasites divide and an ulceration is formed. This may lead to necrosis and perforation of the large intestine may occur. In extreme condition lung or liver may be infected. Besides urinogenital organ, vaginal, uterine and bladder infections may be seen. The parasites do not transmit readily from one species to another because of their food and adaptation in the new hosts. It requires a period of time to adjust to the symbiotic flora of the intestine of new host. Once it is adapted the parasite may become a serious pathogen to human beings.

Balantidiasis infection may take place in immunocompetent persons but the infection never becomes serious.

Usually in the patients with symptoms diarrhoea occurs. The abscess in the liver or in other organs is very very rare. Infection not likely occurs in the malnourished individuals due to low acidity in the stomach or in the immunocompromised individuals.

2.8.1.6 Diagnosis

Saline preparation of the diarrhoeal faeces shows trophozoites in the microscope, if it is present. Cysts can be found also in saline preparation of formed stool.

2.8.1.7 Epidemiology

Balantidium in human beings is found in the people of Philippines though it is cosmopolitan in distribution. The parasites are found among people who are associated with swine. The infection in human beings is rare and only 1%. But it is found that infection rate is very high in pigs and it is seen that in slaughter house in Japan the infection rate was 100%. In middle east countries it is seen that the prevalence of Balantidiasis is connected with the increased population of wild boar.

The ability of the trophozoites to encyst when they come out increases the rate of infection from reservoir host. These cysts remain viable for a long period in the faeces of pig. It is proved that the pigs are common source of infection in human beings. It is seen that infection disappears in healthy person. But they may be carriers of infection.

Some are of opinion that cysts found in pigs cannot infect human beings. The parasites found in pigs are *Balantidium suis*, completely a separate species.

2.9 *Toxoplasma*

2.9.1 *Toxoplasma gondii*

Toxoplasma gondii is a parasitic protozoa, a causative agent of the disease Toxoplasmosis of man and animals.

2.9.1.1 History

In the year 1908 Charles Nicolle and Louis Manceaux discovered a protozoan parasite in a rodent called Gundii, *Ctenodactylus gundi* at the Pasteur Institute in Tunis. They named the parasite *Toxoplasma gundi*.

In the same year Alfanso Splender found the same organism in a rabbit in Brazil. It is like *Plasmodium* an obligate intracellular parasite. The disease is extremely common and asymptomatic but sometimes becomes very dangerous.

But it was for the first time found in human beings in an infant girl delivered full term by caesarean section on May 23, 1938, at Babies' Hospital in New York City.

It was believed by the scientists that transmission of *T. gondii* may be due to consumption of undercooked meat (Weinman and Chandler 1954). To prove this way of transmission in 1960 the cyst wall of oocysts was shown to dissolve in the proteolytic enzymes present in the stomach and as a result infectious bradyzoites are released into the stomach and then pass into the intestine. This type of transmission in human beings was tested in an orphanage in Paris in 1965.

In Bombay in the year 1959 *Toxoplasma gondii* was found in strict vegetarians also. This led to believe that there must be another major route of infection besides meat consumption and in 1970 oocysts were discovered in cat faeces. The faecal–oral route of infection was demonstrated with the help of oocysts experimentally in laboratory.

For 10 long years from 1970 to 1980 quite a number of experiments were done to test the ability of the animals to shed oocysts of *T. gondii*. Seventeen different species of Felids (cat family) have been found to shed oocysts but never the non-felids as the sexual reproduction of *T. gondii* and subsequent formation of oocysts never occur in non-felids.

The disease is asymptomatic although mild flu like symptoms may occur during the first few weeks of infection.

But in immunocompromised persons, AIDS patients and infants the disease may be serious and occasionally fatal.

2.9.1.2 Geographical Distribution

The parasite is cosmopolitan in distribution and can infect all warm blooded animals including human beings. It is believed to be one of the most common parasite and is estimated that one-third of the world population harbour this parasite within them.

2.9.1.3 Morphology

The parasites are crescent shaped or oval measuring 6–12 μm in length and have a distinct central nucleus. They are presented in pairs or groups of pairs.

Electron microscopical picture shows the presence of an organelle at the anterior end which may be a cytostome or a holdfast and fibrils extending two-thirds of the body length which is responsible for gliding movement of the parasite.

The parasites are found swimming in the bloodstream, within tissues and within cells of R.E. system, W.B.C. and epithelial cells.

Toxoplasma has a number of phases: Oocysts or tissue cysts, Tachyzoites, Merozoites, Bradyzoites and Sporozoites.

Oocysts

Oocysts are product of sexual reproduction and are formed only in the intestine of cats, the definitive host. They are spherical in shape measuring 10 × 15 μm × 18–12 μm in size.

Tissue Cysts

Same as oocysts but are product of asexual reproduction in tissues. They are 12–100 μm in size and contain 50 to several thousand bradyzoites.

Sporozoites

Sporozoites are formed within mature oocysts. Mature oocysts contain two sporocysts each with four sporozoites. These are very small in size.

Tachyzoites

They are motile, multiply rapidly and are responsible for increasing population of the parasites within the host. Tachyzoites are crescent shaped, measuring 3–7 μm × 2–4-μm in size and have one end more round than the other.

Merozoites

They also divide quickly and increase population of the parasite within the cat's intestine before sexual reproduction. They are also very small.

Bradyzoites

These are slowly dividing phase of the parasites. When an infected host consumes oocyst, bradyzoites come out from the cyst, infect epithelial cells of intestine of the host and slowly convert into the rapidly dividing stage tachyzoites. After sometimes tachyzoites convert back to bradyzoites. They reproduce inside host cells and form oocysts or tissue cysts in the new host, cysts usually range between 5 and 50 μm in diameters.

2.9.1.4 Life Cycle

Life cycle of *Toxoplasma* is very much complex and interesting.

The cat is the definitive host where sexual cycle is completed. All other warm blooded animals (non-feline) are intermediate host.

There is only one species of *Toxoplasma*, *T. gondii* known still today. The *T. gondii* has no choice about its host, it can infect all kinds of warm blooded animals like birds and mammals. Experimentally it is shown that it can survive even in lizards and turtles. As indicated by antibody reaction, the parasite is known to occur in 59% of dogs, 34% of cats, 48% of goats, 30% of pigs, 20% of rats, 10% in pigeons. But in human beings in the USA it is 17–35% and 0–68% in Eskimos at Tahiti (Chandler and Read 1961).

The human infection is found to be highest in warm, moist areas, but less in warm, dry areas and very few in very cold areas of the globe.

About transmission it is observed that the closest friends of human beings are dogs and cats and they are the reservoirs of the disease. The rats are also very important animals acting as reservoirs of the disease.

The transmission of the disease from person to person is very rare. Laarman in the year 1956 reported mechanical transmission by Stomoxys (Stable fly) and fleas after feeding them with *Toxoplasma* infected animals. These flies and fleas have been reported to harbour *Toxoplasma* for 60 days after experimental feeding of an infected animal (Fig. 2.14).

Life cycle of *Toxoplasma* has been divided into three components for the convenience of study:

1. In definitive host—Feline cycle.
2. In definitive host and intermediate host—feline and non-feline cycle.
3. In two intermediate hosts—non-feline to non-feline cycle.

Definitive Host Cycle

Toxoplasma multiplies through sexual reproduction only in cats or animals of feline groups.

The cycle starts by consuming rat heavily infected with *Toxoplasma*. The product of sexual reproduction is oocyst which is formed in the intestine of cats.

Then the oocysts escape from the body of the definitive host through faeces of the host. The oocysts reach the external environment on soil. The oocysts then mature within 2 days. During maturation the oocysts contain two sporocysts and each sporocyst contains four sporozoites.

Some sporozoites in case of cat only enter into the epithelial cells of gut and produce enteroepithelial cycle. Other sporozoites enter into the mucus membrane and start development process in the lamina propria, lymph nodes, WBC and other organs.

Within the cats the result of the cycle in gut epithelium are the tachyzoites. The dividing stages in acute infection, tachyzoites 8–32 in number, accrued within parasitophorus vacuole in the host cell and ultimately ruptures escaping the parasites which will again infect other cells.

After some such cycle of asexual multiplication gametogony starts: macro and microgamets are produced. They are formed within 3–15 days of infection. Gametocytes are formed all along the small intestine, especially in the ileum. Very small percentage like 2–4% gametocytes are male gametocytes. The microgametocytes develop 12 microgamets from each microgametocyte. The microgamets fertilize the macrogamets and form zygotes. Zygote converts into an

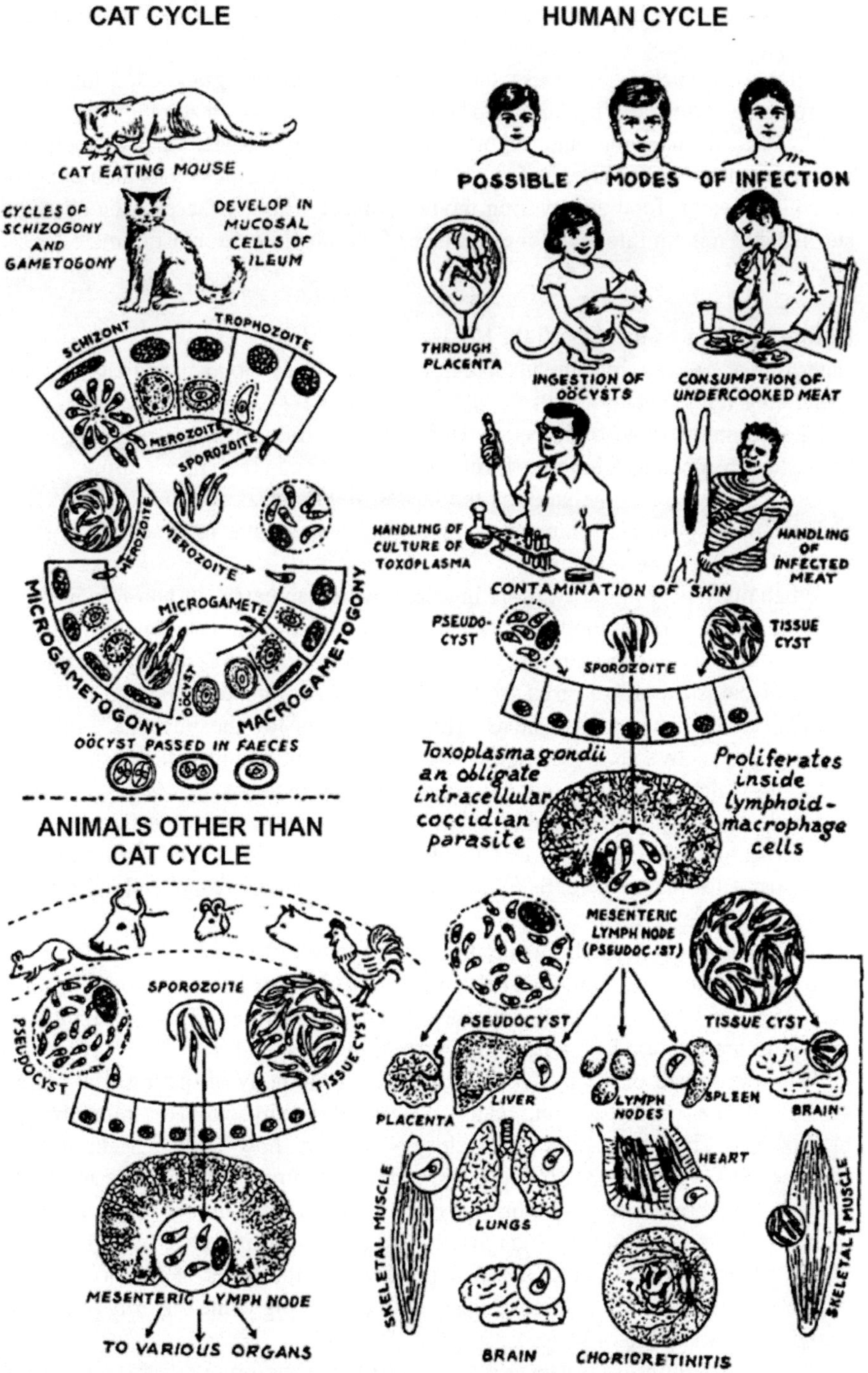

Fig. 2.14 Life cycle of *Toxoplasma gondii* (in different animals)

oocyst. Oocysts then come out in the faeces of cat after 3–5 days of infection by cysts.

After some such cycle of asexual multiplication gametogony starts: macro and microgametocytes are produced. They form micro and macrogamets and are fertilized to form zygote and zygote converts into an oocyst. Oocysts then come out in the faeces of cat after 3–5 days of infection by cysts. Oocysts then deposited on soil, in water, food as anything having contact with the faeces. The cysts may survive and remain infective for a number of months in cold moist climate.

Definitive Host to Intermediate Hosts

Ingestion of oocysts by human beings or other warm blooded animals starts the feline–non-feline cycle.

The common route of infection is faecal–oral route. Human beings consume unwashed vegetables in the form of salad or unknowingly drink oocysts contaminated water or by handling the faeces of a domesticated or pet infected cat.

Though cats are also infected by ingesting oocysts the intermediate hosts are infected by ingesting oocysts.

When oocysts or tissue cysts are ingested by human beings or non-feline animals through contaminated food and drink the cyst wall dissolves by the action of proteolytic enzymes present in the stomach and small intestine. The infective forms of parasites are released from the oocysts and invade epithelial cell of the intestine of the intermediate hosts. The parasites inside the intestinal epithelium convert into tachyzoites, the motile and rapidly multiplying cellular stage of the parasites, which infect the other cells and organs of the host.

Inside the intestinal epithelial cells the tachyzoites multiply inside a vacuole formed within the cell known as parasitophorous vacuole created during the entry of the parasites inside the cells. Now the parasites remain as intracellular but extra cytoplasmic one.

The tachyzoites multiply inside the parasitophorous vacuole of the infected cell until the cell dies and ruptures. Then the tachyzoites are released which find their way into the bloodstream and ultimately reach organs and tissues of the body including brain of the host where they form tissue cysts.

With the spread of the tachyzoites throughout the body of the host the immune system of the host react and causes *T. gondii* tachyzoites to transform into bradyzoites. The bradyzoites form clusters and are now known as tissue cysts because the cyst wall is produced from the membrane of the parasitophorous vacuole. The bradyzoites containing tissue cysts can form in any organ and found mainly in brain, eyes and striated muscles including heart muscles. It is seen that the choice of organ by the bradyzoites depends upon the hosts like majority of the parasites form tissue cysts in muscle tissues in case of pig while majority of cysts are found in brain in case of mice.

The cysts within the tissues are formed to avoid the immunity reactions of the hosts.

The transmission in case of human beings takes place through ingestion of undercooked meat, unwashed vegetables and contaminated drink. Transplacental transmission also occurs from mother to foetus.

Intermediate (Non-feline) to Intermediate (Non-feline) Cycle
Bradyzoites in the tissues cysts of non-feline animals, i.e. intermediate hosts undergo only asexual multiplication. Bradyzoites infect the intestinal cells of the intermediate host and starts endodyogeny, i.e. two daughter cells develop inside a parent cell. The parent cell then rupture and two daughter cells are released. The individual parasite then again undergoes the same process forming four parasites. These tachyzoites infect new cells and can invade all cells except RBCs.

T. gondii theoretically may be passed to intermediate hosts through consumption of tissue cysts in meat. But the life cycle of parasite begins and completes only when the parasite enters into a feline host (from rat to cat) where the parasite undergoes sexual reproduction.

2.9.1.5 Transmission

1. Consumption of raw or undercooked meat.
2. Drinking of unpasteurized goat milk.
3. Contact with contaminated soil.
4. Consumption of unwashed and unboiled raw vegetables or fruits.
5. Handling of infected cat and dogs.
6. Blood transfusion or organ transplant.
7. Transplacental transmission when mother is infected during pregnancy.

2.9.1.6 Precaution

As suggested by United States Centre for disease control and prevention.

From food:

1. Peeling and washing of consumable fruits and vegetables thoroughly.
2. Freezing meat for several days at subzero temperature before consumption.
3. Avoid eating undercooked meat.
4. Washing of hands with hot soapy water for the dog or cat handlers.
5. To drink only pasteurized milk.

From environment:

1. Changing and disposing of cat and dog litter daily.
2. Wearing of gloves during gardening.
3. Drinking of only filtered water.

2.9.1.7 Pathogenicity

In immunocompetent patients most of the time the Toxoplasmosis is overlooked as it is asymptomatic.

Toxoplasmosis if symptomatic is mild and remains only for a few days.

In acute Toxoplasmosis the symptoms are cervical lymphadenopathy, headache, illness and fever. Then the symptoms subside and the patients become carrier for the rest of their life as the tissue cysts remain in their organs.

Immunocompromised patients may be patients with AIDS or the patients on which immunosuppressive drugs are regimented develop severe illness. The common symptoms in AIDS patient if infected with *T. gondii* is encephalitis. This condition arises when the immunity developing WBC, CD4 cell count decrease below 100 cells/cubic millimetre. If the patients in this condition remain untreated the disease may be fatal.

It is found that transplacental transmission in pregnant mother does not occur if the infection took place more than 6 months before conception. The risk of infection of the foetus by transplacental transmission during pregnancy depends upon the trimester. The risk of infection by *T. gondii* is lowest in the first trimester and highest during third trimester.

The typical symptoms noticeable in an infected offspring are hydrocephaly, microcephaly, intra-cerebral calcifications, convulsions, chorioretinitis, optic nerve atrophy, hepato-spleenomegaly and jaundice.

In acute infection, the infection takes place in intestine at the onset.

In subacute infection, the tachyzoites start destroying cells of the liver, lung, heart, brain and eyes. It is found that CNS is damaged more than other organs.

Chronic infection takes place when the patients develop immunity. In this condition tachyzoites formation is suppressed. Now bradyzoites are formed from tachyzoites. The bradyzoites are also called zoitocysts. These bradyzoites remain intact for years but the disease is now asymptomatic. It is seen if the infection reach retinal cells of the eyes then it cause blindness. In chronic infection also myocarditis, heart damage, pneumonia may develop. Spontaneous abortions like serious phenomenon may occur.

2.9.1.8 Diagnosis

Clinical diagnosis can be done from the symptoms like cerebral calcification, chorioretinitis, hydrocephalus, symptoms like pneumonia, myocarditis, spleenomegaly, hepatomegaly, etc.

Direct methods may be applied like presence of parasites in the blood smears, bone marrow puncture, centrifuged deposit of CSF by staining with Giemsa and examined under microscope.

By serological tests like Sabin–Feldman dye test, indirect immunofluorescence, indirect haemagglutination, complement fixation test and ELISA.

2.9.1.9 Epidemiology

The toxoplasmosis, a disease is distributed worldwide. The disease is associated with cats mainly. Though birds and other mammals may be infected. The complete natural cycle remains within cats and mice. The mice devour materials contaminated with faeces of cats where there may be oocysts of *Toxoplasma*. If the cats consume such infected mice, they are infected.

Human toxoplasmosis is a zoonotic disease. Human infection takes place through food and drink contaminated by mature oocysts of *Toxoplasma gondii* and/or undercooked meat having tissue cysts. Over and above cockroaches and flies act as mechanical vector.

Infection may be transmitted through blood or WBC transfusion. The incubation period of Toxoplasmosis is ordinarily 1–3 weeks.

The result of infection depends upon the immunity status of the persons concerned. The important of all is the incident of congenital toxoplasmosis. Out of all it is such a disease that the stage of a parasite known as tachyzoites have been found in the nasal, vaginal and eye secretions. Not only that breast milk, saliva, urine, seminal fluid may also contain the infective forms of the parasite.

2.9.1.10 Host Defence Mechanism

Toxoplasmosis is an opportunistic disease. The disease manifests in two states: an acute stage when early spread of tachyzoites take place by the lymphocytes to the tissues another chronic stage when cysts are formed.

During the acute stage IFN-γ is produced by the NK cells which are induced by IL-12 coming from macrophage along with TNF-α.

IFN-γ and TNF-α activate anti-Toxoplasma action in macrophages. During chronic stage Th2 type response occurs. Here CD8+ cells and IFN-γ play major roles.

2.10 Systematic of the Studied Protozoan Parasites

2.10.1 *Entamoeba histolytica*

Kingdom	Protista
Subkingdom	Protozoa
Phylum	Sarcomastigophora
Subphylum	Sarcodina
Class	Lobosea
Order	Amoebida
Family	Entamoebidae
Genus	*Entamoeba*
Species	*histolytica*

2.10.1.1 General Characters of Kingdom Protista

1. They are mostly unicellular organisms.
2. These eukaryotes have a nucleus and organelles.
3. They are mostly heterotrophic organisms.
4. Symbiosis or parasitism is observed in the member of Protista.
5. Locomotory organs are cilia or flagella or pseudopodia.

2.10.1.2 Subkingdom Protozoa

1. All are microscopical.
2. Body composed of single cell.
3. Single cell performs all the biological functions.
4. Body is naked or covered by pellicle.
5. Single cell usually contains single nucleus.
6. Presence of contractile vacuole in free living forms.
7. Free living, symbiotic or parasitic.
8. Reproduction asexual or sexual.

2.10.1.3 Phylum Sarcomastigophora

1. Locomotion with the help of pseudopodia or flagella.
2. Reproduction by binary fusion or syngamy.

2.10.1.4 Subphylum Sarcodina

1. Body is naked.
2. Mostly pseudopodia are present.
3. Presence of a single nucleus.
4. Most of the members are free living, some are parasitic.

2.10.1.5 Class Lobosea

1. Pseudopodia are lobose type.
2. Presence of single or more nucleus.

2.10.1.6 Order Amoebida

1. Presence of a cell membrane which encloses the cytoplasm and cell organelles.
2. As there is no cell wall, so cellular structure is not definite.
3. Presence of a single nucleus.

2.10.1.7 Family Entamoebidae

1. Cells are small.
2. Presence of a simple life cycle.
3. Trophozoites feeds on bacteria.

2.10.1.8 Genus *Entamoeba*

1. It is found as internal parasite of G.I. tract of human beings or may be commensal.

2.10.1.9 Species *histolytica*

1. An anaerobic parasitic protozoa.
2. Infects digestive tract of human beings and other primates.
3. Trophozoetes ingest RBC.
4. Trophozoetes transform into cysts.
5. Occasionally penetrate the intestinal wall by a type of secretion and enters into blood vessels and form abscess in different organs.

2.10.2 *Giardia intestinalis*

Kingdom	Protista
Subkingdom	Protozoa
Phylum	Sarcomastigophora
Subphylum	Mastigophora
Class	Zoomastigophora
Order	Diplomonidida
Family	Hexamitidae
Genus	*Giardia*
Species	*intestinalis*

2.10.2.1 Kingdom Protista
Same as Entamoeba.

2.10.2.2 Subkingdom Protozoa
Same as Entamoeba.

2.10.2.3 Phylum Sarcomastigophora
Same as Entamoeba.

2.10.2.4 Subphylum Mastogophora
1. Presence of single or more than one flagella.
2. Asexual reproduction by binary fission.

2.10.2.5 Class Zoomastigophora
1. Absence of chloroplast.
2. Sexual reproduction may be seen in some cases.

2.10.2.6 Order Diplomonidida
1. Presence of two nuclei in the body.
2. Presence of four flagella.
3. They are zooflagellates.
4. Live as parasites or commensals in the digestive system of host.

2.10.2.7 Family Hexamitidae
1. Intestinal parasites of invertebrate and vertebrate hosts.
2. Bilaterally symmetrical.
3. Oval in shape with two nuclei.

2.10.2.8 Genus *Giardia*
1. Anaerobic flagellated protozoan parasite.
2. Colonize and reproduce in the small intestine of the hosts.

2.10.2.9 Species *intestinalis*

1. A flagellated parasitic microorganism.
2. Colonize and reproduce in small intestine of man and other vertebrates.

2.10.3 *Trichomonas vaginalis*

Kingdom	Protista
Subkingdom	Protozoa
Phylum	Sarcomastigophora
Subphylum	Mastigophora
Class	Zoomastigophora
Order	Trichomonadida
Family	Trichomonadidae
Genus	*Trichomonas*
Species	*vaginalis*

Kingdom, Subkingdom, Phylum, Subphylum and Class same as *Giardia*.

2.10.3.1 Order Trichomonadida

1. They are anaerobic protista.
2. They are zooflagellate.
3. Presence of three to six flagella.
4. One trails or borders an undulating membrane.
5. Resides within digestive system of vertebrates.
6. They may be uninucleate or multinucleate.

2.10.3.2 Genus *Trichomonas*

1. They are anaerobic excavate parasites of vertebrates.
2. Presence of four flagella.
3. One recurrent, along the outer margin of the undulating membrane.
4. Costa originate in the kinetosomal complex at the anterior of the parasite.
5. Presence of axostyle.

2.10.3.3 Species *vaginalis*

1. Causes sexually transmitted infection.
2. It is an anaerobic flagellated protozoan parasite.
3. Usually resides in vagina and urethra of female human beings.

2.10.4 *Trichomonas hominis*

Up to Genus same as *T. vaginalis*.

2.10.4.1 Species *hominis*

1. Flagellate parasitic protozoa.
2. They are nonpathogenic.
3. Sometimes causes diarrhoeic stools.
4. Resides in large intestine of the host.

2.10.5 *Nagleria fowleri*

Up to Subphylum same as *Entamoeba histolytica.*

Class	Heterolobosea
Order	Schizopyrenida
Family	Vahlkamppfidae
Genus	*Nagleria*
Species	*fowleri*

2.10.5.1 Class Heterolobosea

1. Free living heterotrophs.
2. Many are amoeboflagellates.
3. Some of these can reversibly transform into flagellates.

2.10.5.2 Order Schizopyrenida

1. They are small amoebas.
2. Many of them can transform into flagellates.

2.10.5.3 Family Vahlkamppfidae

1. They are uninucleate.
2. Can transform from amoeba to flagellate.
3. Forms polar masses in mitosis.

2.10.5.4 Genus *Nagleria*

1. Presence of DNA in the nucleus.
2. Presence of membrane bound organelles.

2.10.5.5 Species *fowleri*

1. Presence of two flagella in their flagellate stage.
2. May be parasite and causes primary amoebic meningitis.

2.10.6 *Trypanosoma cruzi*

Kingdom	Protista
Subkingdom	Protozoa
Phylum	Sarcomastigophora
Subphylum	Mastigophora
Class	Zoomastigophora
Order	Kinetoplastida
Family	Trypanosomatidae
Genus	*Trypanosoma*
Species	*cruzi*

Up to Class same as *T. vaginalis.*

2.10.6.1 Order Kinetoplastida

1. Parasitic life style.
2. Cell body is elongated.
3. Locomotion by flagellum.
4. Presence of kinetoplast containing DNA.
5. Kinetoplast is a special compartment of mitochondria.

2.10.6.2 Family Trypanosomatidae

1. A type of protozoan parasite.
2. They are dixenous.
3. Infection through insect vector.

2.10.6.3 Genus *Trypanosoma*

1. A monophylectic group of unicellular organisms.
2. Parasitic flagellate protozoa.
3. Locomotion like corkscrew.

2.10.6.4 Species *cruzi*

1. Parasitic to human beings,
2. Metacyclic trypomastigotes and amastigotes stages found in the life cycle of human hosts.

2.10.7 *Trypanosoma gambiense*

Kingdom	Protista
Subkingdom	Protozoa
Phylum	Sarcomastigophora
Subphylum	Mastigophora
Class	Zoomastigophora

(continued)

Order	Kinetoplastida
Family	Trypanosomatidae
Genus	*Trypanosoma*
Species	*gambiense, rhodesiense, brucei*

Up to Genus same as *T. cruzi.*

2.10.8 *Leishmania donovani*

Kingdom	Protista
Subkingdom	Protozoa
Phylum	Sarcomastigophora
Subphylum	Mastigophora
Class	Zoomastigophora
Order	Kinetoplastida
Family	Trypanosomatidae
Genus	*Leishmania*
Species	*donovani*

Up to Family same as *T. cruzi.*

2.10.8.1 Genus *Leishmania*

1. Presence a well-defined nucleus and other cell organelles.
2. Presence of kinetoplasts and flagella.
3. Presence of amastigote form in the mononuclear phagocytes and circulatory systems of human beings.

2.10.8.2 Species *donovani*

1. Causes the disease leishmaniasis.
2. Responsible for visceral leishmaniasis or kala-azar.
3. Infection is transmitted by sandfly.

2.10.9 *Plasmodium vivax*

Kingdom	Protista
Subkingdom	Protozoa
Phylum	Apicomplexa
Class	Sporozoa
Order	Haemosporida
Family	Haemosporidae
Genus	*Plasmodium*
Species	*vivax, falciparum, malariae, ovale*

Up to Subkingdom same as *Entamoeba.*

2.10.9.1 Phylum Apicomplexa

1. Presence of apical complex, only seen under E.M.
2. Presence of one or more polar ring.
3. Locomotory organ is microfibril.
4. Flagella may be present in gonad.
5. All members are intracellular parasite.
6. Sporozoites are formed within spore.

2.10.9.2 Class sporozoa

1. Colloid forms complete ring.
2. Sporozoites are formed from oocytes.
3. Sexual and asexual reproduction both are seen.

2.10.9.3 Order Haemosporida

1. They are blood parasites.
2. The zygote is capable of movement called ookinite.
3. Sporozoites are free not enclosed by sporocysts.

2.10.9.4 Family Haemosporidia

1. Parasite produces pigment.
2. Asexual cycle takes place in peripheral blood.

2.10.9.5 Genus *Plasmodium*

1. Parasites are responsible for the disease malaria.

There are four species vivax, falciparum, malariae, ovale.

2.10.10 *Cryptosporidium parvum*

Kingdom	Protista
Subkingdom	Protozoa
Phylum	Apicomplexa
Subphylum	Sporozoa
Class	Sporozoasida
Subclass	Coccidiasina
Order	Eucoccidiida
Suborder	Eumerii
Family	Cryptosporidiidae
Genus	*Cryptosporidium*
Species	*parvum*

Up to Phylum same as *Plasmodium*.

2.10.10.1 Subphylum Sporozoa

1. Absence of any special organ for locomotion.
2. Reproduce asexually by fission followed by sexual union or syngamy.
3. Presence of alternation of generations.

2.10.10.2 Class Sporozoasida

Reproduce by sexual and asexual process oocysts are produced.

2.10.10.3 Subclass Coccidiasina

Life cycle usually involves mesogony, gametogony or sporogony. Gamonts are small.

2.10.10.4 Order Eucoccidiida

Mesogony or schizogony present.

2.10.10.5 Suborder Eimeriine

Micro and macro gamonts produce independently. Zygote is non-motile.

2.10.10.6 Family Cryptosporidiidae

1. Oocysts contain four naked sporozoites.
2. Absence of sporocysts.
3. Endogenous stages with attachment organelle.
4. Life cycle monoxenous.

2.10.10.7 Genus *Cryptosporidium*

1. Important gastrointestinal agents.
2. Presence of two types of oocysts, a thin walled and a thick walled.
3. Both contain four sporozoites.
4. Thin walled oocyst reinfects the same host.
5. Thick walled oocysts are exit forms.

2.10.10.8 Species *parvum*

1. Life cycle in a single host.
2. Sporozoites 4 in number are liberated in the intestine of host.
3. Infection remains restricted in the brush border cells of intestine.
4. Oocysts are resistant to most disinfectants.

2.10.11 *Babesia bigemina*

Kingdom	Protista
Subkingdom	Protozoa
Phyllum	Apicomplexa
Class	Sporozo

(continued)

Order	Haemosporida
Family	Piroplasmida
Genus	*Babesia*
Species	*bigemina*

Up to Order same as *Plasmodium*.

2.10.11.1 Family Piroplasmida

1. Piroplasm parasitize RBCs of the host.
2. Presence of apical complex with polar ring.
3. They lack flagella.
4. Do not form oocysts or spores.
5. Sporogony occurs within invertebrate vector.

2.10.11.2 Genus *Babesia*

1. Pear shaped appearance in blood cells.
2. Sometimes may be spherical or amoeboid in shape.
3. Ticks are vector.

2.10.11.3 Species *bigemina*

1. Presence of intraerythrocytic merogony.
2. Sporogony forms sporonts which produce sporoblasts.

2.10.12 *Balantidium coli*

Kingdom	Protista
Subkingdom	Protozoa
Phylum	Ciliophora
Subphylum	Mastigophora
Class	Zoomastigophora
Order	Spirotrichida
Family	Heterotrichina
Genus	*Balantidium*
Species	*coli*

Up to Class Zoomastigophora is same as *T. vaginalis*.

2.10.12.1 Order Spirotrichida

1. They are a group of ciliate protozoa.
2. They have prominent oral cilia.
3. The body of cilia is fused to form cirri in some.

2.10.12.2 Family Heterotrichina

1. Presence of a distinct mouth, cytostome and a cytopyge (anal pore).
2. The cilia cover the entire body arranged in longitudinal or obliquely spiral parallel row.

2.10.12.3 Genus *Balantidium*

1. Presence of two nuclei macro and micro.
2. Macronucleus large, slightly curved and vegetative in function.
3. Micro-nucleus, vesicular type, acts as generative nucleus.

2.10.12.4 Species *coli*

1. Parasite is able to form cyst.
2. All are parasitic to vertebrate and invertebrate hosts.
3. Only species of medical importance.

2.10.13 *Toxoplasma gondii*

Up to Order same as Cryptosporidium.

2.10.13.1 Family Sarcocystidae

1. Parasites of the carnivorous hosts.
2. They are characterized by an obligatory two hosts life cycle.
3. Infection takes place through predator relationship.
4. Cysts are not found in raccoon dogs.

2.10.13.2 Genus *Toxoplasma*

1. Occurs in three morphological form—Trophozoite.
2. Produces cysts in the muscles of intermediate hosts.
3. The cysts contain numerous bradyzoites.

2.10.13.3 Species *gondii*

1. Tissue cysts occur in muscles and other tissue.
2. Parasites multiply slowly within the host cell and develop a cyst wall.

References

Bray RS (1957) Studies on the exo-crythocytic cycle in the genus Plasmodium. Mem Lon Sch Hyg Trop Med 12:192

Chandler AC, Read CP (1961) Introduction to parasitology. Wiley and Toppan Co Ltd, Tokyo

Das N, Jha D (1967) On a new flagellate, *Giardia panthera* n sp, from the gut of a lion (*Panthera leo*). Ind J Anim Health VI(II):305–307

Das N, Ray HN (1967) PA Schiff positive ectocystic envelop in *Entamoeba coli* type cyst from Mandrill (*Cynocephalus mormon*). J Protozool USA 14(Suppl):29 Art 131

Das N, Ray HN (1970) Cytomorphological studies of some parasitic protozoa. Thesis submitted for Ph. D. degree of N.Das at University of Calcutta
Dennis EW (1932) The life cycle of *Babesia bigemina* of Texas cattle fever in the tick *Margaropus annulatus*. Univ Calif Publ Zool 36:263–298
Garnham PCC (1966) Malarial parasites and other Haemosporidia. Blackwell Publication, Oxford
Hoare CA (1959) Amoebic infection in animals. Vet Rev Annot 5:91
Howard L (1960) Studies on the mechanism of infection of mosquitoes by malarial parasites. DPH thesis, John Hopkins University
Huff CG (1946) The relation and acquired immunity of various avian hosts to the cryptozpites and metacryptozoites of *Plasmodium gallineaceum* and *Plasmodium relictum*. J Inf Dis 78:99–117
Jha D, Das N, Ray HN (1968) Giardial infection in Indian goats. Ind J Anim Health VII(I):39–42
Karyakarta R, Damle A (2003) Medical parasitoloogy. Book Allied, Kolkata
Levine ND (1961) Protozoan parasites of domestic animals and man. Burges Publishing, Minneapolis, p 129
Levine ND (1980) Journal of protozoology. Society of Protozologists, Athens
Markel EK, Voge H (1958) Diagnostic medical parasitology. WB Saunders, Philadelphia
Noble ER, Noble GA (1961) Parasitology: the biology of animal parasites. Lea & Febiger, Philadelphia
Schimidt GD, Roberts LS (2013) Foundations of parasitology. McGraw-Hill International, New York
Simitch T et al (1959) Contribution à la connaissance de la fauna des protozooire intertinaux de pore de yogoslavia. Ard Inst Pasteur, D'Algeric 37:401
Weinman D, Chandler AH (1954) Toxoplasmosis in swine and rodents: reciprocal oral infection and potential human hazard. Proc Soc Exp Biol 87:211–216
Wenrich DH (1937) Studies on Iodoamoeba butschlii (Protozoa) with special reference to nuclear structure. Proc Am Philosop Soc 77:183

3 Introduction to Parasitic Helminthes and Cestoda

Subkingdom Metazoa belongs to the Kingdom Animalia. The members of the metazoa are multicellular animals and consist of tissues, organs and systems. Subkingdom metazoa includes two phyla of medical importance and contains multicellular parasitic organisms Phylum Nemathelminthes and Phylum Platyhelminthes.

All the members of Phylum Nemathelminthes are also termed nematodes. All nematodes are roundworms. They are cylindrical and have tapering ends.

Nematodes are unsegmented and possess a tough cuticle over the body. Digestive system is complete. It consists of mouth, oesophagus, intestine and anus. They exhibit sexual dimorphism. The size of the male is smaller than that of females. The posterior end of the male is curved ventrally. Males have testes, vas deferens, seminal vesicle, ejaculatory duct which open to cloaca. There are accessory sex organs like spicules, copulatory bursa, caudal alae or genital papillae.

Females have one or two tubular ovaries, uterus or uteri united to open to the exterior through genital pore or vulva. They are either viviparous, oviparous or ovoviviparous. During the development moulting takes place in larvae. Most of the parasitic nematodes live in the intestine of the host.

Some of the parasitic Nematodes are:

Ascaris lumbricoides	*Dracunculus medinensis*
Enterobius vermicularis	*Wuchereria bancrofti*
Ancylostoma duodenale	*Onchocerca volvulus*
Trichuris trichiura	*Loa loa*
Strongyloides stercoralis	*Trichinella spiralis*

P. K. Bandyopadhyay et al., *Biochemical, Immunological and Epidemiological Analysis of Parasitic Diseases*, https://doi.org/10.1007/978-981-16-4384-2_3

3.1 Platyhelminthes

All the members of Phylum Platyhelminthes are called flatworms. The parasitic Platyhelminthes of human beings belong to two classes: Cestoda and Trematoda. The parasitic organisms under Class Cestoda which parasitize human beings are known as tapeworms.

3.1.1 Cestoda

The tapeworms or cestoda are dorsoventrally flattened worms and have a segmented body. As the body looks like measuring tape they are called tapeworms. Adult tapeworms live in the intestine of human beings and other animals. Surprisingly the length of the different tapeworms of different species varies from a few millimetres to several metres.

All tapeworms have three regions: head, neck and body. Head is known as scolex. In the scolex, suckers and in some hooks are present on the scolex. They serve as organs of attachment. Behind the head or scolex there is neck, this is the region of growth. The portion after neck is called strobila. Strobila is made up of proglottids, the segments.

During lifetime of the parasite the proglottids are formed continuously. The newly formed proglottids are called immature. Behind them are mature proglottids which contain developed reproductive organs. These proglottids contain both testes and ovaries so they are hermaphrodites. Self-fertilization or cross-fertilization leads to production of eggs. Next to mature proglottids are gravid proglottids which are loaded with eggs.

Tapeworms do not have body cavity. Tapeworms have no mouth or digestive system, so food is absorbed through body surface. They possess a very simple excretory system.

The reproductive system is very well developed. Each mature proglottid possesses several testes, a bilobed ovary and a uterus.

Some of the helminthic parasites under Class Cestoda are:

Taenia solium	*Hymenolepis nana*
Echinococcus granulosus	*Diphyllobothrium latum*

3.1.2 Trematoda

The parasitic organisms belong to the Phylum Platyhelminthes and Class Trematoda are called flukes and they parasitize human beings.

The characteristic features of the parasites are: body covered with cuticle. Flukes are flattened dorsoventrally and somewhat oval in shape. They are very thick also. Flukes have a powerful sucker around the mouth and most of them possess a ventral sucker situated midventrally called acetabulum. They are termed distome, monostome and amphistome on the basis of their presence of suckers. When they possess a single oral sucker, they are known as monostome. They are called amphistome, when they have an oral sucker and an acetabulum at the posterior end of the body. The flukes are known as distome when the ventral sucker is placed elsewhere on the ventral surface. Most of the members of trematoda are parasitic organisms. These parasites complete their life cycle in two or three hosts and that is the reason they are also known as digenea. The intestine of the parasite is divided into two branches. The reproduction is sexual and parasites are unisexual or bisexual. Most of the organism under trematoda have two testes and single ovary. All of them are endoparasites of human beings. Many of them are hermaphrodite but some are capable of self-fertilization. But to produce viable offspring they practice cross-fertilization.

Digenea is a subclass of Class Trematoda, Phylum Platyhelminthes. The parasite of subclass Digenea includes orders like Paramphistomiformes, Echinostomatiformes, Hemiuriformes, Strizeiformes, Opisthorchiformes and Plagiorchiformes.

Digeneans are parasitic flatworms having a syncytial integument with two suckers: one ventral and one dorsal. Adults usually reside within the alimentary canal and occur throughout the organ, systems of all classes of vertebrates. There are nearly 6000 species known still today.

Out of these only about 12 species are infective to human beings. Near about 200 million people are affected by these parasites. They are grouped into two infective parasites: schistosomes and non-schistosomes. The schistosomes are *S. mansoni*, *S. haematobium*, *S. japonicum*, *S. intercalatum*.

The invertebrate intermediate host of *S. mansoni* is *Biomphalaria* sp., *S. haematobium* is *Bulinus* sp., *S. japonicum* is *Onchomelania* sp. and *S. intercalatum* is *Bulinus* sp.

Seven species of non-schistosomes that infect human beings and reside in the G.I. tract of human beings are:

Name of the species of the parasites	Name of the intermediate host
Fasciolopsis buski	*Segmentina* sp.
Heterophes heterophyses	*Prinella* sp.
Melagonimus yokogawaii	*Semisulcospira* sp.
Gastrodiscoides hominis	*Helicorbis* sp.
Clonorchis sinensis	*Bulinus* sp.
Fasciola hepatica	*Lymnaea truncatula*
Paragonimus westermani	*Oncomelanii* sp.

3.2 Cestoda

3.2.1 *Taenia solium*

3.2.1.1 History

It has a long history of discovery. In 5000 to 15,000 BC Indian Ayurveda the name of tapeworms was mentioned. From 500 to 200 BC Theophrastus named the Taenia. AD 20 to 200 Roman physicians detected tapeworms. In 1300 AD Annaldo de Villanueva coined the term solium. In 1850 to 1900 life cycle of *T. solium* and development of cysticercosis were described.

3.2.1.2 Habitat

Taenia solium, a type of parasitic tapeworm resides in the intestine of mammals including man. This type of tapeworms remain anchored to the intestinal wall of its host by its small pin-like head called scolex and the rest of the body is segmented and sustains through absorption of the digested food of hosts.

3.2.1.3 Geographical Distribution

They are cosmopolitan in distribution. The infection is common among pork eating people. But as Muslims and Jews do not eat pork for religious reason the infection is very uncommon among them.

The distribution of *T. solium* is cosmopolitan because pig is domesticated and consumed almost all over the world but in Jewish and Muslim countries it is rare as the eating of pork is religiously forbidden.

3.2.1.4 Morphology

The pork tapeworm as it is called measures 6–10 ft. in length. The colour of the fresh living specimen is opaque white. The anterior end of the parasite is called scolex. The scolex is very very small and about 2 mm in diameter. The scolex has a rostellum having 22–32 nonretractable curved chitinous hooks arranged in two rows: one row of comparatively long hooks measuring 0.14–0.18 mm and one row of small hooks measuring 0.11–0.14 mm. Hooks of the inner row alternate with those of the outer. Each hook possesses a base by which it is fixed, a blunt projection called handle directed towards the apex and a conical blade. The scolex also bears four hemispherical, highly muscular suckers placed at four sides of the head as the head is quadrangular.

The scolex with the help of hooks and suckers remain fixed in the outer mucosal layer of the intestine. This highly fixative organ prevents the parasite to dislodge from the intestinal wall during peristalsis.

The next portion is called neck. It is short, narrow and unsegmented. Behind the neck there are segmentation and 800–1000 segments called proglottids arranged in a linear fashion. This portion of the body is also called strobila. The younger segments or proglottids are broader than long, in the middle part the proglottids are square and the ripe proglottids are about twice the length of breadth measuring 12 mm long. The

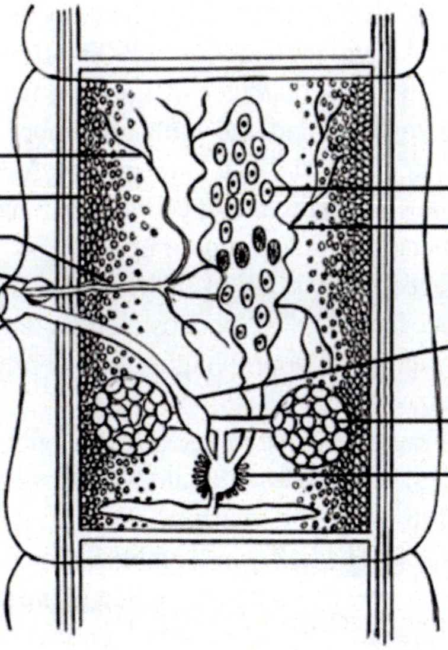

Fig. 3.1 Gravid proglottid of *T. solium*

sexually mature proglottids nearer the neck are young and far from the neck are ripe or gravid (Fig. 3.1).

Both the reproductive systems, male and female, remain in the same proglottid. It is quite interesting that the anterior proglottids 100–150 possess only male reproductive system then after that towards the posterior region the proglottids contain both male and female reproductive systems. After a prolonged period of reproductive activity, the mature proglottids have only the fertilized egg loaded uterus practically spreading throughout the space of the proglottid and all other organs degenerate. This type of proglottid is called ripe or gravid proglottid.

Male reproductive system: There are a number of testes which are small round bodies spread throughout the space of the proglottid and located close to the dorsal surface. There is a controversy about the number of testes. Some workers believe that testis is single but divided into so many parts. From each testis arises vas efferens which is a very slender, thin ductule. The all vasa efferentia unite to form vas deferens. This is a thick much convoluted tube. The terminal end of which forms a copulatory organ the cirrus enclosed in a sheath. Cirrus opens through genital pore.

Female reproductive system: A single bilobed ovary is located in the posterior part of the proglottid. The ovary is dorsoventrally flattened and consists of a number of radially arranged germinal follicles. Oviduct arises from the middle part. The oviduct runs backward and joins with another tube called vagina. There is a swollen structure called ootype which is surrounded by Mehlis's gland. The vagina is a bent tube one end of which joins oviduct to form ootype and the opposite end opens into genital pore. The female genital pore is located behind the male genital pore.

The uterus is highly branched 7–13 in number. Inside of the uterus remains filled up with thousands of fertilized eggs. This is the characteristic of a gravid proglottid.

The gravid segments are separated from the body and are expelled in chain of 5–6 segments. Each gravid segment contains 30,000–50,000 eggs. The eggs are spherical and measure 31–43 μm in diameter. A thick, brown wall surrounds the embryo. This is called embryophore and the embryo proper is called onchosphere. The onchosphere has six hooklets and that is the reason for calling it hexacanth embryo.

Usually a single tapeworm is developed within human intestine but more than one may form in some person which is quite rare.

The eggs hatch outside the body of the intestine in the degenerating proglottids present in the human faeces. Now onchospheres come out within the faeces. The intermediate host, pig, as a coprophagus animal ingests the onchospheres and acquires the infection.

Sometimes man himself may be infected by the onchosphere with the ingestion of improperly cooked vegetables contaminated with onchospheres.

In the stomach of the pig, the intermediate host, the embryophore is lost due to the action of acid pepsin. The hexacanth embryo then passes into the small intestine. The hexacanth embryo now becomes active due to the presence of bile salts and bore the intestinal wall to reach the blood vessels. The secretion of penetration glands and the six hooks are used to bore the intestinal wall. Then the hooks are lost. The entire process takes nearly about 10 min. The larva is being carried to the liver via hepatic portal vein. From here they go to the heart and finally enter into arterial circulation.

From the arterial circulation they enter into the tissues of the intermediate hosts. Here in the voluntary muscles like tongue, shoulder, neck, thigh, etc. they transform into bladder worm or cysticercus. Cysticercus may also develop in brain, heart, lungs, kidney, etc.

The hexacanth embryo loses its hook, absorbs nourishment from the host's tissue and grows in size of about 4.5 mm in length and 6 mm in width. a central cavity appears. The cavity grows in size and filled up with a fluid. The fluid filled vesicle is known as bladder. The scolex is formed which is invaginated having ring of hooks and suckers. There it stays for 8–10 months called bladder worms. Finally the cysticercus dies and becomes calcified.

3.2.1.5 Life Cycle

Life cycle of *T. solium* is completed in two hosts: man, the definitive host, and pig, the intermediate host.

When a human being gets an infection with pork tapeworm expels ripe proglottids of *T. solium* in his stool. The proglottids are discharged in a chain of 5–6 segments with full of eggs. The embryophore are almost round and 35–42 μm in diameter. These are ingested by pigs as they are coprophagus in habit. The eggs survive in the moist faeces for a considerable time. Young pigs are highly susceptible. Not only pigs other animals like camels, dogs, monkeys and man also act as intermediate host.

After ingestion of embryophore oncospheres are liberated in the intestine of intermediate host. From the oncosphere hexacanth embryo comes out. Hexacanth embryo with its six hooks and secretion of penetration gland bore through the wall of the intestine of the intermediate host and comes to swim in the bloodstream when

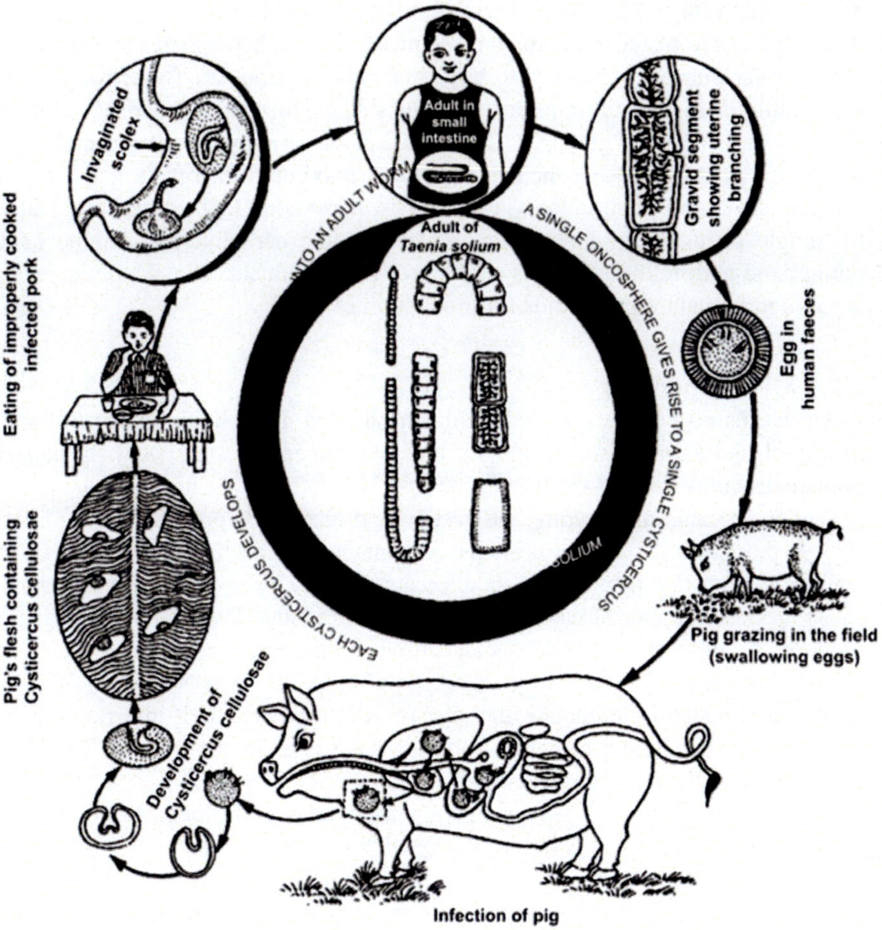

Fig. 3.2 Life cycle of *Taenia solium*

they lose their hooks. The embryos then find passage in hepatic portal vein. From the hepatic portal vein they enter into the liver. With the help of the bloodstream they ultimately go to the arterial circulation, reach heart and from heart to the different muscles and organs. Within the muscles and organs they transform into bladder worm or cysticercus. These are small, oval, white bodies because of calcification and 6–18 mm long with a scolex. The meat or flesh with the cysticercus is called measly pork (Fig. 3.2).

When this infected meat is ingested by human beings in undercooked condition all but the scolex is digested. Now they invaginate, i.e. they turn inside out and with the help of hooks and suckers attach with the wall of the intestine of final host. Within a short time they grow to maturity in 2–3 months.

3.2.1.6 Pathways Associated with Energy Production

Adult tapeworms reside in the small intestine of the host, here oxygen tension is very low so organisms are compelled to have anaerobic metabolism. By active transport and simple diffusion, the nutrient molecules like glucose polymerise within the parasites and stored as glycogen. The other major carbohydrate is galactase from where the parasites derive their energy. The maximum energy is derived from substrate level phosphorylation via glycolysis. Most adult tapeworms absorb lipids by simple diffusion. Other nutrient molecules like carbohydrates, amino acids, purines and pyrimidines are transported through integument. The maximum energy is required in mature proglottids for production of eggs.

3.2.1.7 Pathogenicity

Adult worms do not always produce much harm to the definitive host. But some degree of abdominal discomfort with malnourishment of the hosts may happen. Sometimes congestion in small intestine may occur. The symptoms include pain in the abdomen, nausea, vomiting, indigestion and increased appetite. The hooks and suckers present in the scolex of the parasite may cause mechanical irritation on the wall of the intestine resulting in small ulcers. The toxins produced by the parasite cause increase number of eosinophil count in the blood. The continuous irritation produced by the hooks and suckers sometimes produces antiperistalsis which leads to autoinfection.

The autoinfection produces the disease cysticercosis. Self-infection due to contaminated hands and sometimes the liberated eggs in the intestine hatch and enter back into the stomach by antiperistalsis.

The effects of cysticercosis depend upon the location of the lodging of cysticercus in the body.

Cysticercus may be produced in the muscles or in sub-cutaneous tissues. If the number of cysticercus is few in the muscles, then nothing serious happens. But if they are produced in the eye, heart, spinal cord, brain or other important organs serious condition of the host appears. This type of cysticercosis in human beings needs surgical removal. But if the cysticercus are formed in brain, then serious complications arise. Symptoms of brain cysticercosis include epileptic convulsions, giddiness, violent headache, vomiting and local paralysis.

There are three types of cysticerci. The first common one is the ‘cellulose cysticercus’. In this type a fluid filled bladder measuring about 0.5–1.5 cm in diameter with an invaginated scolex is found.

The second one is called ‘intermediate form’ where there is a scolex and another type called ‘recemose’ without scolex are larger and dangerous. They may measure 20 cm and may contain 60 ml fluid.

3.2.1.8 Host Immune Response

The infection with the cysticercus of *T. solium* causes the disease neurocystisticercosis occurring from the migration of the oncosphere larva to the brain. Within the brain tissue it develops into a cysticercus or metacestode.

As the parasite dies it is calcified there. When the larva remains alive, it has the ability to suppress local immune responses. It is believed that the parasite covers its outer surface with proteins derived from the host. Some are of opinion that the parasite shows molecular mimicry by synthesizing proteins which resemble those of the host. When the parasite starts to die, localized immunoresponse exhibits. At first, Th1 response occurs with the production of IFN-γ, IL-4 and IL-18 cytokines, these result in chronic mixed Th1 and Th2 responses with mature granuloma formation. The granuloma along with the fibrosis produce both good and bad results for the patients. On the one hand, it protects adjacent neural tissue from injury, on the other hand, it results in permanent damage of the nervous tissue which surrounds the cysticercus.

3.2.1.9 Diagnosis

T. solium is detected by the presence of proglottids in faeces with naked eye or presence of eggs in faeces by making saline preparation and examined under microscope.

Faeces was examined on three consecutive days for eggs. In case of *T. solium* perianal swab (cellophane swab) method is used to detect eggs that are deposited in perianal region.

To become more sure, faces are examined under microscope after administration of antihelminthic drugs in which unarmed scolex may be seen.

3.2.1.10 Epidemiology

Cysticercosis can be prevented by detecting early the presence of cysticercus and removal of adult worm. One should avoid consumption of food and drink contaminated with faeces of infected persons. Some suggest ELISA based tests in faeces for *T. solium* antigens are more reliable than faecal examination. In USA it is a major health problem. The disease cysticercosis is a dangerous parasitic disease in Indonesia, Papua New Guinea, etc.

3.2.2 *Echinococcus granulosus*

3.2.2.1 History

Rudolf Virchow in 1855 first identified the parasitic cestode *Echinococcus granulosus*. The first two cases were observed in Southern Germany. The parasite was successively recognized in Switzerland then Russia, Austria and France. The disease was also disclosed in Turkey in 1939. In 1991 the situation was totally changed when Chinese endemic areas were discovered.

3.2.2.2 Habitat

Echinococcus granulosus is commonly called Hyper Tapeworm or Hydatid worm included in the class Cestoda under Phylum Platyhelminthes that parasitizes canines and feline animals usually dog and resides in the small intestine of human beings accidentally.

Besides dogs the definitive hosts are wolves, coyotes, etc. The intermediate hosts are herbivorous such as sheep, deer, moose, kangaroos and wallabies. Human beings sometimes fall victim as intermediate host of this tapeworm accidentally but life cycle of the parasite ends here. It is widespread in rural areas having lots of grazing areas. The human infection is very much frequent in areas where large number of sheep and cattle are raised for commercial purpose.

The cestode parasite produced hydatid cysts in man and domestic animals. There is another species *E. multicularis* which forms malignant alveolar form of hydatid cysts.

The infection usually passed back and forth between dogs and sheep or cattle but unfortunately to human beings also.

3.2.2.3 Geographical Distribution

The disease is prevalent in North and South Africa, the middle East, Australia, New Zealand, Southern South America and Iceland. The disease is so epidemic that 1/4th of dogs and ½ of the domestic animals are infected in these areas. The eggs of this particular parasite are very much resistant to environmental stresses and are present wherever their hosts are available. They can survive freezing and damp conditions also. But vulnerable to heat and desiccation.

3.2.2.4 Morphology

The adult worm measures 2–7 mm and has three to four proglottids. The scolex possesses four suckers, a protrusible rostellum with two rows of 28–50 hooks, usually 30–36. Next to scolex is the narrow neck followed by proglottids, the first is immature, the second is mature and the third is gravid. The ripe or gravid segment contains 500–800 eggs. The eggs are ovoid and measure 32–36 by 25–32 μm. The eggs contain a hexacanth embryo surrounded by a thick radially striated embryophore. The eggs usually infect sheep, cattle and man.

Though the adult is small in size yet they require 4–6 weeks to mature in a dog.

Development of Hydatid Cysts

The eggs enter into the intermediate hosts with contaminated vegetables or water. Human infection results from too intimate association with dogs especially children as the eggs remain attached with the fur of the dog. Unclean habit of dogs is an efficient means of transfer of tapeworm eggs to the man.

The larval forms are called the hydatid cysts. Hydatid cysts develop in the tissues of intermediate hosts. The common organ of development of the cysts is the liver. In man 60–75% cysts are formed in the liver whereas 20% only in the lungs and lesser number in kidneys, spleen, muscles, bone, heart, brain and other organs. The growth rate of cysts is 1–5 cm in a year.

The development of cysts is very slow. The young larva changes into a hollow bladder like structure filled with a fluid. The cysts have double covering, outer hyaline, non-nucleated, the ectocyst white in colour and 1 mm thick. The inner covering, endocyst nucleated and 10–20 μm thick. The endocyst, i.e. the inner covering produces ectocyst, hydatid fluid and brood capsules. Brood capsules remain attached by slender stalks and fall free into the fluid filled cavity of the mother cyst. With the growth of the hydatid more and more brood capsules are formed. Several protoscoleces develop within the brood capsules. Each of them may produce an adult worm. In the fluid of the cysts granular deposit of liberated brood capsules and free scolex is called hydatid sand. A cyst of about 2.25 l capacity with more than two million scoleces may be formed.

The hydatid fluid is slightly acidic pH 6.7, colourless and contains sodium and calcium salts having specific gravity of 1.005–1.010. It is antigenic and toxic.

The size of the cysts may be up to a size of orange and after 10–20 years they may contain 15–22 l of fluid. Sometimes by pressure they may form a daughter cyst and detach.

The hydatid cysts may rupture and the toxic hydatid fluid leads to anaphylactic shock resulting in death.

It is interesting to note that a fully developed protoscolex reverts to a hydatid cyst instead of developing into an adult.

3.2.2.5 Life Cycle

This tapeworm performs both sexual and asexual reproduction. Asexual reproduction takes place by the process of budding in the intermediate hosts and sexual reproduction takes place by fusion of gametes in the definitive host. *E. granulosus* is a hermaphrodite organism having both male and female reproductive organs in the same organism rather same proglottid.

The fertilized eggs come out from the detached gravid proglottid along with the faeces of the definitive host, like dog. These eggs enter into the alimentary canal of intermediate hosts through grazing where the grasses are contaminated with infected dog's faeces. Human beings are infected from mainly dogs either directly from unclean fur or from contaminated vegetables or vector.

In the small intestine of the intermediate hosts including man the fertilized eggs hatch to form oncosphere. The hexacanth embryo present within the oncosphere burrows the wall of the small intestine of their host by using their hooks. Here they come into contact with the bloodstream and travel to many organs where they form hydatid cysts. Daughter cysts are formed from the mother cyst by the process of asexual budding. In the organs they grow in size and number and are transformed into the definitive hosts after the canines ingest the infected tissue of the intermediate host (Fig. 3.3).

After ingestion the organism travels to the small intestine where it attaches itself and develops into an adult form. As it is a hermaphrodite complete set of both male and female reproductive organs remain in the proglottid. Now the ripe or gravid proglottid starts producing eggs. The proglottid having fertilized eggs detaches from the body and ruptures. The eggs are released into the small intestine of definitive host and the eggs escape through the faeces of definitive hosts only to be taken by the intermediate hosts. But in case of human beings the cycle comes to a dead end.

It is really astonishing that such a small tapeworm may sometimes become responsible for the disease and death of many animals even human beings.

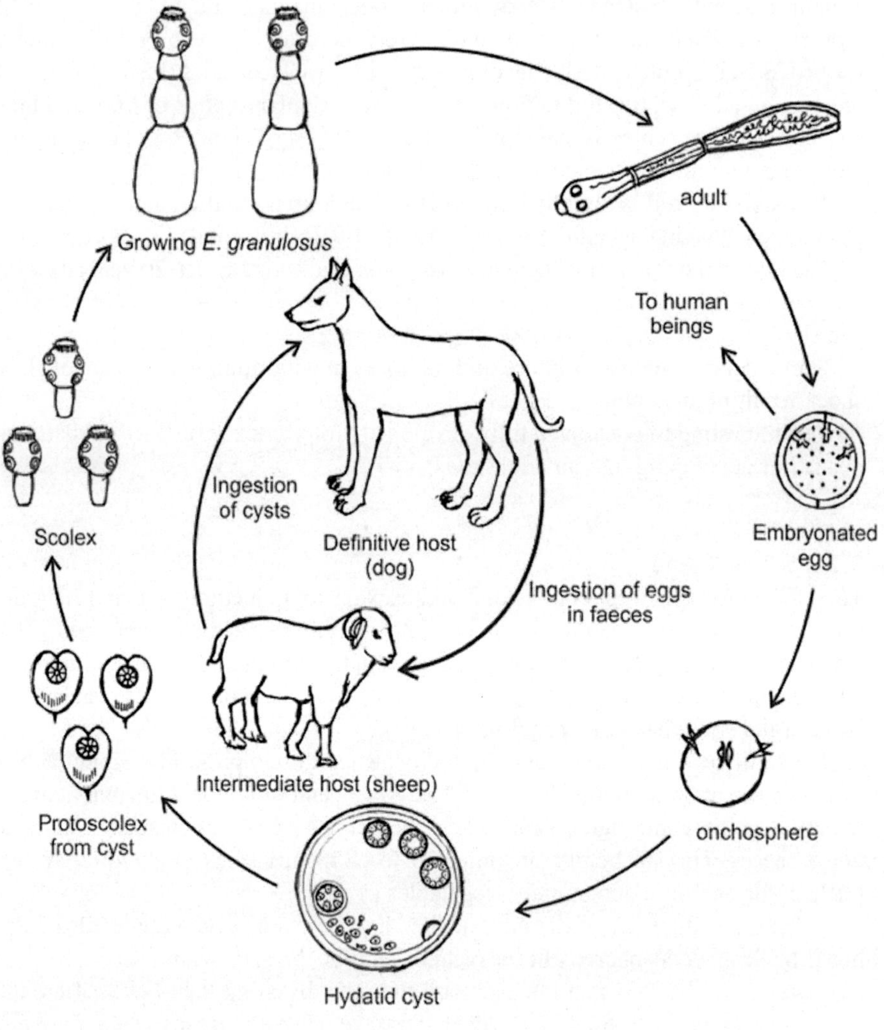

Fig. 3.3 Life cycle of *Echinococcus granulosus* (*Advanced Parasitology*, Das)

The parasite takes its nutrition from the digested food of the host passing through the small intestine of the host. They absorb the food from the immediate environment or the fluid in which they are submerged. All nutrients are absorbed by the surface. The very small body ensures the nutrients reach every cell by diffusion. The maturity of the worm takes 56 days and the lifespan is 5–20 months.

3.2.2.6 Pathogenicity

E. granulosus causes formation of hydatid cysts in the organs of definitive host. The common site is liver and lungs. If they are formed in liver abdominal pain and palpable swelling are the manifestations. Compression of bile duct and as a result obstructive jaundice may also take place.

Hydatid cysts in lungs cause cough, chest pain and haemoptysis. In bones cause erosion, in heart conduction defect and pericarditis may occur. Rupture or leak of hydatid cyst may cause fever, pruritis, urticaria, eosinophilia or fatal anaphylaxis. The location of hydatid cysts in the body causes different types of pathology. The cyst in the nervous system causes very early manifestations of the symptoms of disease. In some organs and sites the hydatid cysts grow and become enormous which may contain near about four gallons of fluid along with millions of protoscolexes. When these cysts rupture it causes sudden death of the host. When the infection is very old the host becomes sensitized to the parasitic antigens. The brood capsules and protoscolexes constitute hydatid sands.

3.2.2.7 Diagnosis

Sometimes accidentally the infection is discovered on routine X-ray or ultrasound studies. In symptomatic cases plain X-ray, ultrasound, CT-scan for hydatid cysts are important. Only when cysts are present in lungs plain X-ray can reveal the infection. But until the cysts are calcified they are not detected by X-ray. CT scan and USG are important for the detection of cysts in organs.

Serological test is done by indirect haemagglutination test, fluorescent antibody test, immunoelectrophoresis test and ELISA.

Antibodies against antigen of *E. granulosus* are detected and are the most specific test for hydatid cysts.

Casoni's test is widely used for diagnosis of hydatid cysts. The Casoni's antigen is Seitz filtered hydatid fluid. In Casoni's test, 0.2 ml of antigen is injected intradermally on forearm. A wheel with multiple projections develops within half an hour and disappears in an hour. This is a positive test for hydatid cyst disease.

The brood capsules can be surgically removed and surgery is the main line of treatment in hydatid cyst disease.

Certain tests are very useful to differentiate hydatid cyst disease with malignant tumour in liver.

Serum—glycoprotein
Serum—aminotransferase
SGOT and SGPT
ESR

In hydatid cyst disease the above test result will be more or less normal. But in cancer of liver all the test value will be abnormal. ESR value is moderately raised in hydatid cyst disease but in cancer of liver it is considerably raised.

3.2.2.8 Prophylaxis

It consists of (a) prevention of infection in dogs in the endemic zones, (b) deworming of dogs by administrating antihelminth drugs, (c) personal prophylaxis by washing of hands after handling dogs.

3.2.3 *Hymenolepis nana*

The cestode parasite is commonly called dwarf tapeworm as it is the smallest 'parasitic tapeworm' of man.

The parasite is found throughout the world and infects a large number of children. Though, cosmopolitan in distribution but found generally in temperate zones. In the Southern United States 1–2% of the population of children are infected.

3.2.3.1 History

In the year 1852 Van Siebold for the first time reported *H. nana* as a human cestode parasite residing in the lumen of the intestine causing the disease hymenolepiasis.

The number of species found to parasitize man and other mammals are:

Hymenolepis apodermi—in rodents
Hymenolepis asymetrica—in rodents
Hymenolepis diminuta—in human
Hymenolepis horrida—in rodents
Hymenolepis rymzhanovi—in rodents
Hymenolepis microstoma—in rodents
Hymenolepis nana—in human

H. nana infection is seen highest among children in warm countries having poor sanitation. It is found in Australia, Cambodia, Turkey, Mexico, etc.

3.2.3.2 Symptoms

The Hymenolepiasis exhibits symptoms like diarrhoea, muscle cramps, irritability, anorexia, etc. in children. Hymenolepis also destroy intestinal villi of the host where they grow and develop. Sometimes with heavy infection symptoms manifest like anorexia, vomiting, nausea, blood diarrhoea, headache, dizziness and behavioural changes, sometimes also epilepsy.

3.2.3.3 Geographical Distribution

H. nana infection is seen highest among children in warm countries having poor sanitation. It is found in Australia, Cambodia, Turkey, Mexico, etc.

3.2.3.4 Habitat

It lives in the villi of the small intestine and lymph channels of the villi of human beings.

3.2.3.5 Morphology

The adult tapeworm is flat, segmented having a very slender neck that varies 7–100 mm in length but mostly 40 mm long and 1 mm wide. The most interesting fact is that the length of the adult worm depends upon the population density of the worm. The scolex bears a well-developed retractable rostellum with a single circle of 20–30 hooks. The neck is long and slender, the proglottids are broader than length.

Each mature segment possesses three testes. The uterus is a sac like structure that spreads between the excretory vessels and mostly filled up the segment. The portion between the proglottids and the wall of the uterus breaks down to permit the eggs to spread from the segment to segment and come out due to degeneration or separation of the proglottids from the posterior end. The proglottids contain both male and female reproductive organs, so the animal is hermaphrodite. Genital pores are unilateral.

The gravid proglottids in which the eggs are fertilized either separate or disintegrate at the posterior portion releasing the fertilized eggs.

The eggs are ovoid and have striated appearance with an outer thin shell practically colourless measuring 40 × 50 μm. The embryophore measures 16 × 20 μm with two little knob like structures at both ends, from which arises a number of long delicate wavy filaments which are located between the embryophore and the outer shell. In onchosphere six embryonic hooks lie parallel as organ of attachment.

The dwarf tapeworm does not have digestive system and feeds by absorption of the nutrient by their body surface. They absorb the nutrients available in the lumen of the intestine of the host. When it attains the sexual maturity in the lumen of the intestine it gets attached to the wall of the intestine with its scolex and rostellum to develop into an adult worm.

3.2.3.6 Life Cycle

H. nana the cestode parasite unlike others does not require any intermediate host. They are capable to complete their life cycle within a single host and auto infects the human beings.

Surprisingly, it is seen that they may develop within invertebrate arthropodan host like fleas or grain beetles before they establish themselves into human beings.

The gravid proglottid of *Hymenolepis* containing fertilized eggs separates one by one from the posterior end and then disintegrates and fertilized eggs are escaped with the faeces of the infected vertebrate like man, rat or mice. Besides, the eggs may also settle in the microvilli of the intestine of the host where they hatch and larvae develop to sexual maturity without the involvement of the intermediate host (Fig. 3.4).

Infection is acquired by the faecal–oral route from another infected man. A healthy person is infected by consuming food and drink contaminated with fertilized eggs of *Hymenolepis* coming out along with the faeces of infected persons. The eggs hatch in the duodenum of the new host after ingestion. The eggs develop into spherical embryos in which there are six hook like structures present in three pairs of claw like hooks, they are known as oncosphere. There develop one or two membranes around the developing embryo inside the eggshell proper, the inner one is called embryophore. The onchospheres within the lumen of the intestine burrow inside the villi of the intestine and come to lie in the lymph channels of the villi here they develop into cysticercus with a tail and a scolex. The time takes about 4 days. It has now became spade shaped and the rest of the parasite still remains inside the eggshell. After 5–6 days the cysticercus stage mature and comes out again into the lumen of the intestine. Now the scolex with the rostellum attaches itself to

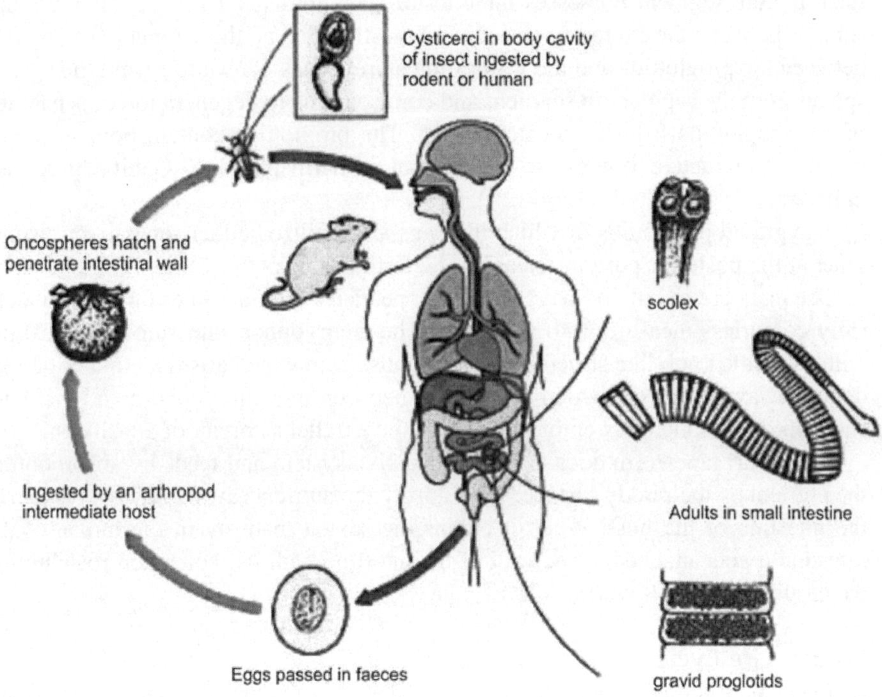

Fig. 3.4 Life cycle of *Hymenolepis nana* (*Advanced Parasitology*, Das)

the wall of the intestine and grow to maturity. The time taken by them to become adult is about 15–20 days. But when they choose to develop in the invertebrate host like fleas or grain beetle they develop into cysticercus with tail in about 12–14 days if the rat or mice ingests the contaminated faeces.

It is very interesting that one host life cycle of *H. nana* instead of two is a recent adaptation in relation to evolution. The most probable reason is that the cysticercus of *Hymenolepis nana* can develop at higher temperature, but accidentally they can still develop within invertebrate host like larval fleas and beetles.

Infection of human beings spread from human to human by faecal–oral route but also the sudden and accidental infection by an infected grain beetle or flea cannot be discarded. The spread of infection from person to person of *H. nana* without the involvement of intermediate host rules out dependence on the intermediate host.

3.2.3.7 Pathogenicity

H. nana causes hymenolepiasis. The infection is usually asymptomatic. If it is symptomatic patients suffer from anorexia, abdominal pain and diarrhoea.

3.2.3.8 Diagnosis

Diagnosis is done by examining the faeces under microscope. The eggs are visible. The eggs float in strong saline solution. The concentration technique may be applied after floating the eggs and then by centrifugation.

3.2.3.9 Host Defence Mechanism

Adult tapeworm infection stimulates Thl/Th2 response. During early infection transient Thl response takes place which is partly or totally replaced by Th2 response as the infection progress. The Thl response is evident by increased secretion of IFN-γ and IL-2, it is associated with protective immunity. The Th2 response is manifested by release of IL-4 and B cell activation and is related to susceptibility.

3.2.3.10 Epidemiology

The transmission of the disease occurs through ingestion of eggs, by faecal contamination of hands and food of the cestode parasite. Eggs do not survive outside the body of the host long after their passage in the stools. Occasional infection may occur from rodent sources.

3.2.4 *Diphyllobothrium latum*

3.2.4.1 History

This cestode parasite was discovered by Luha in the year 1910. *Diphyllobothrium latum* is called fish tapeworm. It is the largest tapeworm found in human beings. *D. latum* develops its maturity in some domestic and wild animals that are fish eating carnivores.

3.2.4.2 Geographical Distribution

It has almost worldwide distribution. It is found in Central Europe, Finland, Ireland, Siberia, Palestine, Japan, Central Africa, Chile, Michigan, Minnesota, Manitoba, Alaska, Florida.

3.2.4.3 Habitat

The adult worm resides in the jejunum of human beings where the worm attaches itself by the suctorial grooves.

3.2.4.4 Morphology

The organisms are found in four forms: adult worm, coracidium, procercoid larva and plerocercoid larva.

D. latum is also called a monster tapeworm as the adult is more than 10 m in length and its width is near about 20 mm. The adult worm resides in small intestine of man. Adult worm contains 3000–4000 proglottids. In the most part, the proglottids are more in breadth than length but the terminal one is approximately square.

The scolex is elongated measuring 2–3 mm in length and 1 mm in breadth. It has two slit like sucking grooves called bothria from which the genus name is given. There is no hook in the scolex. The shape of the scolex is more or less finger shaped. In the proglottids, there are numerous testes and vitelline follicles. The male and female genital pores are midventral in position. At the posterior position of the segment the bilobed ovary is located.

3.2.4.5 Life Cycle

The adult lays one million immature eggs per day which is liberated in the small intestine of the host and ultimately escape outside along with the faeces of the definitive host. The worm is hermaphrodite and eggs are fertilized within the proglottids.

The eggs are oval and operculated the average size is 60 × 42 μm and brown in colour having a knob like structure at the opposite ends. The egg contains an unsegmented ovum and yolk cells.

The eggs mature slowly in 2 weeks in freshwater depending upon temperature. The eggs hatch into a ciliated spherical embryo or larva called coracidium.

The coracidium is 50–55 μm in diameter and has six embryonic hooklets. The coracidium swim with their cilia or creep on the bottom after coming out of their ciliated coverings. Crustacean larva like cyclops consumes these coracidium in less than 24 h for the continuity of their life. Here within the cyclops they undergo development. The coracidium are very much choosy about their first intermediate host. If within 24 h the suitable intermediate host is not found, they perish.

Soon after the coracidium is ingested by the cyclops it loses its cilia and the onchosphere comes out which are 24 μm in diameter.

The coracidiums bore through the coelomic cavity of the cyclops. In 2 weeks they develop into an enlongated procercoid larva.

The procercoid larva is 500 μm long and has a caudal appendage with three pairs of vestigial hooklets.

Next phase of development takes place in fish when the fish consumes infected cyclops.

The procercoid present in the cyclops reaches intestine of fish and the larva slowly pass through intestine and enters body cavity of the fish. The larvae then develop into elongated plerocercoid larvae and are not encysted in the flesh of the fish (Fig. 3.5).

The plerocercoid larva is white in colour, wrinkled and unsegmented. It measures 10 mm in length and 2 mm in breadth and possesses one head with sucking grooves. These plerocercoid larvae are the infective forms of human beings. The larvae when reach intestine of human beings upon consuming uncooked or undercooked fish they attach itself to the mucous coat of the intestine and develop into adult. The time taken by the worm to become an adult is 2–6 weeks. The mature worm starts laying eggs and completes the life cycle. The lifespan of an adult worm within human beings may be up to 20 years.

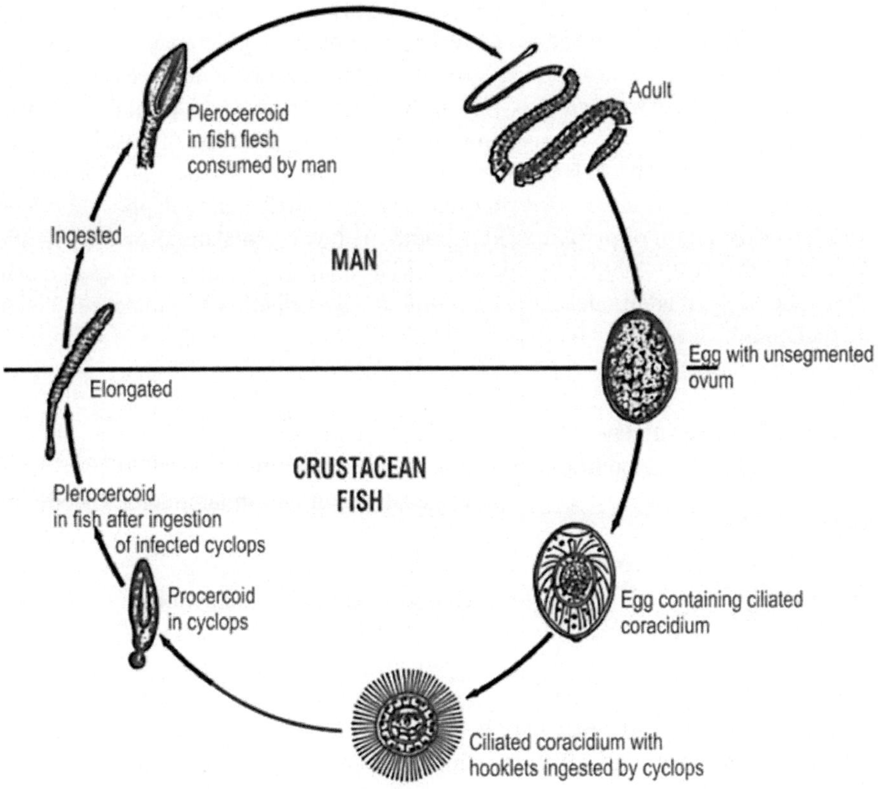

Fig. 3.5 Life cycle of *Diphyllobothrium latum*

3.2.4.6 Energy Metabolism

The adaptive morphology of the adult tapeworm is that these parasites derive nutrient materials from the host and nutrient molecules must cross the integument. The parasites reside in the small intestine of the host where oxygen tension is very low. So they are compelled to practice anaerobic metabolism. The nutrient molecules cross the integument by active transport. The most important nutrient molecules are glucose which after polymerization within the parasite transformed into glycogen. The other carbohydrate is galactose. Here in the intestine the parasites derive energy from substrate level phosphorylation via glycolysis. Adult tapeworms absorb carbohydrates, amino acids, purines and pyrimidines from different sites of the tegument. They also absorb lipids by simple diffusion.

It is observed that the neck and immature proglottids have much higher rate of metabolism than mature and gravid proglottid. The energy is required by mature proglottid for egg production only.

Energy metabolism is more or less same in all the cestodes. The adult cestode parasites are facultative anaerobic organisms. They deduce energy from glucose and glycogen through catabolic process. They oxidize one molecule of glucose and excrete short chain organic acids which are highly reduced end-products. The adult cestodes and larvae are used to store large amount of glycogen. This glycogen supply food when the host is not feeding i.e. the periods between feeding. When stored glucose is exhausted it is replenished when it is again available. Glucose coming from glycogen or coming from the food of the intestine of host is degraded by the pathway of glycolysis. *In* this process lactate is formed by dephosphoenolpyruvate (PEP). After that lactate is produced by dephosphorylation of PEP or reduction of pyruvic acid. Another pathway is malate formed by fixation of carbon-di-oxide to produce oxaloacetate. The oxaloacetate then reduces to malate. Over and above if additional energy is required it is available when malate enters into mitochondria.

3.2.4.7 Pathogenicity

Usually the Diphyllobothrium infection is asymptomatic. But sometimes gastric disturbance with abdominal pain, rarely obstruction and inflammation of bile duct occur.

D. latum absorbs large quantity of Vit. B_{12} from the intestine and so interrupts absorption of digested material through intestinal wall of the definitive host. Due to the lack of Vit. B_{12} the patient suffers from megaloblastic anaemia.

3.2.4.8 Host Immune Response

Usually adult tapeworm infections stimulate chronic Th1/Th2 responses. In early infection Th1 response is transient and it is replaced by Th2 response. Th1 response is marked by increased level of IFN-γ and IL-2 secretions for protective immunity while the Th2 response is characterized by release of IL-4 and B cell activation which is related to susceptibility.

3.2.4.9 Diagnosis

Saline preparation of faeces is examined under microscope and eggs are seen. The eggs have characteristic features like operculum at one end and a knob like structure at the other end.

3.2.4.10 Epidemiology

This is a fish tapeworm so human beings are infected if they consume raw or undercooked fish. The infection is highest among the persons who are in the habit of taking undercooked fish. The intermediate hosts are crustacean larva, Cyclops and freshwater fish mostly Salmon as the people of the USA are very much crazy about consuming Salmon as sushimi.

3.3 Nematoda

3.3.1 *Ascaris lumbricoides*

3.3.1.1 History

In 1758 Linnaeus named them Ascaris lumbricoides. For many centuries they are believed to arise by spontaneous generation. In 1855, Ascaris eggs were found in the faces of human beings by Henry Ransom in England. It was described by Casimir-Joseph Davaine in France in the year 1857.

3.3.1.2 Geographical Distribution

The most common nematode parasite of man is the roundworm. They used to infect the children from time immemorial. It is cosmopolitan in distribution but mainly thrives in moist warm climate. They are prevalent in the tropical countries like China, India, South East Asia. The parasite is particularly prevalent in the regions where sanitation standard is very poor and children are not trained to defecate in a particular place not here and there. Children are prone to infection than adults. It is observed that 500–1000 Ascaris may live within single human being.

3.3.1.3 Habitat

This intestinal parasitic nematode of human beings reside within the small intestine of man and they complete their life cycle within a single host. They do not require any intermediate host.

3.3.1.4 Morphology

The Ascaris in living condition is light yellow to pink in colour. The semi-transparency of the body wall makes some of the internal organs to be seen from outside. The body is elongated and cylindrical. The body is covered by a cuticle which prevents the parasite to be digested in the small intestine of the host.

The adult males measure 15–20 cm in length and 2–4 mm in diameter. The posterior tail end is curved having two copulatory setae measuring 2.0–3.5 nm long. The female is 20–40 cm in length and 4–6 mm in diameter. The tail end is blunt and straight. The female is oviparous. The ovaries are long and extensive.

Along the length of the body there are four longitudinal lines: one mid-dorsal, one midventral and two lateral.

At the anterior terminal end of the body there is the mouth aperture. The mouth is guarded by three lips. Each lip has a forked fleshy core having denticles along its inner margin. In the ventro-latteral lips there is a round groove like structure called amphid. This is a chemo-receptor. At midventral position of the body there is another pore, the excretory pore.

In female the vulva and anus are separate apertures. The vulva is located midventrally at about 1/3rd distance from the anterior end. The anus is present a little behind terminal end.

In male there is cloaca where the genital duct and anus open within the same chamber. The cloacal chamber opens outside through an aperture known as cloacal

Table 3.1 Differences in sexual characteristics

	Male	Female
1.	Adult male is comparatively small	Adult female is larger than male
2.	Adult male possesses cloaca and cloacal aperture	Adult female has anal aperture
3.	Presence of a pair of penial setae at the posterior end	No such structure is present
4.	Tail end is curved ventrally in the form of a hook	Tail end is straight
5.	No such structure is present	Vulva is present at about 1/3rd from anterior side on the midventral line

aperture. There are two penial setae seen repeatedly protruding through the cloacal aperture in living condition.

In both the sexes there are special reduced papilla, chemoreceptors on the lips called amphids.

Male and female copulate in the small intestine. At the time of copulation male orients its body at right angle to that of the female so that the cloacal aperture of male apposed to the vulva of female. The penial setae of male help to open the mouth of the vulva and sperms are transferred to the female body. In the uteri the eggs are fertilized by the sperms coming from the vagina.

The females are oviparous. The sexually mature female lays 2 lacs eggs per day. These eggs are not always fertilized, there are some unfertilized one also. The oval or round fertilized eggs measure 45–75 μm in length and 35–50 μm wide. The outer shell of the eggs has a thick proteinaceous lumpy layer. It is estimated that a sexually mature female contains 27,000,000 eggs in their two uteri. The released eggs come out through the vulva and reach the intestine of the host and these are expelled out, through the faeces of the host.

Outside the body under optimum temperature, moisture and oxygen the larvae develop within the shell of the fertilized eggs. The fertilized eggs are bile stained, yellow brown in colour and unsegmented within eggshell. At both poles of the eggs there is clear crescent shaped space. The eggs develop in soil and moult to form infective second stage larva called rhabditiform larva. The larvae possess a nerve ring and oesophagus at the end of which there is a bulb like structure. The first stage larva is not infective to human beings and it takes 30–40 days to become infective second stage larva.

The unfertilized eggs measure 90 mm × 40 mm and are bile stained. There is no polar space and the cytoplasm is disorganized having highly refractile granules of different sizes.

The chemoreceptors are present in many nematodes. They are present in *Ascaris lumbricoides*. These chemoreceptors are called Amphids. These are a pair of sensile organ present on each side of the head. The opening of the Amphids goes to deep cuticular pit. In these pits there are several nerve fibres. In some species of nematodes they may act as thermoreceptors.

Differences in sexual characteristics of *Ascaris lumbricoides* are given in Table 3.1.

3.3.1.5 Life Cycle

The life cycle is completed within a single host. The fertilized eggs form first stage larva within 9–13 days. The second stage infective rhabditiform larvae are formed after 20 days which is viable for 6 years outside the body of the host in proper condition. The second stage larvae enter human beings with food and drink. The eggshell dissolves in the secretions of small intestine. The presence of CO_2 concentration in the intestine stimulates the larva to come out from the egg cover. The larva now secretes chitinase, esterase and protease enzymes which dissolve a portion of the inside membrane and larva comes out through this dissolved portion. The second stage larva is 0.2–0.3 mm in length and 1–15 μm in breadth. The larvae contain alimentary canal, excretory duct, nerve ring and reproductive system.

Within the intestine they bore through the wall of the intestine and reach blood vessels of hepatic portal vein. The larvae swim and carried via bloodstream to liver from liver to right auricle through post caval vein. Then they float from right auricle to right ventricle and to the lungs via pulmonary circulation. The larvae reach alveoli by breaking the pulmonary capillaries. Within the alveoli they remain for 10 days and moult twice here. first moulting takes place after 6 days and transforms into third stage larva then after another 4 days to fourth stage larva. These larvae now crawl up taking the route of alveolar duct, bronchioles, bronchus, trachea and epiglottis. On reaching pharynx they cause irritation and cough. The patient swallow the mucous along with the living larvae. Within 10 days they reach intestine again after making a long journey throughout the body. The fourth stage larvae in the meantime grow to 2–3 mm in length. After 25–27 days again another moulting takes place and fifth stage larvae take permanent residence in the intestine where they sexually mature after 60–70 days. The time taken by the larvae to migrate is 15 days (Fig. 3.6).

The sexually matured male and female copulate within the lumen of the intestine and the cycle is completed.

Moulting: Nematodes undergo four moultings in their life cycle involves: (1) formation of new cuticle, (2) loosening of the old cuticle, (3) rupturing of the old cuticle and (4) escape of larva.

3.3.1.6 Pathogenicity

The disease caused by Ascaris is commonly called Ascariasis. The infection of Ascaris is greater in children than adults.

Due to the migration of larvae of Ascaris patient mainly suffers from lung disease. Blood tinge sputum, fever, eosinophilia may happen. The disease caused by migration of larvae in lungs is called Loeffler's syndrome. When the larvae break up capillaries and enter into alveoli haemorrhage takes place. Due to haemorrhage the air space of the lungs may be choked. WBC and dead epithelial cells accumulate in that space and secondary infection takes place by bacteria. This condition is known as Loeffler's syndrome. Sometimes this pathological condition of the lungs may lead to death.

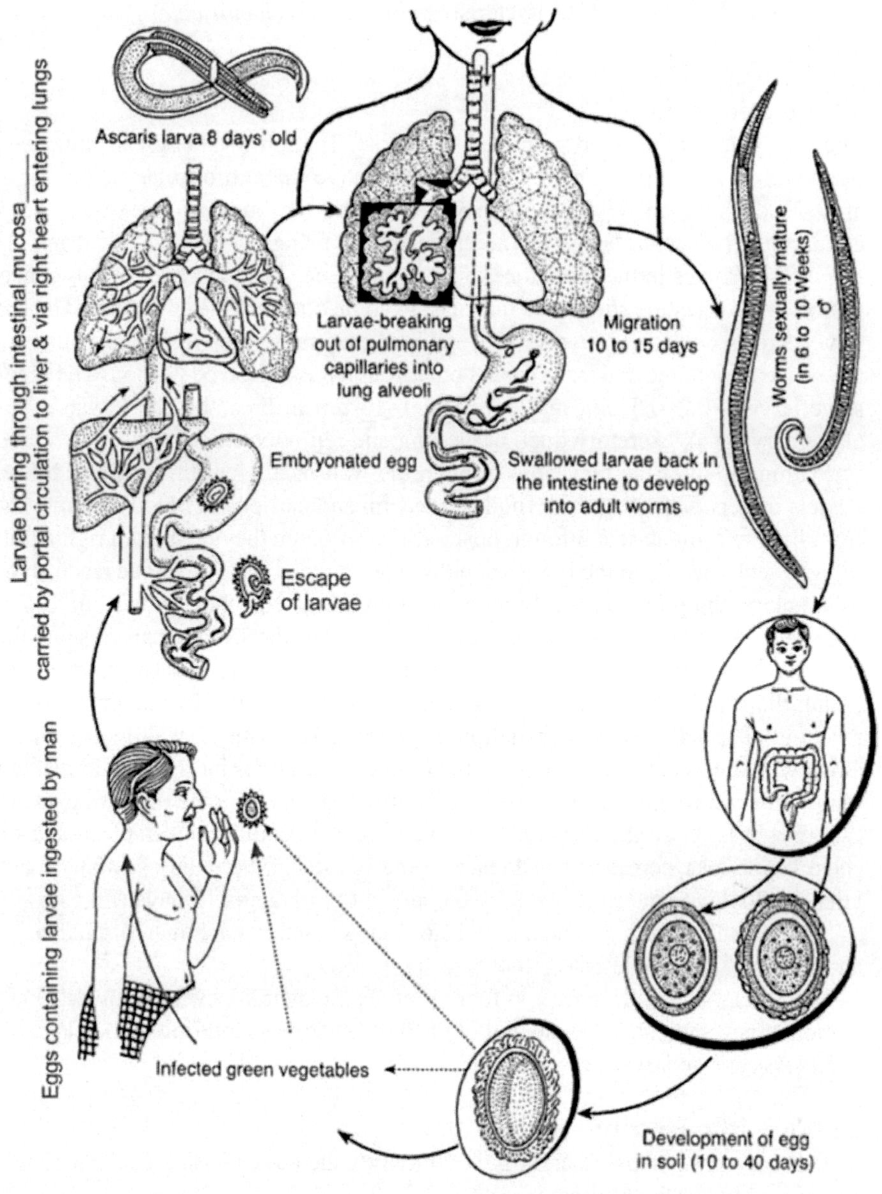

Fig. 3.6 Life cycle of *Ascaris lumbricoides*

The second stage larva while wandering through different organs with the help of blood circulation may sometimes enter into spleen, liver, lymph nodes, brain, etc. and cause inflammation due to the death of the larva in those places.

The lipid binding protein of Ascaris expresses IgE antibody response. This may help to create Loeffler's syndrome.

Ascaris while in small intestine depend upon the digested food material of the host. So they cause malnutrition and underdevelopment in children. The production of metabolites of the Ascaris creates allergic responses. Sometimes abdominal pain may occur due to heavy infection. Rashes, asthma, insomnia, eye pain, restlessness may happen due to allergic responses. A massive infection of *A. lumbricoides* may cause intestinal blockage.

The adult worms if present in large number especially in children malnutrition interfere in absorption, Vit A deficiency may occur. If strangulation happens then surgical interference is required. The worms sometimes enter appendix and appendicitis may then happen. The adult worm may enter ampulla of vatar causing biliary colic, cholicystitis, pancreatitis and abscess. Adult worm may come out through mouth and nose leading to sometimes nausea and vomiting.

3.3.1.7 Diagnosis

Direct method: Presence of fertilized and unfertilized eggs in the saline and iodine preparation of stool. The eggs will be seen under light microscope, plain X-ray, barium X-ray and USG can reveal the presence of adult worms in the intestine.

3.3.1.8 Host Immune Response

The individuals who are chronically infected with *Ascaris* exhibit reduced Th1 responses with dramatically lower level of TNF-y and IL-2 secretions. Th-2 and IgE (stimulated by IL-4 and IL-5) immune responses are increased. The lung stages (L3/L4) parasites cause remarkable and highly polarized Th2 responses which stay throughout the life cycle. Th2 cytokine release causes protective immunity, suggesting a host's adaptation to control the population stress of parasites.

There is another method discovered that migrating larvae evoke the production of a number of types of antibodies. These antibodies involve themselves in ADCC mechanisms for which the larvae move back to the gut again to escape themselves from the reaction. This reaction prevents later invasion of larvae and as a result overcrowding does not take place.

3.3.1.9 Energy Metabolism

In *Ascaris* sp. malate dismutation takes place. In this case glucose is oxidized up to PEP. Then CO_2 fixation occurs which is catalysed by PEPCK. This process form oxaloacetate which is then reduced to malate. There are three enzymes of TCA cycle which are present in the mitochondria of the Ascaris. These enzymes are called MDH, fumarate reductase and fumarate hydratase. They catalyse reductive reaction-

$$\text{Oxaloacetate} \underset{\text{MDH}}{\longrightarrow} \text{malate} \underset{\text{fumerate hydrase}}{\longrightarrow} \text{fumerate} \underset{\text{fumerate reductase}}{\longrightarrow} \text{succinate}$$

Malate enters the mitochondria and dismutation occurs. Here one molecule of a malate is oxidized and the second is reduced. This is called malate dismutation. But here one molecule of malate oxidized and two molecules are reduced. The resulting products of dismutation are acetate and succinate which are excreted as free acids. As a result more ATP is produced than usually obtained from homolactate fermentation.

Electron transport: The electron transport system of the helminth is anaerobic. Electrons are transported through cytochrome oxidase or fumarate reductase system. Helminths possess rhodoquinone an anaerobic adaptation. The redox potential is less in rhodoquinone or fumarate succinate couple. So electron is transported in the fumarate direction.

3.3.1.10 Epidemiology

The transmission of this parasite is usually through the hands of the host which is contaminated with embryonated eggs. To prevent this infection personal cleanliness and better nutrition are very much necessary. According to parasitologists contaminated vegetables and water is not so common way of transmission. But the use of the human stool as manure, the handling of this manure may be an important method of infection.

3.3.2 *Enterobius vermicularis*

3.3.2.1 History

Enterobius vermicularis or pin worm was discovered by Ferreira and Hugot.

3.3.2.2 Geographical Distribution

It is worldwide in distribution found almost all over the world but more prevalent in temperate zones than tropical countries. Most members of the white people are infected with pin worm but infection remains in children mostly. Sample survey of white children in the USA and Canada shows that 30–60% children are infected.

3.3.2.3 Habitat

The adults reside in caecum, appendix and adjacent regions of the ascending colon of the GI tract.

3.3.2.4 Morphology

The parasites are white in colour, very small in size and show sexual dimorphism. This nematode is covered by finely striated cuticle. But both male and female have cervical alae at the anterior end of the body. They have no buccal cavity but has three lips and a bladder like expansion of the cuticle (Beaver et al. 1952). The most important feature of this nematode is the presence of a dilated posterior end of the oesophagus.

Males are 2–5 mm having 0.1–0.2 mm in diameter. It has a coiled tail with a single penial seta which is 100–140 μm in length. The cuticle is semi-transparent and oesophagus with a bulb like structure at the posterior end can be seen. The double-bubbled oesophagus of both the sexes is a characteristic feature of the pin worm.

The female worm measures 8–13 mm long with a diameter of 0.03–0.05 mm. The posterior end of the body is drawn into a long point. The female is oviparous. The gravid worm has an uterus which is swollen because of the presence of large number of eggs. The average number of eggs is about 11,000.

Reproductive organs: The males are equipped with single testis having two seminal vesicles, ejaculatory ducts which open into the cloaca. Male possesses sharply curved prominent copulatory spicule or penial seta nearly 70 μm long.

The female has anterior and posterior oviducts and uteri opening into the vagina. Vagina extends some distance and opens to the vulva. Vulva opens outside nearly one-third of the anterior portion of the body.

The infective forms are embryonated eggs. The features of the egg are as follows:

(a) Eggs are colourless and not bile stained.
(b) Plano-convex in shape and asymmetrical.
(c) Measuring 50–60 μm in length and 20–30 μm in breadth.
(d) Egg is covered by transparent shell, shall consist of a hyaline membrane, two layers of chitin and an inner most layer of lipid. The eggs remain on the perianal region of the host where they become embryonated having a coiled tadpole like larva. These embryonated eggs become infective after 36 h. These larvae survive outside the body of the hosts for 4–5 weeks.

3.3.2.5 Life Cycle

Enterobius vermicularis requires a single host to complete its life cycle. The host is human beings.

The infective forms are embryonated eggs. The transmission of infection is faecal–oral. The embryonated eggs hatch in the intestine to liberate the larvae. They are found to be attached to the mucosa of intestine wall. The larva transforms into adult in the large intestine of human beings.

The male worm fertilize female and the average number of eggs are about 11,000. After hatching in the intestine they temporarily burrow into the mucus membrane of the intestine in the caecum and tadpole larvae 140–150 μm by 10 μm hatch out and set free. They transform into adult when they come out and reside in lumen of the intestine. The larvae pass down the small intestine and moult twice in the crypts of jejunum and upper part of ileum (Fig. 3.7).

The adolescent worms remain attached to the mucus membrane and become adult in 15–43 days after ingestion of the embryonated eggs.

The gravid female now moves freely in the lumen of the intestine. They pass colon and crawl to the anus to lay eggs in the perianal surface at night. The eggs are with albuminous coating by which they adhere to the surface of the skin. Due to albuminous attachment the patient feels irritation and itching. Many worms are seen to be out of the anus and others pass via faeces. Their movement in the rectum and perianal region at night causes intense itching. When the adult female worm is stimulated in contact with air deposits eggs. The worm sometimes becomes dry in contact with air and then explodes liberating thousands of embryonated eggs. The

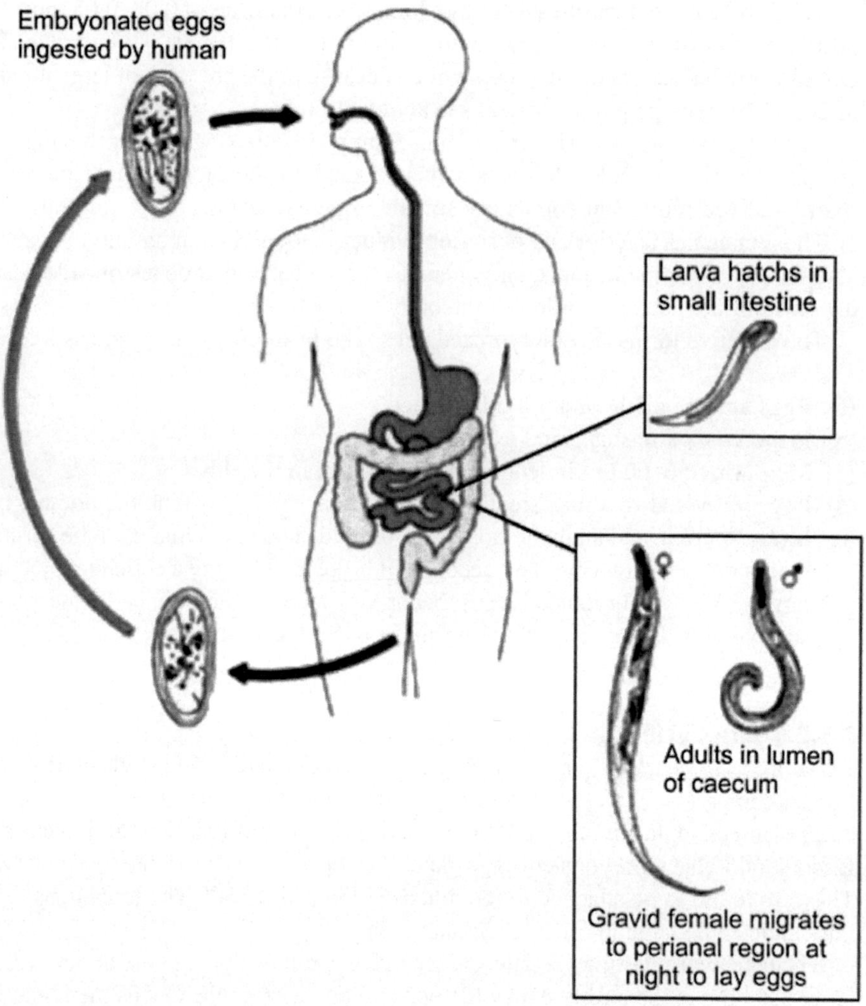

Fig. 3.7 Life cycle of *Enterobius vermicularis* (*Advanced Parasitology*, Das)

embryonated eggs when freshly laid contain partially developed larvae in the 'tadpole stage' but they transform into infective forms within 6 h.

The eggs and larvae can be detected in the scrapings of perianal region or using cello tape may be used to retrieve the eggs.

3.3.2.6 Transmission

The infective forms of larvae either may infect the same person, i.e. autoinfection especially in children and when they unknowingly scratch the perianal region for itching sensation by hands the larvae get access in the gap of the nail and when the same children due to their usual habit put the fingers into the mouth autoinfection takes place. Sometimes it is observed that the larvae move retrogressively to the anus and rectum from there to the caecum region. The liberated eggs are deposited on bed sheets, clothing, towel, etc. and shaking of these cause the embryonated eggs to be in the air and then these eggs are inhaled from the air. They hatch in the nasal mucus and get access into the G.I. tract. The cycle is completed in 15–43 days.

According to Cram the eggs of *E. vermicularis* survive 2–6 days under cool humid climate. But in dry and temperature above 25 °C they may survive as long as 12 h. But in dry climate at 36–37 °C less than 10% eggs survive for 3 h and more up to 16 h.

The lifespan of the adult female is 37–100 days and males are more short living organisms.

3.3.2.7 Pathogenicity

Most pin worm infection is asymptomatic. Minute ulcerations may occur in the perianal area of the patients and in the intestine inflammation may occur along with secondary bacterial infection. The itching caused by migration of the parasites in the anal region causes perianal pruritus. The allergic irritation of the skin around anal region often causes loss of sleep, restlessness, nervousness and even sexual disorders. In girls the worm may cause vulvovaginitis if invade female genital tract. They may even reach fallopian tube or in the peritoneal cavity where they become encysted.

Immature burrowing worms may create inflammation in the caecal region with abdominal pain and digestive disturbances.

Moreover in case of female patients the migration of the adult worms may take them up to the vulva, vagina, Uterus, oviducts, etc., and may form granulomatous tissue in the peritoneum. Sometimes it has been observed to become encapsulated in the graafian follicles. Surprisingly granulomatous tissue may form sometimes even in the vulva. The other symptoms observed are restlessness, nervousness, grinding of teeth, loss of appetite, insomnia, nausea, vomiting, bed wetting due to irritation, etc.

3.3.2.8 Diagnosis

Direct observation of adult worms and/or eggs under light microscope. The best method is administration of laxative or application of enema and then examination of stool under microscope from smear preparation, NIH swab from the perianal region of the skin may also help to detect the eggs under light microscope.

3.3.2.9 Epidemiology

The eggs of *E. vermicularis* may be found in the dust, cloths and lavatories. Eggs survive in the dust for 3 weeks and such dust with eggs may cause mild infection. But heavy infection occurs in children when they put their fingers in the mouth after scratching the perianal region with their hands. By this method they may ingest 100–1000 eggs at a time.

3.3.3 *Ancylostoma duodenale*

The group of hookworms under the sub-order Strongylata, the parasitic nematodes cause tremendous injury to man and economic loss of the society through attack on his domestic animals. Many of them are blood suckers and sometimes haemorrhages take place due to blood sucking habit of the parasite as a result the hosts suffer from anaemia, loss of vitality and loss of economy, through death of the domestic cattle.

3.3.3.1 History

The parasite is a roundworm and was first discovered by Italian Scientist Angelo Dubini in the year 1838. The eggs were identified by Grami and Parona from human faeces in 1878. The development of free living forms from eggs was observed by Perroncits in 1880. The structure of adult worm was described by Looss in 1897 and the life cycle was recorded by Chandler in 1929.

3.3.3.2 Habitat

It is primarily a parasite of human beings. The adult worm resides in the jejunum of small intestine less often in duodenum and rarely in ileum. They suck blood, lymph and bites mucous membrane of the lining of intestinal wall with a suctorial pharynx.

3.3.3.3 Geographical Distribution

It is found in all tropical and subtropical countries wherever humidity and temperature are favourable for the development of larva in soil. It is quite abundant in South East Asia, India, North Africa, Sri Lanka, China and Caribbean Islands, Japan, Europe, Western Asia and some states of the USA.

It affects seriously the miners and causes anaemia. Mines are an ideal place for the development of eggs and juvenile of this nematode parasite.

3.3.3.4 Morphology

The body is more or less cylindrical with curved anterior end covered by tough cuticle. They are reddish brown in colour. The adult male is 8–11 mm long and 0.4 mm in diameter. The females are 10–13 mm in length and 0.6 mm in diameter. The anterior end of the parasite possesses mouth and a buccal capsule. The buccal capsule is lined by hard substance which bears six teeth and four hooks. There are four ventral pointed teeth and a pair of blunt dorsal teeth. There are five digestive glands secretions of which help in digestion. The buccal capsule has a pair of amphid. The amphids are chemoreceptors. The oesophageal gland present in the oesophagus secretes anticoagulant of blood which prevents coagulation of blood during feeding of the parasite.

The buccal capsule opens into oesophagus which is lined by cuticle and has a triradiant lumen. The oesophagus of the parasite continues into a straight intestine which opens into the rectum.

The posterior portion of the parasite differs in male and female. In females mouth leads into buccal capsule. The buccal capsule opens into oesophagus. Oesophagus in turn opens into intestine. Intestine continues into rectum and rectum opens into anus present at the posterior end.

But in males the rectum opens into cloaca where the ejaculatory duct with protrussible and retractile copulatory spicules 1 mm long lie in a sac lateral to ejaculatory duct. Here also lies the copulatory bursa. The copulatory bursa is an umbrella like structure with 13 rays. The copulatory bursa is applied on the female genital pore of the parasite so that both the sexes remain firmly attached during copulation. A short distance behind the head excretory pore exists where a pair of excretory glands open.

There is a circumoesophageal nerve ring present slightly above the excretory pore.

3.3.3.5 Reproductive System

Reproductive system: The sexes are separate. The genital openings in male and female are at different levels of the body so during copulation a Y-shaped figure is formed between male and female.

The male reproductive system has a single tubular thread like testes and coils around the intestine in the middle portion of the body of the parasite. The testis opens into vas deferens followed by an elliptical seminal vesicle which at the end tapers and forms a muscular ejaculatory duct which opens into cloaca. A pair of spicules about 1 mm long located in a sac lateral to ejaculatory duct have the capability of protraction and retraction by muscles.

The female reproductive system has a pair of thread like ovaries coiled around the intestine. Each ovary leads into a short oviduct which is dilated backward to form seminal receptacle. Then there is a muscular uterus at the end of the oviduct. The two uteri join with one another at their posterior side to form vagina. Vagina ultimately opens outside through an aperture called vulva.

3.3.3.6 Life Cycle

The sperms are ejaculated by the ejaculatory duct and smooth and continuous discharge of sperms is helped by the secretions of cement gland. The sperms are collected into the seminal receptacle of the female where the eggs coming from uterus are fertilized (Fig. 3.8).

The sperms are 2 μm in length and look like bacteria.

Eggs

The female worm is oviparous and lays 10,000–25,000 eggs per day which go outside in the environment along with the faeces of the host.

The eggs are oval or elliptical in shape having a hyaline hard covering measuring about 60 × 40 μm. Each egg is colourless and contains embryo which has four blastomeres, i.e. two cleavages are performed. There is a clear space between the eggshell and segmented zygote.

The eggs with eggshell are released into the intestine of infected person which are deposited on soil along with faeces. The segmented zygote under favourable condition of moisture, oxygen and temperature first stage larva hatches out and remains

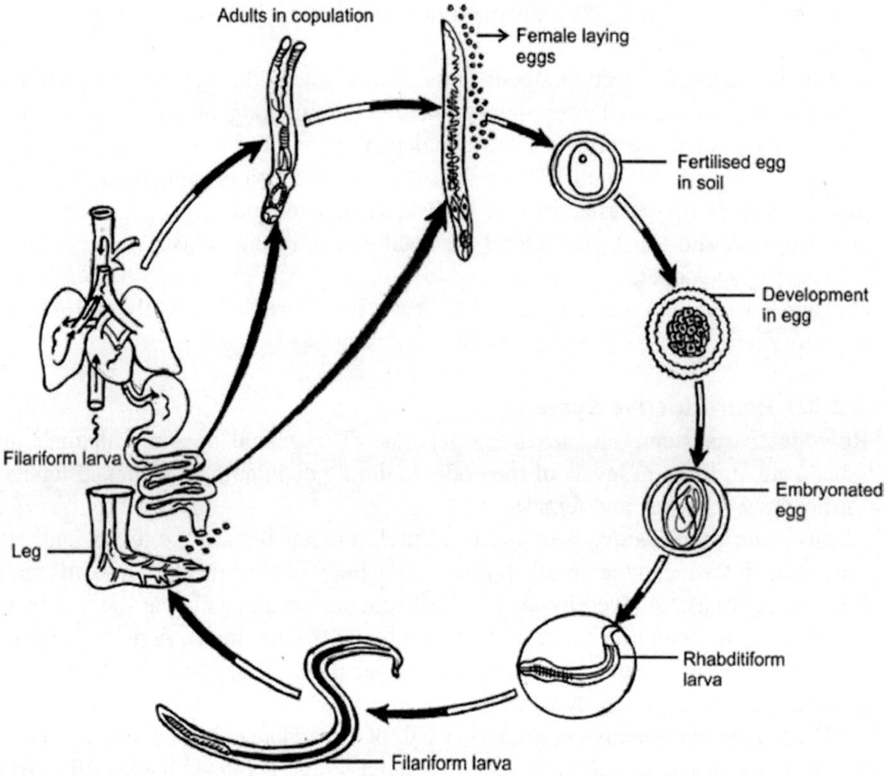

Fig. 3.8 Life cycle of Hookworm, *A. duodenale*

within the eggshell. The first stage larva then converts into second stage larva called Rhabditiform larva within 1–2 days and comes out on the soil along with the faeces of host. Rhabditiform larva is the infective form. The larva is now 0.25–0.30 mm long and 17 μm in diameter. Newly hatched larva starts feeding on bacteria and organic debris. The free living Rhabditiform larva has a mouth, buccal cavity, a flask shaped muscular oesophagus with an intestine. It also possesses a nerve ring at their anterior end.

After 3–4 days larva moults twice and after first moult Rhabditiform larva grows to 0.5–0.6 mm in length and undergoes several moulting. After second moulting the third stage larva called filariform larva becomes 500–600 μm in length. This slender filariform larva is the infective form of the hookworms.

The filariform larva casts off its sheath and penetrates the skin on the sides of the foot where the skin is soft in the case of bare footed human beings. The penetration of larva is accompanied by severe itching called 'Ground itch'.

On reaching the subcutaneous tissue of the human host it gains entry into general circulation and reaches right auricle from right auricle to right ventricle from there to pulmonary capillaries. They now break through the capillary wall and obviously drop into the alveolus. From alveolus the larvae crawl up and ascend via trachea from trachea to pharynx. On reaching pharynx the larvae create irritation and cause cough. The larvae are now swallowed along with some amount of mucous without the knowledge of the host. Ultimately the larvae reach intestine. In the intestine after 3–4 weeks they are sexually matured and undergo fourth and last moulting.

The sexually matured male and female copulate and after copulation the female starts laying eggs. The cycle is thus repeated. The total time taken by the hookworm from penetration of the skin to the appearance of eggs in the host's faeces of infected person is 40–45 days.

An interesting story has been told that in 1896 Arthur Looss in his laboratory has been working with Ankylostoma and infecting guinea pigs putting juveniles in the mouth of them. Accidentally he spilled some amount of culture onto his hand. He found that he has been infected and found that after some days he was passing hookworm eggs in his faeces. He found that juveniles of hookworm may develop into adults without migration through lungs. It is an extreme case.

3.3.3.7 Pathogenicity

A. dudenale causes ancylostomiasis which is also known as hookworm disease. Most of the times the disease is asymptomatic. In symptomatic infection larvae cause pruritic maculo-papular dermatitis (ground itch) at the site of the penetration of the skin by the hookworms. This is experienced only when quite a large number of larvae of parasite penetrate the skin. The ground itch lasts for 2–4 weeks. Mild pneumonitis may occur during the migration of the larvae through lungs. Adult worm in intestine creates abdominal pain, diarrhoea, nausea and vomiting during early stage of the infection.

The most important expression of the disease is microcytic hypochromic anaemia.

A single adult sucks 0.2 ml of blood per day. The worm consumes plasma as food. But RBCs are not digested in the intestine of parasites. The worm frequently changes their feeding sites. The earlier site continues to bleed due to the action of anticoagulant. So iron deficiency anaemia develops in the hosts. Malnourished patients also suffer from hypoproteinemia. These conditions result in weakness and shortness of breath in the hosts.

3.3.3.8 Diagnosis

The infection is diagnosed by detecting eggs in the faecal sample of the suspected patients. In mild infection concentration technique by centrifugation is very useful.

If the stool is kept for more than 24 h in the laboratory, rhabditiform larvae may be seen under microscope from the faecal sample of the patient. Occult blood test will be positive.

3.3.3.9 Host Immune Response

It is evident that hookworm infection produces strong immune responses but these responses are not protective in nature. Information on T cell activity in hookworm infectionis quite lacking. Available data indicate that Th2 responses predominantly generate IgE and eosinophils. Eosinophils can kill infective L3 larval stages, but not the adults. The dominant Th2 cytokines are IL-4, IL-5 and IL-13 with IL-4 promoting IgE synthesis. Immunity is expressed and effective against migratory larvae. The immune response against nematodes is capable of preventing the infection. Firstly the immunity process is against the adult worm and secondly against larval forms. The lymphocytes of the alimentary canal first recognize the antigens of the worm which help in the T-lymphocytes to release a number of different types of cytokines that make active the other immunity cells to secrete mediators like prostaglandins, etc. to make the environment of the alimentary canal hostile to the parasites so that they may be expelled from the gut.

3.3.4 *Trichuris trichiura*

3.3.4.1 History

It was first described by Blanchard and called them *Trichocephalus trichuris* in 1895, then Siles in 1901 called them *Trichuris trichiura* or whipworm.

3.3.4.2 Geographical Distribution

This parasitic nematode of human beings is called whipworm because of its whip like structure. This nematode ranks third in the roundworm infection in man after Ascaris and hookworm. About 50 crore people are parasitized by this nematode worldwide. It is found in moist warm part of the world like South East Asia, Caribbean and tropical Africa. It is also common in India.

3.3.4.3 Habitat

The adult worm resides in the large intestine of man mainly the caecum and vermiform appendix of the definitive hosts.

3.3.4.4 Morphology

The adult female whipworm measures 30–50 mm in length having a thick fleshy posterior part containing the intestine and reproductive organs and a long three-fifth elongated coiled anterior portion contains capillary oesophagus. The oesophagus is surrounded by stichocytes, a large unicellular gland. The oesophagus at its anterior end is more or less muscular and have no stichocytes.

The adult male is slightly shorter and measures 30–40 mm long. At the terminal end of the narrow portion there is mouth. The mouth has no lips but is provided with a minute spear. The male is recognized by its curled tail end of the body just like *Ascaris*. Males have a single long spicule which can be withdrawn into the body in a sheath with a spiny, bulbous end.

The ejaculatory duct, the distant part of the sperm duct runs parallel to intestine side by side. At the end is the cloaca where penial sac, ejaculatory duct and anus open.

In female the anus is at the posterior terminal end and the vulva is present at the junction of the narrow and broad portion of the body. The two long uteri containing barrel shaped eggs measure about 50 × 22 μm and are in unsegmented condition when they are expelled from the host.

The female is oviparous and lays about 50,000 eggs per day. The eggs are brown in colour and have projecting mucus plug on the both sides.

The further development of the eggs takes place outside on soil which are coming out with the faeces of the host.

Furthermore, according to Desmommier on the ventral surface of the oesophageal region there is a band of minute pores. These pores connect with underlying glandular and nonglandular cells. This is called a bacillary band. The function of this band is not known. But the ultrastructure of this band indicates that gland cells may have functions of osmotic regulations and secretion.

The larva hatches within the egg and takes about 3–6 weeks under favourable condition of moisture and temperature. The infective larva remains within the eggshell.

3.3.4.5 Life Cycle

T. trichura requires only single host to complete its life cycle.

Infective forms of *T. trichura* are the larvae remaining within the eggs called embryonated eggs. Human beings are infected by ingesting the embryonated eggs along with food and drink. The eggs develop very slowly, they need 3–6 weeks for the embryo to hatch within the embryonated eggs. The eggs hatch in the small intestine and then the larvae enter into the crypts of Lieberkuhn present in the large intestine. After passing through stomach the larvae within eggshell is dissolved in the alkaline medium of small intestine and hatch here. Here in the base of the crypts the juvenile worms grow and form tunnel within the epithelium. In this way the juveniles enter into so many places of the gut mucosa. According Bundy and Cooper, the larvae which penetrate the mucosa of large intestine only they develop further. When they start to mature the swollen posterior portion of the larvae come out of the gut epithelium and project into the lumen of the intestine and anterior end which is slender remain within the mucosa of the gut (Fig. 3.9). The hatched larvae transform into adult in 2 months and migrate to the caecum and appendix.

Then the adults become sexually matured. The male and female copulate here and the female is inseminated. The female with fertilized eggs pushes their anterior portion of the body within the mucosa from where they suck blood. The posterior portion remains projected in the lumen of the intestine. The fertilized embryonated eggs are expelled from the uterus and discharge outside along with the faeces of the host. It takes about 3 months from the date of infection to the eggs found in faeces. The eggs mature on soil and become ready to infect human beings.

The adult worms live for a number of years and infection builds up gradually.

3.3.4.6 Pathogenicity

The clinical symptoms of the disease trichiuriasis caused by *Trichuris trichiura* are not seen if the infection is done by less than 100 worms. Most of the infections are symptomless.

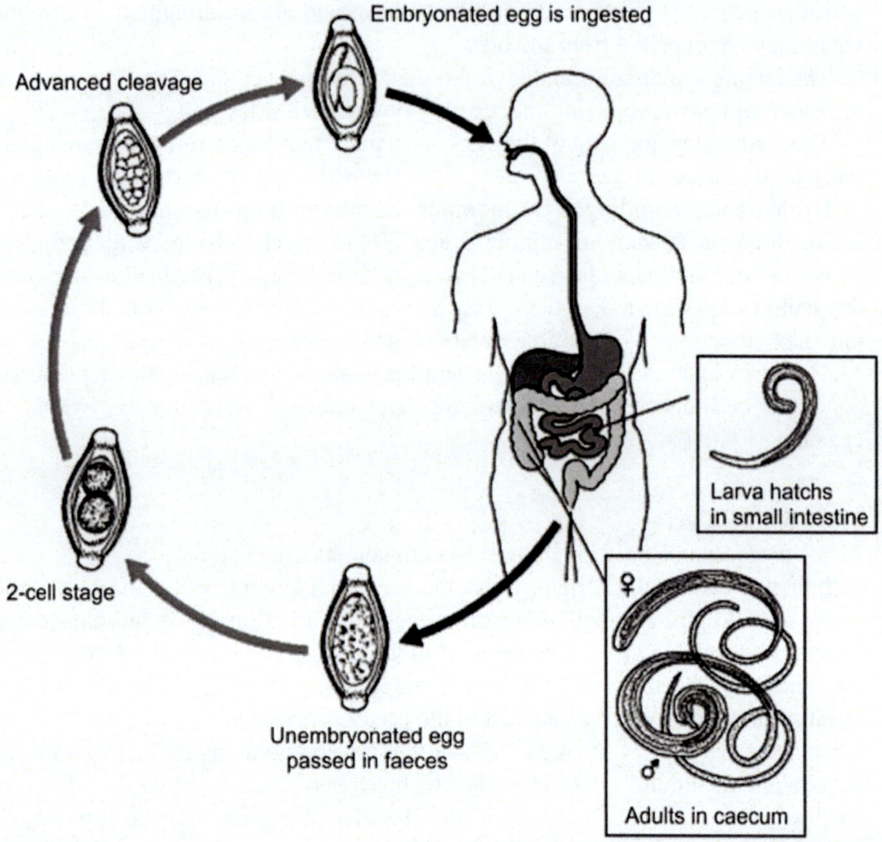

Fig. 3.9 Life cycle of *Trichuris trichiura* (*Advanced Parasitology*, Das)

But severe infection causes abdominal pain, mucoid diarrhoea, anaemia, malnutrition and prolapse of the rectum found in chronic cases.

When the worms remain partly buried in the mucus membrane they feed on blood or cell debris. This may cause chronic haemorrhage which may cause anaemia.

Kouri and Valdey Dias (1952) reported massive infestations in children of 1–5 years old in Cuba due to severe diarrhoea, vomiting and emaciation quite a number of children have to face death.

3.3.4.7 Diagnosis

Direct method of demonstrating worm or eggs in the stool smear preparation and to find them under light microscope is the best one. The other way is colonoscopy to see adult worms. The saline or iodine preparation of patients faeces may reveal the characteristic barrel shaped eggs.

3.3.4.8 Epidemiology

Infection takes place by ingesting embryonated eggs contaminated vegetables, food and water. Poor standard of sanitation may lead to trichuriasis problem. It is observed that eggs survive and develop in warm climate, humidity, rainfall, moisture in soil, etc. A recent survey described that throughout the world the yearly infection is more or less 795 million with 10,000 deaths per year in human population.

3.3.4.9 Host Immune Response

In *T. trichura* infection, IgE level of protection is seen along with increased level of IL-10 secretion. The presence of IL-10 cytokine, produced by monocytes and B and T cells, is known to downregulate Th2 cell response. In *T. trichura* infection a negative correlation is found between the cytokine IL-10 secretion and age of the infected patient. Conversely Th1 pro-inflammatory responses with IFN-y production increase in older patients.

3.3.5 *Strongyloides stercoralis*

3.3.5.1 History

It was first reported in French soldiers returning home from Indo-China in nineteenth century. Vietnam, Cambodia and Laos are endemic to the parasite. Some areas of Japan were also endemic but central programme have eradicated the disease completely from there. But still some areas of Brazil and Central America have prevalence. In some African countries like Zaire *S. fulleborni* was very common up to 1970. Some species of Strongyloides infect children in New Guinea highlands and western province.

3.3.5.2 Geographical Distribution

They are found in Central Africa and Papua New Guinea. It is found that in developed countries where there is well-defined sanitation, i.e. where faecal contamination of soil and water is remote, there the distribution of this parasite is rare. It is found in rural areas where sanitation standard is poor. Tropical and subtropical moist climate is best suited for the parasite.

3.3.5.3 Habitat

They are very small parasitic nematodes those parasitize human beings. The adult resides in the mucosa of the small intestine burrowing tunnels. The genus Strongyloides contains near about 53 species still known today. *S. stercoralis* not only parasitizes human beings but also they are found in the cats and dogs. Nonhuman primates are reported to be infected by another species called *S. fuelleborni* and *S. cebus*.

3.3.5.4 Morphology

S. stercoralis exists in two forms: Adult and larva. The adults may be parasitic or free living. The parasitic females are parthenogenetic. But some scientists have found a few male worms of the free living forms in the lungs of dogs.

The parasitic females are very very slender and measure 2.5 mm × 0.04–0.05 mm and possess a very small buccal cavity surrounded by four lips. They have a bulbless oesophagus about 1/4th of the length of the body. The vulva is located in the posterior one-third of the body and two uteri filled with 10–20 eggs at a time. The eggs are 50 × 32 μm. The eggs are laid in the tissues. The eggs hatch and Rhabditiform larvae come out from the mucus membrane and come to lie in the lumen of the small intestine.

Free Living Adult

They live in moist soil. The females are 1 mm long and males are 0.7 mm in length. The free living male and female copulate in soil. The inseminated female then lays fertilized eggs. These eggs hatch to form Rhabditiform larvae.

Larva

In case of parasitic worm Rhabditiform larvae come out with the faeces of the host. They measure 200–300 μm in length and 16 μm in diameter. They are actively motile. The larvae possess a small mouth and oesophagus with two bulb like structure.

3.3.5.5 Life Cycle

The life cycle of the parasite is homogenic but for the free living it is heterogenic.

In parasitic condition it requires only one host. Human infection is acquired with the contact of contaminated soil or by autoinfection.

This parasite depends on a chemical signal for action, i.e. to find the definitive host. They use sensor neurons of class AFD to identify the chemical signal excreted by the host. *S. stercoralis* is attracted to nonspecific attractants of warmth, carbon dioxide and sodium chloride. Urocanic acid, a component of skin secretions in mammals is a major chemoattractant. Larvae of *S. stercoralis* are strongly attracted to this compound.

The infective forms are filariform larvae which penetrate the skin of the foot like hookworm. After penetration the larvae enter into the bloodstream and through bloodstream they are carried to the lungs via heart. In the lungs they break into the alveoli, from alveoli they ascend taking the route alveolar duct → bronchiole → bronchus → trachea → epiglottis. Here they create irritation and cause cough to the host. Then they are swallowed and reach small intestine. Within the intestine they mature to form adults (Fig. 3.10).

The adult females burrow tunnel in the mucous membrane of the intestine in any place from just behind the stomach up to the rectum. But upper part of the intestine is their choice. It is reported that some may mature even in bronchial tubes. The female lays eggs measuring 50 × 32 μm and are deposited in the mucous membrane of the intestine of the host. Here the eggs hatch and larvae come out into the lumen of the

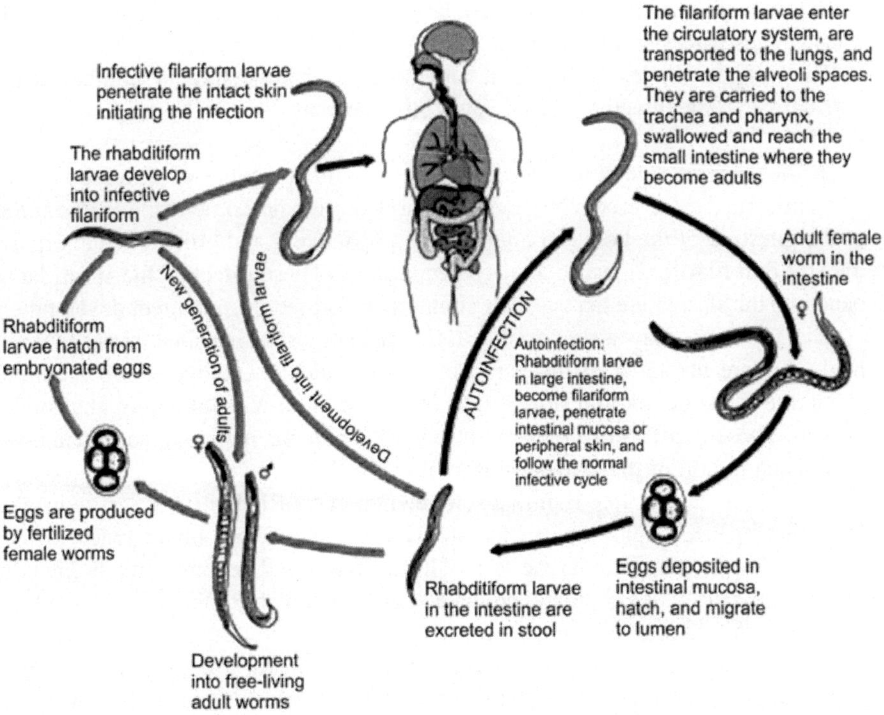

Fig. 3.10 Life cycle of *Strongyloides stercoralis* (*Advanced Parasitology*, Das)

intestine. They are then expelled along with the faeces of the host. The egg output is less than 50 per day.

The larvae in the faeces are Rhabditiform larvae and measure 300–380 μm long and are very slender. They are first stage juvenile (J1) and are passed with stool. These juveniles have two way of development either develop into free living organisms or infective one. The infective form is J3 measuring 490–630 μm in length. These parasites are called filariform juveniles and development is arrested. Then these J3 stages of the worm is infective and wait for the chance to penetrate into the skin of the human host or by ingestion. Hookworm larvae but they can be distinguished by their very small buccal cavity whereas the buccal cavity of hookworm larvae is comparatively large.

The development of the larvae may be direct homogenic or indirect heterogenic.

In the indirect heterogenic development the Rhabditiform larvae grow in 36 h and after four moulting transform into adult male and female. They are 1 mm in length and 40–60 μm broad. These adults copulate on the soil and produce fertilized eggs which hatch Rhabditiform larvae like parasitic females reside in intestine of the host.

These Rhabditiform larvae result in heterogenic development after two moultings transform into filariform larvae. They are slender and long having an oesophagus and a long tail notched at the tip. These larvae 600–700 μm long and have a cuticular sheath. These are infective forms.

They appear in less than 48 h and become innumerable in number within 5–6 days. These infective forms enter into the body of the host by penetrating the skin.

In the case of free living Strongyloides in direct course of development the Rhabditiform larvae produced from the parasitic female come to lie in the lumen of the intestine of the host and after a period of feeding and growth metamorphose directly into filariform larvae after two moultings. These infective filariform larvae penetrate the skin of the host as their counterpart formed from indirect development.

The infection may also possible through another path called hyper-infective method. Here the larva of the parasitic female quickly undergoes two moultings inside the intestine and transforms into filariform larva without taking any food or growth. These filariform larvae bore through the mucous membrane of the intestine or perianal region of the skin causing re-infection.

In short, in case of free living developmental cycle Rhabditiform larvae hatched from the fertilized eggs pass in the stool. They may either moult twice to form infective forms, filariform larvae or moult four times and becomes free living adult male and female. They copulate, produce eggs from which Rhabditiform larva hatch which transforms into infective form filariform larva. They penetrate the skin of the host. After penetration some larvae remain in the skin for a long time. The larvae appear in the lungs after 3 days of infection. The larvae develop to adolescence in the lungs and then migrate to the intestine through trachea and throat. The larvae begin to appear in the faeces after about 17 days of infection in human beings.

3.3.5.6 Pathogenicity

Firstly, the skin penetration by filariform (J3) larvae cause slight haemorrhage and at the site of penetration there may be slight swelling and very strong feeling of itching. Most of the cases are asymptomatic. During migration of larva cutaneous symptom like larva currens is seen. Migrating larva may create lesions in lungs.

During burrowing in the mucus layer of intestinal epithelium abdominal discomfort with pain, nausea, vomiting, diarrhoea, peptic ulcer are the symptoms.

In immune-compromised persons may be due to immune suppression drugs, malignancy, malnutrition, pregnancy conditions hyper-infection with *S. stercoralis* happen. In hyper-infection large number of larvae invade gastrointestinal tissues causing colitis and dehydration. The larvae also may attack CNS, peritoneum, kidney and liver.

In severe cases secondary infection may take place resulting in septicemia, meningitis, loeffler's syndrome, etc.

3.3.5.7 Host Immune Response

S. stercoralis shows an autoinfection cycle permitting the infection to stay for a long time in an immunocompetent host also. In such a host the population of the parasites generally remains under control due to increased level of IL-5 secretion along with high level of IgE antibodies against *S. stercoralis*.

In immunocompromised patients the parasitic population of adult and filariform larva increase tremendously. This proliferation in number finally causes injury to almost all organs and systems making them an important opportunistic parasite. In such patients, there is an increased level of IFN-y and IL-10 secretions, a decreased level of IL-4, IL-5 and IgE secretions and responses change from Th2 to Th1. Usually Th2 response secretes IL-10.

3.3.5.8 Epidemiology

Human beings are infected with this parasite when the person contact filariform larvae present in the soil. It is observed that the infection happens in the tropical countries but may extends up to the temperate regions. The poor sanitation condition in the countries may cause infection with *S. stercoralis*. In one survey it is estimated that infection may be up to 10% by examining the stool of the people concerned. But by ELISA 72% people are positive. The infection rate also varies with the age group. Increasing age coincides with increasing infection rate.

3.3.5.9 Diagnosis

For detecting Strongyloides infection repeated stool examination is necessary to find rhabditiform larvae. Concentration techniques like formol-ether and Baermann method can also be applied.

The laboratory technician or pathologist should take care to differentiate the rhabditiform larvae of hookworm and that of Strongyloides.

3.3.6 *Trichinella spiralis*

3.3.6.1 History

The disease trichinosis is caused by a smallest nematode called *Trichinella spiralis*. The parasite was first discovered by Owen in 1835 and in the year 1895 it is called *Trichinella spiralis* by Railliet.

The disease occurs in carnivorous and omnivorous animals like rats, pigs and human beings. The nematode is also known as ‘pork worm’ as they are mostly found in pork products.

3.3.6.2 Geographical Distribution

The parasite is common in the USA, Europe and in Arctic regions. It also reported from parts of Africa, China and Syria. But fortunately it is still not been reported from India.

3.3.6.3 Habitat

The disease Trichinosis occurs in Carnivorous and omnivorous animals like rats, pigs and human beings. The nematode is also known as ‘pork worm’ as they are mostly found in pork products.

Trichinella are the smallest parasitic nematode of human beings which has an unusual and complicated life cycle. The adult worms sexually mature in the intestine of an intermediate host like pig. Female worm produces about 1500 larvae, which bore through the intestinal wall and hide themselves in the villi. Then they enter blood vessels and are carried to the voluntary muscles on reaching the muscles they encyst.

3.3.6.4 Morphology

The adult male measures 1.4–1.6 mm in length with 0.04 mm in diameter. At the tail end there are two conical structures called copulatory lobes.

The body of male is long and slender anteriorly. The anus is located at the terminal end. On each side of the anus there is a large copulatory pseudobursa. Copulatory spicule is absent. The characteristic feature of the order Trichinellida is the presence of stichocytes arranged in a row below the level of muscular oesophagus. The females are 3–6 mm long and 0.06 mm in diameter, whitish, slender and tapering from the middle of the body towards the anterior end. In both sexes, the anterior end is pointed and the posterior end blunt and fleshy. The outer covering cuticle has transverse striations. The long capillary oesophagus occupies 1/3rd of the length of the body.

The vulva opens near the middle of the oesophageal region. The anterior part of the uterus remains crowded with larvae, whereas the posterior part contains developing eggs. The females are viviparous and give birth to larvae instead of laying eggs. The anus is present at the terminal end of the body.

Larva

The active larvae measure 10 μm by 6 μm in dimensions. They enter into the voluntary muscles via blood and remain encysted there. Inside the cyst, the larvae continue to develop up to the stage of sexual differentiation and when fully grown increase ten times its original size, i.e. from 100 μm to 1000 μm (1 mm) in length. The maximum size is attained by 35th day. Commonly one larva remains in a single cyst. The formation of cyst is completed within about 9 weeks. Larva within the cyst remains viable for 10–30 years, but finally dies and undergoes calcification.

Encystations of the larvae begin in about 21st day and are completed within 3 months of infection. A blunt elliptical sheath (0.4 by 0.25 mm) develops due to host–parasite reaction around the tightly coiled larva. The long axis of the cyst remains parallel to that of striated muscle fibre. Calcification occurs in 6–18 months. It appears in naked eye as a fine, opaque granule. The muscles which become heavily parasitized are diaphragm, intercostal muscles, deltoid muscles, biceps and gastrocnemius. It is seen that there may be 1000 trichenella cysts per 1 g of muscle.

The larvae finally roll themselves into a spiral structure and become infective after about 17 or 18 days.

After encystment in the muscles further development does not take place until the flesh is consumed by an animal or man. There the worms mature and begin reproducing within a few days.

3.3.6.5 Life Cycle

The life cycle is very unusual the same individual acts as both definitive and intermediate host. Lifespan of the adult worm is very short. Man usually becomes infected accidently by consuming imperfectly or undercooked infected pork meat.

The sexually mature male copulates with female within the mucus membrane of the intestine after 30–32 h of infection. Within the intestinal epithelium the gravid female gives birth within a period of 4–16 weeks thousands of larvae.

The larvae or juveniles go to the hepatic portal system and through blood go to the liver, heart, lungs and arteries from where they are distributed throughout the body. When they reach voluntary muscles they enter into the muscle fibres. There they are nourished, the myofilaments are degenerated. The nucleus of the cell enlarges and SER increases in number. Apoptosis starts with the replacement of mitochondria. Then the cyst is formed around the parasite.

The whole life cycle is passed in one animal like pig, rat, etc. but for the sustenance of the species, they must transfer themselves from one host to other. In case of human beings pigs act as intermediate host. But cycle ends here in case of human beings as human flesh is rarely consumed by others.

Infection usually passes from pig to pig and rat and from rat to rat. Infection to a new host in man is always brought about by ingestion of undercooked flesh of the trichinosis infected animal containing viable encysted larvae (Fig. 3.11).

Larvae encysted in the striated muscle remain viable for sometimes. When the infected flesh is consumed by human beings, the cyst wall is digested in the acidity of the stomach. The larvae are released and migrate to the villi of the intestine to start a new life cycle. The larvae from here carried to muscles via bloodstream where they encyst themselves.

The tissue affected are mostly eyes, tongue and muscles of mastication. Moreover diaphragm, muscles of anus and legs and intercostal muscles are prone to infection.

3.3.6.6 Pathogenicity

Clinical features develop during enteric invasion, larval migration and larval encystations.

When the larvae bore through the intestine of the host pain in the abdomen, nausea, vomiting and diarrhoea develop.

During larval migration through the body and different organs fever, eosinophilia, oedema in the eyes, facial oedema and splinter haemorrhages take place. The migration of worms in the intestinal epithelium can cause traumatic damage of the host tissue and their waste products, their excretion can cause immunological reaction in the body of the host. The resulting inflammation may produce symptoms like nausea, vomiting, sweating and diarrhoea. Five to 7 days after the appearance of symptoms of facial oedema fever may appear. After 10 days intense muscular pain, difficulty in breathing due to formation of cyst in lungs, weakening of pulse, low blood pressure and heart damage may occur due to formation of cyst in heart muscles. At this time extra eosinophilia may occur and body temperature may rise up to 40 °C. Various nervous disorders may occur due to the formation of cyst in the

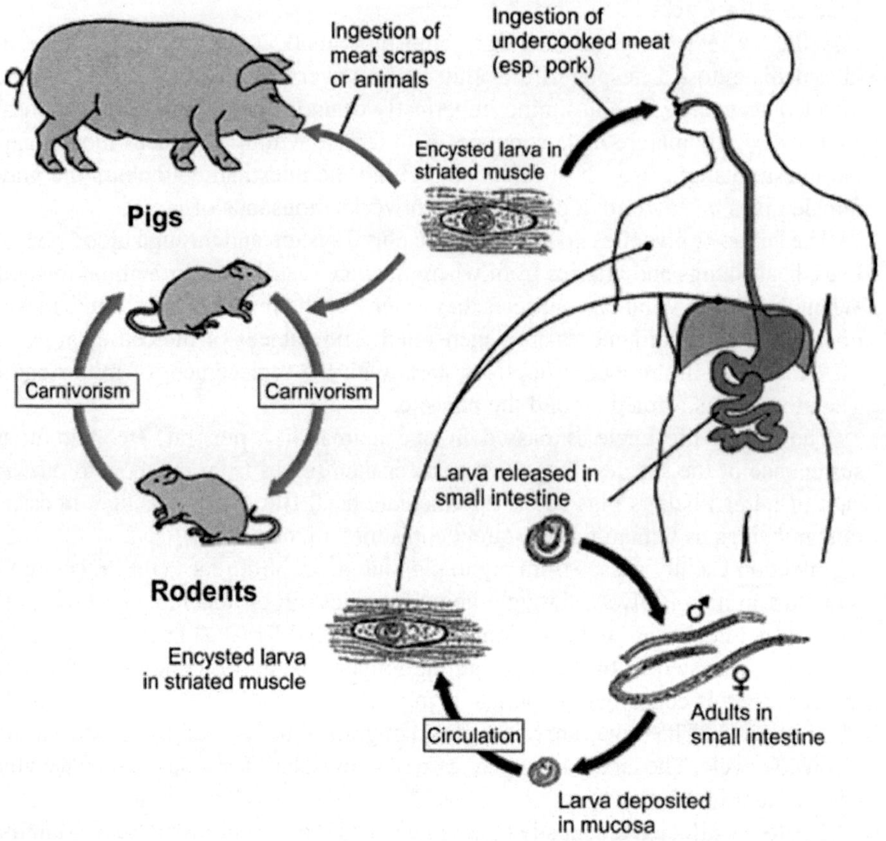

Fig. 3.11 Life cycle of *Trichinelia spiralis* (*Advanced Parasitology*, Das)

brain eventually leading to death. One of the classic signs of *Trichinella spiralis* infection is splinter haemorrhage on the stool.

The first symptom may appear between 12 h and 2 days after ingestion of infected meat.

3.3.6.7 Source of Energy

Both larvae and adults of *T. spiralis* have a functional TCA cycle and electron transport system with oxygen being the final electron acceptor. The energy is generated primarily through oxidative phosphorylation.

3.3.6.8 Host Immune Response

During early infection an increase in lymphocyte, macrophage and eosinophil occurs along with transient increase in the Th2 response. In later stages a limited response is evident. It is reported that the eviction of adult worms from the alimentary canal is stimulated by IL-9, IL-4 and IL-13 secretions.

3.3.6.9 Diagnosis

For definite diagnosis of *Trichinella* infection, biopsy of at least 1 g of suspected muscle is required.

The fresh muscle tissue is compressed between two glass slides and examined under microscope to detect encysted larvae.

Routine histopathology may be done in case of suspected *Trichinella* infection.

Certain tests are necessary like Eosinophilia which develops in 90% of the cases between 2 and 4 weeks after infection. Serum IgE and certain enzymes like creatine phosphatase, lactate dehydrogenase and aspirate aminotransferase level are raised. To detect antibodies against *T. spiralis* Bentonite flocculation test and latex agglutination test may be performed.

3.3.6.10 Epidemiology

The disease trichonisis today is perceived as a zoonotic disease. In the life cycle of the parasite now human as host is not so important. The encysted parasite in the muscles of human beings cannot infect others as the consumption of human flesh by the carnivores is very very remote these days. So the parasites in the infected human host meet dead end.

There is two prominent cycle in the life of Trichenella one is called domestic cycle, i.e. cycle within pigs and rats in human inhabitation.

Another cycle is sylvatic cycle which is between wild animals.

Domestic cycle of *T. spiralis* is very important because of the close approximity of the human beings with rats and pigs. Pigs are infected by devouring the infected rats. Rats are again infected due to their cannibalism habit. Infection spread in the piggery from pigs to pigs. Human infection takes place from eating undercooked pork or sausage.

Sylvatic cycle of trichinosis remains among wild animals. Human infection from sylvatic cycle is very remote. But when people consume game animals may be infected.

3.3.7 *Dracunculus medinensis*

3.3.7.1 History

Dracunculus medinensis is the longest nematode tissue parasite of human beings. This parasite was once abundant in the Muslim dominant population of Africa, India and the middle east.

It is known that at one time it was estimated that 3.5 million cases were found in 20 countries of Africa and Asia in 1986.

The first detailed account of the parasite along with its morphology and life cycle was described by the Russian scientist Aleksej Fedchenko between 1869 and 1870.

The disease caused by this parasite is known as dracunculiasis and the parasite is called Guinea worm. Actually the prime agent of the disease is the female nematode that parasitizes human beings.

Dracunculus medinensis meaning 'little dragon of Medina' inhabiting western Asia, 'fiery serpent' as it is known from the time of Bible is one of the important causes of great sufferings of life of human beings from central India to Arabia and is also an important parasite of East India, Egypt and Central Africa. Stoll (1947) estimated guinea worm infection in 48,000,000 human beings of the world. The disease is associated with dry climate because of less water supply and concentration of water reservoir or step wells in some restricted place, where the possibility of drinking water contaminated with copepods (Cyclops) is high.

3.3.7.2 Habitat

In the USA it has been founded in dogs, raccoons, foxes, etc. The adult worms occur in the subcutaneous tissues of legs, arms and shoulder of the definitive hosts.

3.3.7.3 Morphology

The gravid female with eggs resides in the subcutaneous tissue where they can be felt as loose coils or like varicose vein.

The female is 100–400 cm in length and has been recorded up to 800 mm in length. The male is only 12–30 mm in length. The triangular, small mouth sclerotized plate around it, reduced lip at the level of amphids there are outer circle of cephalic papillae. The head end is bluntly rounded and usually rupture to expel embryos. The tail is attenuated and sharply hooked. At the anterior end there is a cuticular ring bearing mouth. They have an inner circle of six labial papillae and an external circle of four double papillae. The posterior end of male possesses ten pairs of papillae: four pairs pre-anal and six pairs post-anal.

The oesophagus is divided into two parts: an anterior narrow muscular region and a posterior broad glandular region. The female bears a didelphic uterus and vulva is present near the middle of the body. But in adults it is degenerated and becomes non-functional. The male has a pair of copulatory spicules. The spicules are not equal and 490–730 μm in length. Females are viviparous. The first stage larvae which are discharged in water measure 500–700 μm in length and have a round anterior end measure 15 μm. The tail is long and pointed.

3.3.7.4 Life Cycle

Human beings are infected by drinking water in which living copepods (Crustacean) are thriving. Copepods also must be infected with larvae of *D. medinensis*.

The definitive host of *D. medinensis* is human beings and the intermediate host is copepods. The type of copepods is Cyclops. These Cyclops are very very small and can only be seen under microscope.

The mature long female parasitic nematode contains quite a large number of embryos in their much enlarged uterus. The inseminated female with fertilized eggs migrates to the subcutaneous tissue mainly towards the surface of the skin of knee, legs and shoulder. They form a blister by the toxic secretion from their posterior end. Here soon an ulcer is formed due to the rupture of the blister. The blisters are formed in that part of the host which regularly comes in contact with cold water. The patient due to irritation and burning sensation in the ulcerated portion seek to relieve the local discomfort by placing the ulcerated portion in the cold water. The female parasitic worm now projects it uterus and releases a milky fluid containing large number of very very small coiled embryos in water. These embryos are called first stage larva. The first stage larvae are 600 μm in length, one-third of which is the long filamentous tail.

It is reported that after 1 h or so again with the contact of cold water fresh ejection of larvae happens. In this way all the supply of embryos is exhausted. After each ejection the protruded portion of the uterus dries up and sealing of the uterus takes place and save the other larvae for next turn. This intermittent releases increase the chance of larvae invading the Cyclops. This is an unique adaptation in behaviour of the female parasite which gives fair chance of her offspring to find cyclops, the intermediate host (Fig. 3.12).

When all the embryos are exhausted the mother dies and dries up in the tissue of the host waiting for absorption.

The released larvae (J1) swim about in the cyclops infested water and remain alive for 4–7 days. But this infective capacity remains for only 3 days. They are now swallowed by cyclops. After entering into the intestine of Cyclops the larvae (L1) go into the haemocoel after 1–6 h where they moult twice and develop into L3 within 12–14 days at 25 °C.

Scientists have never found more than 2 larvae in the coelom of a cyclops. In experimental condition in the laboratory it is seen that if more than 4 or five larvae reach the body cavity the development of the parasite is interfered.

Human beings are affected by drinking water containing such infected Cyclops with third stage larvae. When the Cyclops, the tiny crustaceans, are digested in the stomach of the host, larvae are released.

The third stage larvae of *D. medinensis* are the infective forms. Human beings are affected by drinking water containing such infected cyclops with third stage larvae. When the cyclops, the tiny crustaceans, are digested in the stomach of the host larvae are released. The active motile third stage larvae of *D. medinensis* then reach small intestine and from here they bore through the intestinal wall and reach retroperitoneal space. In this quite space of the host they mature in about 12–20 days after infection. Then they mate and after copulation the male dies and inseminated female with full of fertilized eggs migrate to the subcutaneous tissue beneath the skin of the host to discharge embryos in the water between 10 and 14 months. The cycle is completed.

3.3.7.5 Energy Metabolism

The adult female is seen after filled up with a coloured substance may be blood. The mature females found to have glycogen in their tissues. The glucose is utilized and formation of lactic acid is not interfered by the presence and absence of oxygen.

3.3.7.6 Pathogenicity

In western India especially Rajasthan people use step wells. Here people instead of using buckets and ropes enter into the wells by using steps and fill their containers

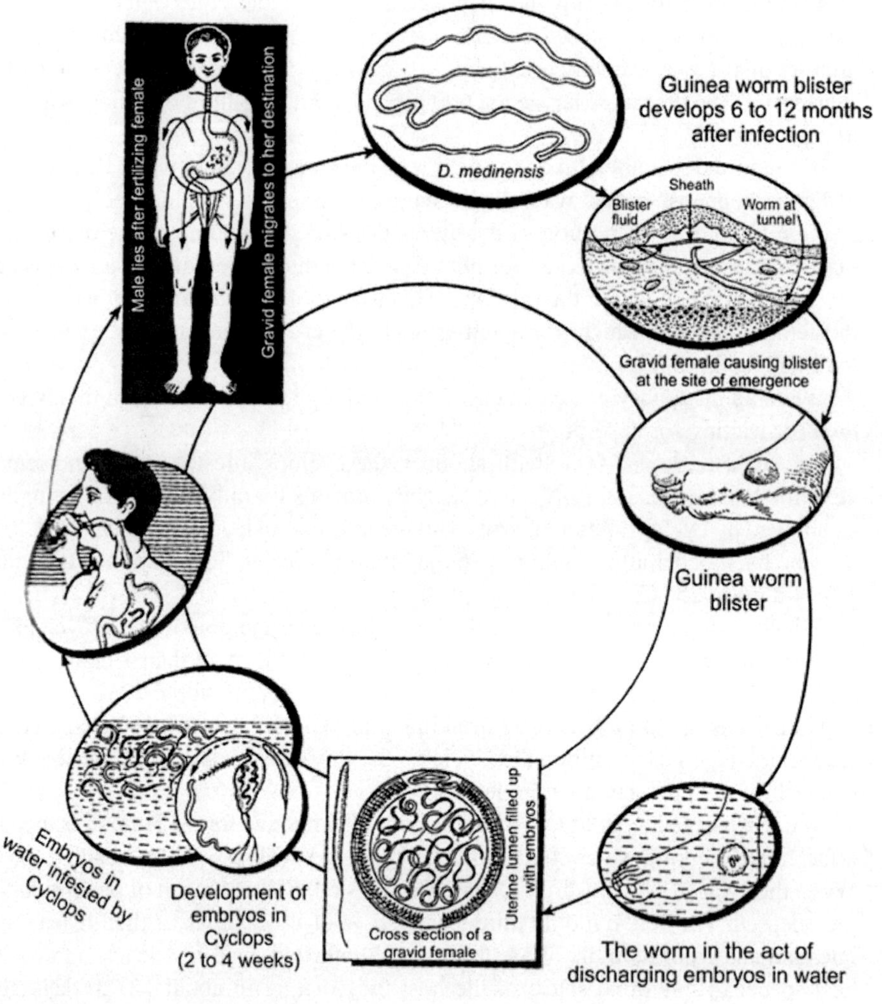

Fig. 3.12 Life cycle of *Dracunculus medinensis*

standing in the knee deep water. This habit is exploited by the parasite to discharge their offspring into the water.

The people at the same time withdraw previously infected cyclops in their containers.

The symptoms of dracunculiasis appear when blister is formed in the host's skin. Due to secretion of toxin by the parasite to make an injury on the skin so that larvae can be discharged from there, nausea, vomiting, diarrhoea, asthma, giddiness and fainting may happen. All these symptoms suggest allergic reaction.

Then in the ulcerated portion secondary infection starts by bacteria which may produce abscesses. This may turn into severe infection and surgical amputation may require or may lead to septicemia. This may result in death of 1% of cases.

Sometimes the gravid female worm cannot reach the skin in that condition the worm dies. The dead worm is encapsulated which is later becomes calcified. This may result in very painful incapacitating arthritis.

3.3.7.7 Diagnosis

Most easy examination is testing of milky fluid released by the worm at the lesion site on exposure to water. In this milky fluid quite a large number of larvae will be seen. Calcified worms sometimes are detected by accidental X-ray examination. ELISA is available in the market and can be used for serological diagnosis in the cases where no other test is positive but physician suspects Dracunculus infection.

3.3.7.8 Epidemiology

Human beings are infected mainly by ingesting infected copepods in the drinking water. The main factors of human infections are the infected persons must come in contact with the water where the larvae will be discharged.

3.3.8 *Wuchereria bancrofti*

3.3.8.1 Geographical Distribution

The distribution is widespread. It occurs almost entirely in coastal areas and islands where there is a fairly long hot season with high humidity (Augustine 1945). The countries where the parasite is prevalent are Africa, east and west coastal areas but not in the interior of central Africa, India (mainly coastal areas), Korea, Japan, China, Arab, East Indies and South Pacific Islands.

In the western hemisphere it is prevalent throughout the West Indies and on the north coast of South America from Brazil to Colombia.

Wuchereria bancrofti causes filaria the disease is also known as elephantiasis in human beings causing permanent bad looking huge swellings in the legs and genital system of both the sexes of human beings and make the person permanently disabled.

3.3.8.2 History

In 1863 the microfilaria, the larva of *Wuchereria bancrofti* was observed by Demanquay in the hydrocoelic fluid of man. The adult female worm was first discovered by Brancroft in human beings in 1876. The adult male was first described by Bourne in the year 1888.

That the insect vector is a female culex mosquito was reported by Manson in the year 1878.

The structure and behaviour of the worm are extensively worked out by a number of scientists the latest is Nelson (1990).

3.3.8.3 Morphology

The adult worms reside in the lymph glands and ducts from where it is very strenuous to extricate.

Table 3.2 Characteristics of sexual differences in Wuchereria

	Male	Female
1.	Measures 2–5 cm in length, 0.1 mm in diameter	Measures 8–10 cm in length, 0.2–0.3 mm in diameter
2.	The tail end of adult is curved and blunt	The tail end of adult is narrow, abruptly pointed and straight
3.	Presence of penial spicules in adult	Absent
4.	Vulva aperture is absent	Presence of vulva with ovijector

The male and female remain coiled together in such a way that it is very difficult to separate them.

The adult male is 35–40 mm in length and 0.1 mm in diameter. The female worm measures 65–100 mm in length and 0.25 mm in diameter. The anterior portion is tapering and ends in a swelling head having an aperture of the mouth. Some portion of the oesophagus is muscular and rest glandular. The posterior tail end of male is curved ventrally and has two unequal penial setae and a number of genital papillae. The tail end of female is abruptly pointed. The vulva of female opens a little behind the middle of the body. Vulva is armed with a pyriform ovijector or ovipositor. The two uteri are long and are present along the entire length of the body. The lifespan of the adult worm is 4–5 years.

The differences in sexual characters in Wuchereria is given in Table 3.2.

3.3.8.4 Life Cycle

The filarial worm needs only two hosts: vertebrate, man and other animals and invertebrate insect vector female culex mosquito.

Sexes are separate and show sexual dimorphism.

They reside in human lymphatic channels. They usually live in different lymphatic vessels near lymph glands. They mainly remain in the lower half of the body. Adults also reside in the lymphatic vessels of scrotum, testes, epididymis.

The copulation takes place in the definitive host like man and females give birth to microfilariae, the larval forms which are surrounded by delicate sheath are active, colourless transparent and possess long, slender bodies having blunt head and pointed tail.

The microfilariae measures 290 μm in length and 6–7 μm in diameter. The covering sheath of the worm is much longer than the actual body about 350 μm in length and larva can move forward and backward. Near the posterior end of the body there is anus. The sheath stains red with dilute Giemsa stain. Tail end tapers uniformly and absence of nuclei in tail. The periodicity is nocturnal.

Microfilariae are born about 1000 in number and they appear in the peripheral blood at night from 10 PM to 4 AM. There are extensive investigations about the nocturnal periodicity of the microfilariae.

One theory is larvae after birth hide in the internal organs of the host and come out at night because their further development will not take place in the definitive host. They wait for the mosquitoes to suck them and that opportunity comes, only at night.

According to Hawking and Thursnton the microfilariae remain concentrated in the capillaries and other blood vessels of lungs but the stimulation that initiates the microfilariae to circulate in the peripheral blood is the resting period of the host that is during sleeping. For examination of blood for microfilariae the patient is asked to lie on the bed for at least 2 h and then the blood sample is taken for pathological test.

Further development of the microfilariae depends upon the sucking of blood by the female Culex mosquito when the blood with microfilariae enters into the stomach of the mosquito.

It is now proved that the insect vector of filarial worm is restricted not only to Culex, the Aedes, Anopheles and other blood sucking insects may also act as vector.

But *C. pipiens* play a leading role as vector and in some places Anopheles may also substitute Culex.

As the mosquito bite and starts sucking infected blood the microfilarie enter into the proboscis and they escape from the proboscis to the G.I. tract.

In the stomach of the insect vector mosquito microfilariae shed their sheath, bore the stomach wall and go to the thoracic muscles or muscles of the wing where they undergo development. On the third day they transform into sausage shaped organism with a short spiky tail. The organisms which are first stage larvae are 125–250 μm in length and 10–17 μm in breadth. The first stage larva has a rudimentary alimentary canal and a nerve ring.

The first stage larva moults and develops into second stage larva measuring 225–350 μm in length and 15–30 μm in breadth.

The second stage larva moults again and third stage larva is formed on tenth or 11th day. At first the tail degenerates to become short in length and alimentary canals, body cavity and genital organs differentiate. The worm now starts growing in length measuring 1500–2000 μm in length and 20–30 μm in diameter. The third stage larva is the infective form which is formed on 10–20th day. The infective microfilaria is a filariform larva. It should be noted that no multiplication of the parasites occurs in the body of the mosquito.

Now when the infected mosquito bite a human being for sucking blood the infective forms of parasites are not injected into the blood of the final host. The filariform larvae are deposited on the skin and the larvae enter into the body of the host through the punctured site. Gradually with the bite of the mosquito the larvae escape from the proboscis and time taken by all the larvae to come out from proboscis is near about 3 weeks.

The third stage larva that enters the human host through the punctured site of the bite of the mosquito invades the lymphatic vessels and reaches inguinal, scrotal and abdominal lymphatic vessels where they grow to maturity. No reproduction or multiplication takes place here. One filariform larva transforms into one adult either male or female. They become sexually matured in about 12–18 months. Then they start copulation and gravid female discharge 50,000 microfilariae per day after fertilization. The microfilariae swim through the thoracic duct or the right lymphatic duct to the venous system and pulmonary capillaries and then pass to the peripheral circulation. In this way the filarial worm completes its life cycle (Fig. 3.13).

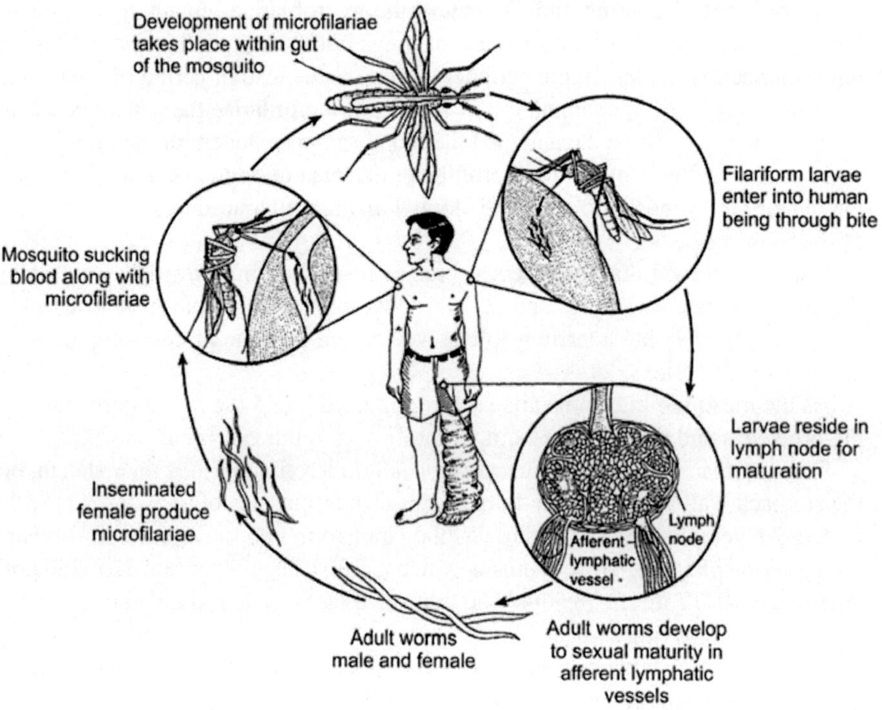

Fig. 3.13 Life cycle of *Wuchereria bancrofti*

3.3.8.5 Pathogenicity

Microfilariae generally do not produce any noticeable symptoms. The symptoms are visible due to the obstruction and inflammation of lymph glands and lymph channels. The inflammation is due to allergic reaction.

In later stages inflammation of lymph glands and lymph channels of male genital organs, legs and arms of host happen. The inflammation sometimes associated with chill, fever and painful localized swellings. This type of filariasis is known as lymphangitis. There is another type of the disease called occult filariasis in which the lesion is caused by microfilariae in the lymph nodes and also in the lungs, spleen, liver, etc.

Lymphangites

The fluid in which the embryos of the worms are discharged often cause allergic symptoms. The allergy may also be due to metabolic waste products of the worms and discharge of protein coming out from dead and phagocytosised worms.

The main symptom of elephantiasis is due to obstruction of lymph channels which arise especially in old infections. The starting of obstruction causes varicose lymph or chyle vessels where the channels are obstructed due to the presence of worms. The varices when burst then divert lymph to the scrotum, bladder, kidney or peritoneum the swellings of scrotum, limbs, breast, vulva become abnormally swollen as a result fibrous tissue increases and the skin becomes hard and dry as the sebaceous and sweat glands degenerate. It is seen that when elephantiasis occurs i.e. when certain portions of the body swells to monstrous proportion number of the micrfilariae in blood decreases. Because the adult worm is dead in the meantime. According to Brown in India the appearance of filarial symptoms after infection is very slow. Brown also reported that he came across with a patient whose scrotum grew 14 lb. in weight from normal size within a year.

In some cases headache, backache, fatigue and nausea are common but fever and uneasiness are unusual, physical and mental depression are very noticeable. Presence of microfilariae in the blood is also very rare. All these were noticed in American troops during Second World War when they were exposed to the non-periodic strain.

Occult Filariasis

When the microfilariae are present in the tissues the disease is called occult filariasis.

Though many symptoms are similar to lymphangitis yet some are suffering from the massive eosinophilia, hepatomegaly, spleenomegaly, enlargement of lymph nodes and pulmonary complications are seen in occult filariasis. It is seen that though adult female gives birth to microfilariae continuously, they are not capable to enter into peripheral circulation as they are continuously degenerate in the tissue due to host's reaction to filarial antigen.

This leads to pulmonary complications and eosinophilia. The symptoms are loss of weight, anorexia, violent dry cough, blood tinged sputum and dysponea.

Differences of characteristics of diseases lymphangitis and occult filariasis are given in Table 3.3.

Adult worm causes dilation of the wall of afferent lymphatics and sinus of lymph nodes. Mechanical blockage and release of toxic substances by the parasites lead to inflammation of lymphatics and lymph nodes resulting obstruction of the lymph channels. Total lymphatic obstruction creates oedema leading to eliphantiasis of the organs affected.

Table 3.3 Difference between lymphangitis and occult filariasis

Characteristics	Lymphangitis	Occult filariasis
Cause	Presence of adult producing microfilariae	Destruction of huge number of microfilariae
Lesion and inflammation	Acute inflammation in the lymph glands and channels surrounding adult worms	Eosinophilic granulose
Organs affected	Lymphatic system	Lymphatic system, lung, liver, spleen
Microfilariae	Present in peripheral blood circulation	Absent in peripheral blood but present in affected tissues
Response to treatment	Absence of any response to any drug	Responds to filarial drug
Complement fixation test	Not sensitive	Highly sensitive

Table 3.4 Lymph nodes are affected due to enlargement of the organs

	Enlargement of organ	Lymph nodes affected
1.	Elephantiasis of leg	Inguinal or iliac
2.	Elephantiasis of arm	Axillary
3.	Elephantiasis of breast (female)	Axillary
4.	Elephantiasis of penis	Superficial and deep inguinal
5.	Elephantiasis of scrotum	Superficial inguinal
6.	Hydrocoele	Para-aortic
7.	Epihidymo-orchitis	Para-aortic
8.	Elephantiasis of vagina	Superficial inguinal
9.	Funiculitis	Para-aortic

Differences in the affected lymph nodes of the enlarge organ are given in Table 3.4.

3.3.8.6 Host Immune Response

In Wuchereria infection there are three stages: (a) The initial infection is called incubation or asymptomatic phase increased Th2 responses take place secreting IL-4, IL-5 and IL-10 secretions; (b) The inflammatory phase known as acute phase, characterized by a change in Th1 cytokine responses, i.e. IFN-γ production; (c) The obstructive phase or chronic phase where increased level of both Th1 and Th2 cytokines responses are found against parasite antigens. The organism that infects human beings is phagocytosed by microphages. After breaking down within macrophage vacuole the broken particles are carried to the surface of the macrophages along with the molecules of class II MHC. The broken particles act as antigen and antigen MHC molecule forms a complex which is detected by T cell receptors. Then the T helper lymphocytes and the complex is bounded tightly. Then the macrophage serves as an APC (antigen presenting cell). APC secretes IL-1 which stimulates Th1 and secretes IL-2, IFN-β, and IFN-γ. This produces immunological specificity which produces memory T and B cells. These memory cells remain in circulation for a long period even if antigen disappears.

3.3.8.7 Diagnosis

The definite diagnosis is made by examining the blood sample in thick smear where microfilaria may be found. The blood should be collected from the capillary at night. The blood collection site should be earlobes than finger. The blood collection site should be earlobes than finger.

Microfilariae can be detected both in wet and stained preparation.

In wet preparation a drop of blood is taken on a glass slide taking from the earlobe of the patient at night and a cover slip is placed over the drop and then is examined under microscope actively moving microfilariae may be seen.

For stained preparation a drop of blood is taken on a glass slide taking from the earlobe at night and streaked then allowed to dry. The slide is then stained with Giemsa stain. After staining the slide is examined under low power of microscope.

If the microfilariae are not detected in conventional method, then concentration technique is followed for cases of strong clinical suspicion.

In concentration technique, either 0.1 ml of capillary blood from earlobe or 10 ml of venous blood is drawn for the test.

Lysed capillary blood technique and capillary tube centrifugation technique are used. In lysed capillary blood 100 μl of earlobe capillary blood is lysed in 1 ml of saponin saline water. The blood is then centrifuged or kept overnight for sedimentation. The sediment is examined under low power of microscope either from wet preparation or Giemsa stained preparation.

In capillary tube technique 50 μl of blood is collected from the capillary tube in two heparinised capillary tubes. The tubes are then sealed and centrifuged in a micro capillary centrifuge machine. After centrifugation the capillary tubes are placed on glass slide and the plasma is examined for microfilariae.

Among other methods serological diagnostic methods are applied.

1. Passive haemagglutination test,
2. Fluorescent antibody test,
3. ELISA.

These tests are done using nonspecific antigens, so exact infection by these tests may not be interpreted correctly.

3.3.8.8 Epidemiology

The vector of *Wuchereria bancrofti*, different species of mosquitoes, all are blood suckers of human beings. Blood sucking mosquitoes are usually night feeders. Because of the factor that a number of species of mosquitoes can transmit the parasite the periodicity has much significance which indicates the particular species of mosquito to be eradicated and what type of control measures to be applied.

During mosquito control it has been observed that there are some sites where mosquitoes may breed but very difficult to control because of their odd sites like tree holes, small unusual places filled with rain water.

It has been observed that human beings are usually infected with the parasite naturally because reservoir hosts have not been found.

3.3.9 *Onchocerca volvulus*

3.3.9.1 History

This nematode parasite was first discovered by Leuckart in 1893 and it was first described by Ralliet and Henry, 1910. In the year 1926 Simulium (blackfly) was described as the vector of *Onchocerca volvulus*.

3.3.9.2 Geographical Distribution

The parasitic tissue nematode *Onchocerca volvulus* causes the disease onchocerciasis in Southern Mexico, Guatemala, Salvador and North West Venezuela, in tropical Africa, Central America and Yemen in Asia.

The disease affects estimated three crore people and cause infectious blindness in human beings due to this *Onchocerca filarial* worm. The infection of this filarial worm was prevalent in Africa but it is introduced in central America not before 1915. In some localities nearly every person is infected and about 5% of them are blind. It is the cause of second most infectious blindness throughout the globe.

3.3.9.3 Habitat

The developing worms within the human beings creep about in the subcutaneous tissue but only when they take rest in the subcutaneous tissue become entangled and form fibrous cysts due to inflammatory reaction. These cysts appear as nodules on the skin. There is a record of nodules found even in a 2-month-old child. It takes longer time to appear nodules on the skin.

In a period of 1 year the nodule grows to a diameter of 1 cm. But usually the growth of the fibrous cyst or nodule is somewhat slower. Within a nodule 3 or 4 worms are seen to remain entangled.

3.3.9.4 Morphology

This tissue nematode exists in three forms: adult, microfilaria and larva.

The adults of this nematode live in the subcutaneous connective tissue of man in knotted condition. The adult worms are long and thin. Their lips are reduced and buccal capsule is very small. The mouth is surrounded by two circles of four papillae.

Males are 20–40 μm in length and 125–200 μm in diameter. There are two ventrally curved spicules at the posterior end of the body. Females are 230–500 mm in length and 0.3–0.5 mm in diameter, the vulva is placed behind the end of oesophagus. The annulation pattern of the cuticle is not same in two sexes. The microfilarae have no sheath around them and measure 250–360 μm in length and 5–9 μm in diameter. A number of nuclei are present at the tail end of the microfilaria and there is a clear space at the tip of the tail.

In Africa in a composite nodule 100 worms were seen knotted together by an investigator. The size of the nodule varies from the size of a pea to a pigeon's egg. Normally one to six nodules found in an infected people. In Africa 25–100 nodules are seen in a person and size is a few millimetres in diameter. Adult worms usually entangle to produce nodule in the subcutaneous tissue. Some investigators found that all the adults do not entangle to produce nodule.

The adult male and female when become sexually matured copulate and the inseminated gravid female releases unsheathed embryos or microfilariae. They measure 250–360 μm in length and 5–9 μm in diameter. The microfilariae escape from the nodules and enter into connective tissue just under the skin where they accumulate in large numbers instead of bloodstream.

3.3.9.5 Life Cycle

O. volvulus needs two hosts to complete its life cycle. Man is the definitive host and Blacklok (1926) discovered that in Africa blackfly, *Simulium damnosum* is the intermediate host. *S. metallicum* also bites man and found to act as intermediate host of *O. volvulus*. It is also discovered that some cattle species of *Onchocerca* is transmitted by *Culicoides*.

The infective forms of *O. volvulus* that infect human beings are the larvae present in the proboscis of the blackfly.

When the fly bites as the mouth parts of the vector are not adapted for deep piercing they suck tissue juices along with blood. So the meal contains a number of microfilaria.

After ingestion the microfilariae reach the gastrointestinal tract and from the G.I. tract of the blackfly microfilariae penetrate the wall of the intestine and enter into thoracic muscles of the fly. They undergo some moulting here and infective larvae are produced in about 6–7 days (Wanson 1950). The infective larvae then find their way into the proboscis of the blackfly from the thoracic muscles and reach there for transfer to another healthy person.

The infective larvae are deposited on the human skin from the proboscis when the fly becomes ready to suck its meal. The deposited larvae find their way into the subcutaneous tissue of the definitive host through the bite wound of the fly. They develop into mature adults in 7 months to 3 years.

The sexually matured adult male and female entangle and mate to inseminate the female within the nodule. The gravid females produce a number of unsheathed microfilariae. They then leave the nodules and creep about in the subcutaneous tissue, dermis and eyes. When they take rest they produce nodules and if in the eyes then blindness occurs. The cycle is completed (Fig. 3.14).

3.3.9.6 Pathogenicity

Both adults and microfilariae produce mainly pathological lesions instead of lymphatic filariasis. The tissue damage of the definitive host is mainly due to the movement of microfilariae and formation of fibrous cyst or nodules that form due to inflammatory reaction and encapsulation of the coiled or entangled adults. These nodules are called onchocercomas.

Onchocercomata is found over coccyx, sacrum and long parts of the lower extremities in Africa. But in Latin America it is common in upper parts of the body. Because here *S. metallicum* usually does not bite below knee. In some parts of Africa the nodules are found on hips, knees, elbows, ribs, etc. and in some parts of Belgian Congo and in parts of Central America nodules are formed on the head even. In Guatemala 95% of nodules are formed on the head around ears. The location of the nodules does not depend on the site of the bite of the intermediate host as the migrating microfilariae roam about and are imprisoned at any place where tissue reaction takes place to encapsulate them.

The nodules on the skin cause no serious disturbances to the patient. But in Africa onchocerciasis leads to prurities, loss of elasticity of the skin, atrophy and fibrosis. The skin is thickened, highly pigmented, scaly and form lizard like skin around the

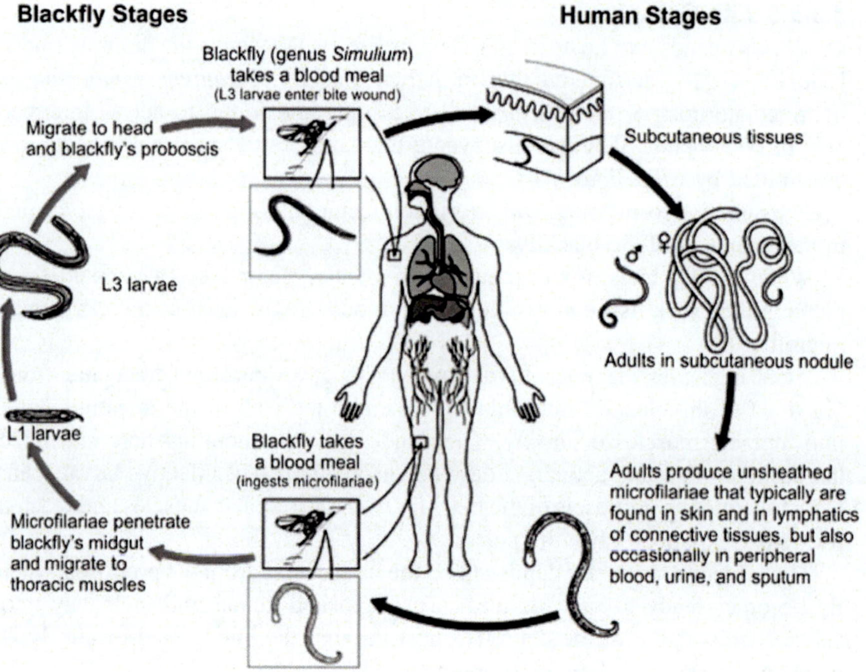

Fig. 3.14 Life cycle of *Onchocerca volvulus* (*Advanced Parasitology*, Das)

middle part of the body of the patient and skin wrinkles appear. Another peculiar symptom is the loose skin hangs containing enlarged lymph nodes around the groin.

In patients with moderate to heavy infections cause blindness due to the entry of the microfilariae in the eye. In cornea they create sclerozing keratites and optic atrophy may also occur.

In Africa lymph channels are more involved than America. Lymph glands become enlarged like filaria leading to elephantiasis of scrotum and legs.

Regarding the interference Strong et al. (1934) investigated the condition in Guatemala where the embryos escaping from the nodules on the head invade the tissues of the eye like conjunctiva, cornea, iris even optic nerves. The blindness occurs among adult patients with 4 or 5 years old nodules.

3.3.9.7 Host Immune Response

In places where onchocerciasis is endemic, three categories of individuals may be identified. One, individuals who are continuously exposed to *Onchocerca volvulus* apparently resistant to infection, they exhibit a dominant Th1 cellular response. The second category consists of individuals who possess microfilariae in their skin and are clinically symptomatic and are characterized by lack of production of Th1 response. Third category exhibits onchocercal skin disease and develops pathogenic immune responses with tremendous reactions to filarial antigens. The last two categories have increased level of IL-10 secretion for protection.

3.3.9.8 Diagnosis

Successful diagnosis is achieved by the demonstration of adult worms in nodules by biopsy or cutting a small portion of the skin of suspected patient by scalpel. This piece of the skin if dipped in saline solution and incubated for 2–4 h in 37 °C temperature the microfilariae will come out and may be seen under microscope.

3.3.9.9 Epidemiology

The vector of this parasite is known as blackfly, Simulium. They lay eggs in the free flowing rivulet and grow in well oxygenated atmosphere. Adults live in high humid atmosphere having adequate vegetation on the sides of the streams. Long ago African people associated the disease onchocerciasis with river and the disease was called 'river blindness'. It was found that the disease occurs in West Africa along the Nile river.

3.3.10 *Loa loa*

3.3.10.1 History

Sanground in the year 1936 for the first time described the adult worm collected from monkeys in Belgian Congo. Duke and Wigers in the year 1958 opined that this eye worm of man has not been transferred from monkeys. He concluded that infected monkeys in the Cameroons are not the reservoir of man as the monkeys live in forest and never have the chance to come in contact with man. At the same time the vectors live in forest canopy do not bite man.

3.3.10.2 Geographical Distribution

It is found in the Central and South Africa mainly in the rain forest areas. This nematode parasite is the cause of the disease loiasis or calabar swellings.

3.3.10.3 Habitat

The adults reside in the subcutaneous tissue of human beings. These parasitic tissue nematodes make journey to different places under the skin causes irritation, itching and a feeling of creeping sensation. It is found that these parasites prefer to move around the eye and at the same time love hot and warm atmospheres. This has been proved by the fact that when an infected person remains in front of fire, the worms become active and go to the exposed part of the body. Their speed of the movement is 2 in. per minute.

3.3.10.4 Morphology

The adults of this worm reside in subcutaneous tissue and tissues of intermuscular connective one. The adults have no lips having a simple head and eight cephalic papillae. Their body is slender with a blunt tail. The body is covered with cuticle of both the sexes having warts along the lateral lines. But the cuticle is absent in head and tail. The adult male is 20–35 mm in length and 0.4 mm in diameter. There are quite asymmetrical pre-anal and post-anal papillae. The body of both the sexes is covered by a cuticle having numerous small warts along the lateral lines. Pre-anal papillae are three pairs and post-anal papillae are five pairs in number. There are two copulatory spicules measuring 123 and 88 μm in length. The adult females measure 20–70 mm in length with 0.5 mm in diameter. The tail measures about 265–300 μm in length. The vulva is located 2.5 mm from anterior end.

The microfilariae are sheathed and move and swim in bloodstream in the day time but disappear at night just opposite to bancroftian filarial microfilariae. They have a diurnal periodicity. They measure 250–300 μm in length and 8–10 μm in diameter. There is a large cephalic space, the nuclei to tip of the tail is 275 μm. The sheath does not take stain in Giemsa, tail short and recurved with nuclei to tip of the tail 300 μm in length. The living specimen stains with methylene blue.

3.3.10.5 Life Cycle

Loa loa needs two hosts to complete its life cycle: the vertebrate host, human beings and intermediate host, or vector insect Chrysops (Deerfly).

When the infected Deerfly bites a healthy person the larvae of *Loa loa* come out of the proboscis of the insect in a row or file and enter into the skin of the definitive host through the site of bite (Fig. 3.15).

The larvae are very much active and remain just under the skin and then with the passage of time take residence in the deeper part of the body. When they come in and

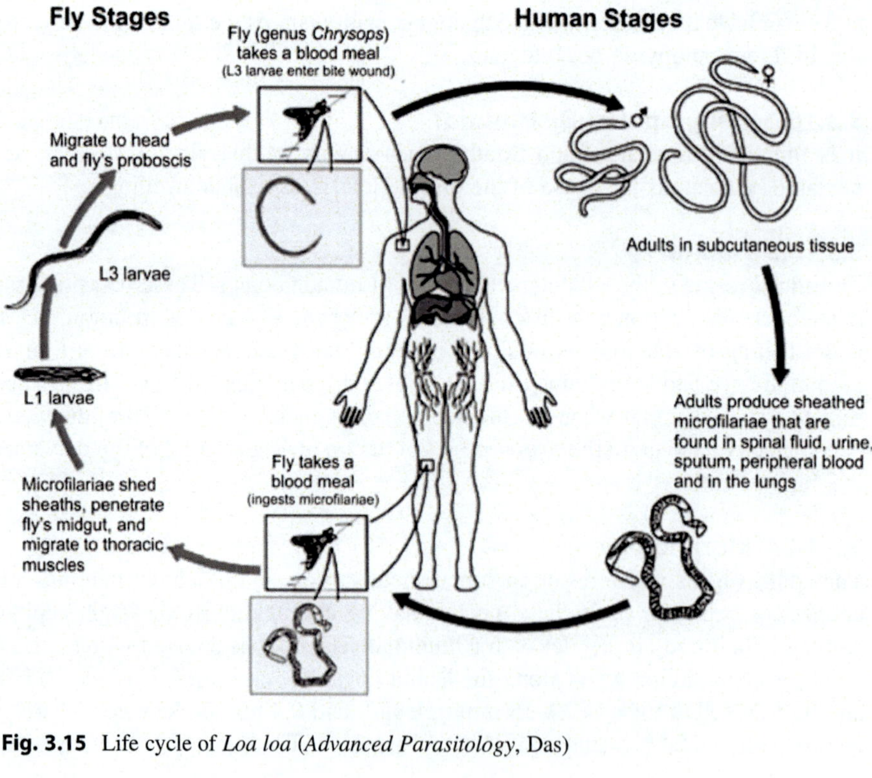

Fig. 3.15 Life cycle of *Loa loa* (*Advanced Parasitology*, Das)

around the eye the patient feel intolerable pain but this is the most convenient place from where they can be easily extracted. The removal of the worms should be done very quickly as the disturbed worm will go deep into the eye for hiding.

The large number of larvae of *Loa loa* enter into the skin of the host and they mature into adults. The adults reside in the tissues of chest, back, groin, penis, axilla, scalp and eyes of human beings. The adults copulate when they remain in subcutaneous tissue. The gravid female is inseminated and release microfilariae. Sheathed microfilarial are found in peripheral blood of human beings in the day time. The development from larvae to adult in human beings takes about 6–12 months. The lifespan of adult worm is 15 years or more.

When the sheathed microfilariae swim in the bloodstream the chrysops come to suck blood. During their blood meal the microfilariae enter into the G.I. tract and from stomach they bore the stomach wall and reach thoracic cavity from here they enter into the thoracic muscles. Worms transform into filariform or larvae (L3) in the fat body of abdomen of the insects. Ultimately goes to the mouth parts of the insect vector where they develop into infective form. Then they wait for transfer to human body. The life cycle is completed.

3.3.10.6 Pathogenicity

Loa loa causes the disease called loiasis. The infections are painless and asymptomatic. But sometimes allergic reactions take place. This reaction leads to the formation of swellings called 'Calaber swellings' as large as the size of pigeon's egg. These swelling are itchy and oedematous. This appears suddenly and last for a few days and then disappear. This allergic reaction is due to metabolic products of the worms or protein coming out from the dead and degenerated worm. Sometimes adult worms enter into the eyes and wander through conjunctiva and cornea.

Microfilariae in rare cases enter into the brain and cerebrospinal fluid. Kivits reported such four cases of fatal encephalitis. Eosinophilia and monocytes are very common.

3.3.10.7 Diagnosis

Diagnosis is done by examining the peripheral blood in wet or stained preparation. The adult worms sometimes are isolated from the eye or from biopsy of subcutaneous tissue. Peripheral blood is collected between 10 AM and 3 PM because the worms are diurnal. Raised level of Eosinophilia and increased level of antifilarial antibodies are detected. Calabar swellings on the skin are also found.

3.3.10.8 Epidemiology

It is prevalent in Africa and sometimes in West Indies for a short time. The control of Deerfly which is very difficult still has reflected some positive results. It is the third most important disease of Africa. It is estimated that 3–13 million people are infected in these areas.

3.4 Digeneans

3.4.1 *Fasciola hepatica*

3.4.1.1 History

Jehan de Brie in the year 1379 first reported the existence of *Fasciola* in the liver of sheep. Pallas in the year 1760 observed the presence of this parasite in human beings. Leukart and Tomas in the year 1883 described the life cycle of this parasitic trematode. Jan Swammerdan described cercaria and rediae of *F. hepatica*. In the year 1816 Prof C.L. Nitzsch detected the similarity between cercariae and adult liver fluke.

In the book alternation of generations published by Johannes in 1844 showed that trematodes have two generations. George Rolleston in the year 1880 indicated that a common snail is the intermediate of liver fluke. Thomas, a 23 year young infected experimentally the snail *Lymnea truncatula* and observed the entire life cycle. It is very interesting that Leuckart described the development of *F. hepatica* in the same species of snail but published his description just 10 days before Thomas. Now credit is given to both the workers.

3.4.1.2 Geographical Distribution

They are cosmopolitan in distribution. *F. hepatica* is the parasite of the far East. The endemic areas include Japan, Korea, Formosa, Southern China, Indo-China and North Vietnam.

3.4.1.3 Habit and Habitat

The parasitic helminth resides as parasites in the herbivorous grazing animal. They reside in the liver and cause the disease liver rot. They live in the liver of sheep, goat, horse, ass and human beings. These parasites require two hosts to complete their life cycle. The second host or the intermediate host is snails of Lymnaea group.

3.4.1.4 Morphology

The body is dorsoventrally flattened and leaf like in appearance. Anterior portion is relatively round and broad. Posterior portion is slender and pointed. They measure 2.5–3 cm in length and 1.2–1.5 cm in breadth. Fresh specimen is pink coloured and margin of the body is black or grey. At the anterior end there is a triangular cone like structure this is known as oral cone. Beyond this cone the body is broadened and at the posterior vent region it becomes slender.

At the anterior there is a sucker some distance away from the anterior terminal end. At the centre of the sucker there is the mouth so the sucker is known as oral sucker. Three to 4 mm below the oral sucker there is a comparatively large sucker called ventral sucker or acetabulum. The suckers are 1 and 1.6 mm in diameter. These suckers are organ of attachment by these suckers they remain attached to the body of the host. At the posterior terminal end there is the excretory pore and on the ventral side between the two suckers the genital aperture. During breeding season there appears another pore called Laurer's canal aperture. The entire body is covered by a thick cuticle with scale like spines which prevent the organism from being digested by the digestive enzymes of the host. The highly branched intestinal caeca extend up to the posterior end of the body.

The large and highly branched male reproductive organ testes placed behind the ovary. The branched, small ovary located on the right side and the uterus is small. It is placed in between ovary and cirrus pouch. Behind the testes there are huge vitelline follicles which fill up the lateral side of the body.

3.4.1.5 Life Cycle

The parasite requires two hosts: definitive and intermediate to complete its life cycle. The vertebrate hosts are definitive hosts as the parasite undergoes sexual reproduction here, asexual reproduction and development of larvae take place within the invertebrate host snail, so the life cycle is digenetic. A single liver fluke can produce 3000–3500 eggs for 11 years. The flukes are hermaphrodite having both male and female reproductive organs in the same individual.

After fertilization within the body the fertilized eggs liberate from the body and escape into the alimentary canal of the host. From here the eggs escape outside in the environment along with the faeces of the definitive host (Fig. 3.16).

Eggs

Eggs are elliptical in shape and grey in colour and 139–140 μm in length and 80–90 μm in breadth. Eggs have an operculum. The eggs remain within a chitinous shell. The eggs develop rapidly within the shell in humid atmosphere. Within 2–3 weeks egg transforms into ciliated miracidium. Miracidium larva comes out of the eggshell and swims in the water.

Miracidium larvae have multicellular body covered with cilia. There is an apical papilla at the anterior end and a pair of cephalic glands. In the body there is a ganglion, two eye spots and two flame cells. Miracidium searches for the snail, *Lymnaea truncate*. If they do not find the suitable intermediate host they ultimately perish. The miracidium larvae with the help of their apical papilla and cephalic glands enter into the mantle cavity of the snail. Within the intermediate host they cast off their cilia and grow in size to form sac like sporocyst.

Sporocyst is sac like and it can divide to form another sporocyst. The sporocyst contains germ cells. These germ cells have the ability to produce redia by asexual reproduction. So sporocysts are known as germinal sac. A sporocyst produces four to eight rediae. The rediae come out of sporocyst and enter into the digestive gland of the snail.

Redia is hollow, cylindrical and 1.3–1.6 mm in length. The body is covered by a thin cuticle. The body has a collar like structure and there is a birth pore behind the collar. The body contains germ cells and flame cells. These germ cells have the power of giving birth to new redia. The daughter redia comes out through the birth pore. In summer several such generations are produced. In winter the germ cells divide again and forms cercaria. Each and every redia can produce 14–20 cercaria.

Cercariae are dorsoventrally flattened and heart shaped. There is a slender tail at the posterior end of the body. The body measures 0.25–0.35 mm and covered by

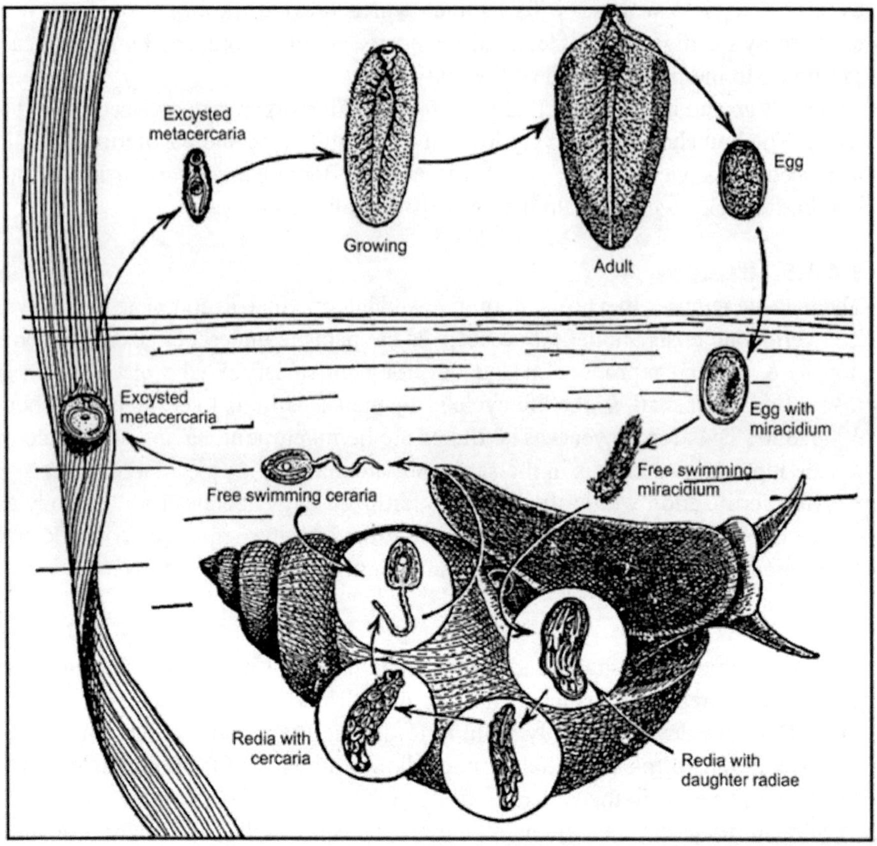

Fig. 3.16 Life cycle of *Fasciola hepatica*

cuticle. Flame cells and dividing germ cells are found in cercaria. The tailed cercaria escapes out of the birth pore of redia. The tailed cercaria comes out in the water and swims for 2–3 days. Generally, at 9–26 °C the cercaria comes out of the redia into the water and swims up to the sides of the water body where they form metacercaria on the leaves of the vegetation. In the meantime they undergo metamorphosis and settles down on the leaves of the aquatic vegetation. The tailed cercaria sheds off its tail and forms a cyst around the body by the secretion of cystogenous glands present in their body and they are now called metacercaria. Metacercaria in cyst condition can live for some months in humid atmosphere.

Metacercaria is round in shape with 0.2 mm in diameter without tail but other characteristics are same with that of cercaria.

When the grazing herbivorous animals like sheep, goat, etc. consume vegetation with metacercaria they enter within the body of the final host. The cyst wall is dissolved in the duodenum by the digestive enzymes of the host and penetrates the wall of the intestine to reach peritoneum cavity. The parasite then enters into the liver and finally gains access into the bile duct. Within the bile duct they are matured. The time taken for the migration of the parasite is about 1 month. After 3–4 months of infection the eggs come out through the faeces of the final host. The cycle is completed. The asexual reproduction taking place in the embryo is called polyembryoni. The lifespan of an adult fluke is about 11 years.

3.4.1.6 Mode of Infection

The vegetation with metacercaria on the leaves if taken by the human beings and consumed uncooked the possibility of infection is high but accidental.

3.4.1.7 Source of Energy

The main source of energy for Fasciola and other digenetic trematodes is the substrate level phosphorylation via glycolysis; glycogen and glucose are the principal carbohydrates metabolized. In *Fasciola hepatica* oxygen is used when available but that oxygen do not satisfy the overall energy requirement. It is controversial whether functional TCA cycle is present in Fasciola or not. Even if the TCA cycle is present, then also production of energy is minimum.

The dependence of the digenetic trematodes on glycolysis for energy helped the pharmacist to design drug for treatment of the patients infected with this parasite.

The miracidia and cercaria of all species are obligate aerobes depending on oxidative phosphorylation for energy. But when remain within the intermediate host, snail, they depend upon substrate phosphorylation for energy just like the adults.

3.4.1.8 Pathogenicity

Primarily the patient suffers from diarrhoea and vomiting. Fever, biliary colic, cough, jaundice are also some of the symptoms of fascioliosis. The parasites in adult condition consume liver cell and drink blood. Sometimes for this reason anaemia may develop. In heavy infection cirrhosis of liver and jaundice occur. It is also observed that wandering larvae may produce ulcer in locations like eyes, brain, skin and lungs. The excretory and secretory products of the parasite are used as antigen and help us in immunodiagnosis. Oedema of the bile duct starts when *Fasciola* resides in the bile duct. Due to their presence in the bile duct fibrous tissues are produced which make the lumen of the duct slender resulting less drainage of bile in the alimentary canal. In severe infection gall bladder, the wall of the bile duct and liver suffer damage and liver rot starts.

3.4.1.9 Host Defence Mechanism

Host immune response to *Fasciola hepatica* excretory/secretory antigens are Th2 type. Humans infected with *F. hepatica* develop specific antibodies of IgM, IgA and IgE class.

3.4.1.10 Epidemiology

It is usually a parasite of sheep and cattle. Sometimes human beings accidentally become infected with the liver flukes if they ingest vegetables deposited with metacercariae or if they drink water contaminated with metacercariae of liver flukes.

It is observed that human infection takes place in the countries like Northern Africa, parts of Europe, Cuba, South America, etc. Rabbits, sheep, cattle are considered as reservoirs of infection.

Besides human beings which are accidental the fascioliasis is a major problem of domestic stock. Huge loss is met with the farmers because of the mortality and reduction in the milk and meat production.

3.4.1.11 Diagnosis

Saline and iodine preparations of faeces are examined under microscope reveal the presence of Fasciola eggs. The eggs are operculated. Concentration technique of faeces is very much important to detect the operculated eggs.

Serological test is useful in confirming the diagnosis of *F. hepatica* infection. ELISA and gel-diffusion test kits are available in the market.

3.4.2 *Fasciolopsis buski*

3.4.2.1 History

This parasite is commonly known as the giant intestinal fluke. They are exceptionally large parasitic fluke. George Busk for the first time discovered Fasciolopsis buski in 1843. But it was unnoticed in the scientific world at that time. He found them in the duodenum of a sailor.

In 1843 English surgeon George Busk found 14 flukes in the duodenum of a sailor from eastern India during autopsy. But the discovery was not published at that time.

In 1852 George Budd noticed them in the liver.

In 1857 E.R. Lankester referred the discovery and named it *Distoma buski*. But Busk objected and suggested the name of the parasite, *Distoma crassum.*

But in 1836 it is reported that *D. crassum* had already been named to describe another parasite.

In 1874 a missionary and his wife consulted an English Physician for their persistent diarrhoea. Doctor found 12 worms in their stool and it was showed to Busk. He at once recognized them as the organisms he had already discovered long before.

In 1887 Poirier described a worm from 35-year-old Chinese women as *D. crassum.*

In 1902 Odhner examined fluke samples from a Chinese boy, he named the worm *Fasciolopsis buski.*

Lastly in 1919 K.W. Goddard at the Christian Hospital in Sheoshing, China studied about 400 samples of the Flukes and finally named and described the Fluke under the name *Fasciolopsis buski* after the name of its discoverer.

3.4.2.2 Geographical Distribution

They are found mainly in Asia, i.e. China, Indonesia, Vietnam, Thailand, Malaysia, etc. in India Bengal, Assam and other eastern states.

3.4.2.3 Habitat

Adult worms reside in the small intestine mainly duodenum of human beings and Pigs. The normal host is Pig which acts as the reservoir of infection for human beings.

The eggs present in the faeces of man and pigs are infective to snails of the Phylum Mollusca, the intermediate host of the genus Segmentina.

3.4.2.4 Morphology

Adult fluke is the largest trematode of human beings. The adults are flesh coloured, flat and somewhat elongated. They vary in size and measure 2–7.5 cm in length and 8–20 mm in breadth and 0.5–3 mm in thickness. The shoulder and cone of the parasite are absent. The body is covered with transverse spines ventrally. The dorsal surface is more or less smooth. The acetabulum is very large and is placed close to the oral sucker. Intestinal caeca of the parasite are simple and with two characteristic curves in the middle. Genital pore is immediately anterior to the ventral sucker. The presence of unbranched caeca, ovary is branched and present anterior to testes. The uterus is small.

3.4.2.5 Life Cycle

This parasite requires two different hosts to complete its life cycle. The definitive hosts are the human beings, pigs and dogs. The intermediate hosts are aquatic snails of the genus Segmentina and Hippeutis.

Eggs are large 130–140 μm by 80–85 μm in size. The eggshell is clear, thick and has a small operculum. The yolk granules are uniformly distributed and not concentrated around nucleus. The eggs are yellow brown in colour and do not float in saturated solution of salt.

The adult worms are found in the small intestine of definitive hosts. Eggs are passed with the stool of the definitive hosts. They hatch and form ciliated miracidia in 2–3 weeks. They develop into sporozoites in 3 days and then into rediae, daughter rediae and cercariae.

Cercariae on coming out from the snail settle on the aquatic vegetation and form metacercariae. When human beings swallow the matacercariae along with the vegetables as salad they excyst in the duodenum. The young worms are liberated and then attach themselves to the wall of the intestine and finally develop into adult worms.

The lifespan is 6 months within human beings.

3.4.2.6 Pathogenicity

Mode of infection is eating of infected plants as raw food stuff. The infective forms are encysted metacercariae. The disease caused by *Fasciolopsis buski* is called fasciolopiasis. The disease causes gastrointestinal irritation, nausea, vomiting, diarrhoea, ulcer in the intestinal wall and haemorrhage. In heavy infection anaemia may result. The absorption of metabolites of the parasites may lead to oedema and eosinophilia.

3.4.2.7 Diagnosis

Finding of eggs in stool of patients by microscope examination of stool smears preparation.

3.4.2.8 Control

To prevent the disease it is very important to immerse the vegetables in hot water for a few seconds to kill the metacercarae. Snail control is also important.

The disease is prevalent among school children. Besides economic factors and lack of sanitation, food habit also are the cause of the disease.

3.4.3 *Clonorchis sinensis*

3.4.3.1 History

McConnel in the year 1875 discovered this trematode parasite in Calcutta Medical College Hospital from the bile passage of a Chinese carpenter. Loose in the year 1907 named the genus Clonorchis because of its branched testes. Faust and Khaw in the year 1927 worked out the detailed life cycle of this parasite.

3.4.3.2 Geographical Distribution

The endemic areas of infection of this parasite are Japan, Formosa, Korea, North Vietnam, Indo-China, Southern part of China and India. These parasites are also found in the countries outside East Asia because people of this country while visiting the endemic areas consumed frozen or dried fish from there. These flukes are the most important human trematode parasite in Asia and actively transmitted in Korea, China, Vietnam and Russia. They are the third most important parasitic platyhelminths. The total estimated infection is 30,000,000 human beings currently suffering from the infection of *Clonorchis sinensis*.

3.4.3.3 Habitat

Clonorchis sinensis is a common parasite of dog, cat, pig, rat and man. The adult worm resides in the biliary tract of the liver and gall bladder of the host.

These flukes are the most important human trematode parasites in Asia and actively transmitted in Korea, China, Vietnam and Russia. The total estimated infection is 30,000,000 human beings currently suffering from the infection of *Clonorchis sinensis*. The most important country in relation to human liver flukes is China and 85% of cases are found there.

Recently it is proved that this organism is capable of causing cancer of liver and bile duct by the International Agency for Research on Cancer. In 2009 this International Organization labelled it as a group of Class I biological carcinogen.

Besides human beings it is found commonly in cats and dogs in those places mentioned above. Human infection is restricted among the persons who are used to consume raw or half cooked fishes as food.

3.4.3.4 Morphology

The adult *Clonorchis sinensis* has oral sucker, pharynx, caecum, ventral sucker, vitalline glands, uterus, ovary, Mehils gland, testes and excretory bladder.

The adult organism is 10–25 mm in length and 3–5 mm in width. Testes are on the posterior side and highly branched. At the anterior terminal end there is an oral sucker. Within the centre of the oral sucker there is a mouth. Below the oral sucker is pharynx then there is a short oesophagus. The oesophagus is bifurcated to form two blind caecum runs straight up to the posterior end bilaterally. Uterus is present at the anterior 1/3rd of the body and extends up to almost middle of the body. The uterus is highly branched. Both male and female reproductive systems are present in the same individual and spread almost entire space of the body. The ovary has a short duct, the oviduct, which is connected with vitelline gland. The genital pore opens in the median portion of the body. Small vitelline follicles are dense and restricted to the level of the uterus. The seminal receptacle is present very near to the branched ovary. There is also a Laurer's canal and an excretory bladder. At the posterior end there is an excretory pore.

The adults reside in the biliary duct leading to the gall bladder and remain there in large numbers.

3.4.3.5 Development

The fertilized eggs are discharged through gonopore in the bile duct or liver from there they escape into the intestine during the discharge of bile into the intestine of the vertebrate host like man.

The eggs are yellow brown in colour and measure 27 μm × 16 μm in size, i.e. very small, ovoid in shape and are like small bowl with a rimmed lid or operculum.

The eggs have a small knob or a curved spine which is the distinguishing character of the eggs.

The eggs when reach water along with the faeces in the villages or cities with poor sanitation arrangement hatch and ciliated miracidum larvae are discharged which find passage due to consumption into the small, conical, operculate freshwater snails, *Parafossarulus manchouricus* also called *Parafossarulus striatus* of the sub-family Buliminae and family Hydrobicidae.

The freshwater snails are the first intermediate host in China, Japan, Korea and Russia. Other species of freshwater snails like:

Bithynia longicomis, B. fuchsiana, B. misella, Parafossarulus anomalosiralis, Melanoides tuberculata, Semisulcorpira libertine, Assiminea lute, Terebia grnifera, Bulimus fuchsianus are the important snail hosts.

Miracidia transform into round sporocysts which develop into rediae.

This development takes about 17 days. Rediae undergo asexual reproduction and produce free swimming cercarae 5–50 in number. A pair of eye spots, bristles and very small spines are present on the cercariae. The colour of the cercaria is brown. The tail possesses dorsal and ventral fins and is called pleurolophocercus cercaria. This process of development is called polyembryonae.

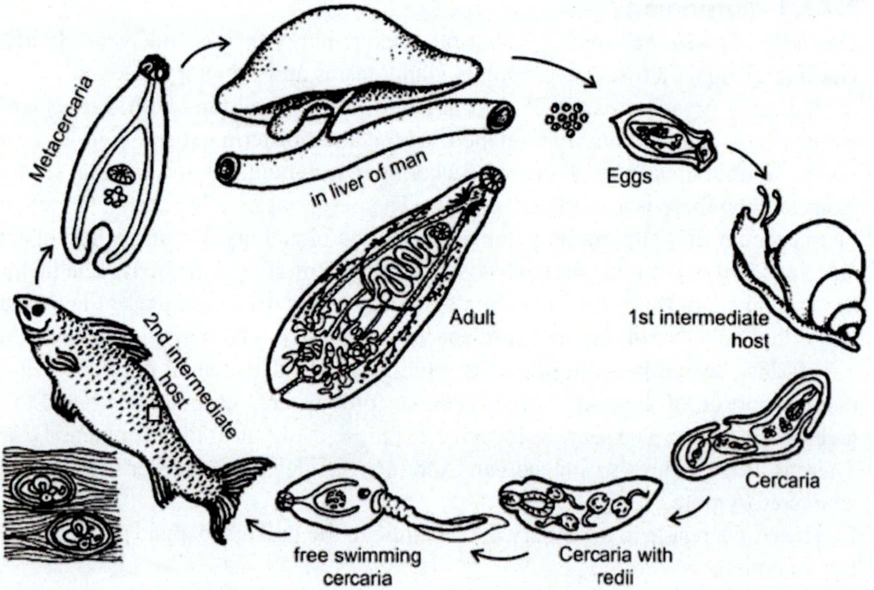

Fig. 3.17 Different developmental forms of *Clonorchis* (*Advanced Parasitology*, Das)

The free swimming cercaria actively bore out of the body of the snail again into the freshwater environment. The process of polyembryonae helps the parasite in reproduction and large number of cercaria come out of a single miracidium which hatches out from the egg. It is observed that cercaria in the freshwater hang upside down and go to the bottom slowly. In this way when a fish swims cercariae are stimulated and touch the surface of the fish. The free swimming cercaria attaches the body of the fish with its suckers, shed its tail and actively bore into the body of the fish.

The cercaria develops into next stage within the fish, the second intermediate host. Fish does not consume the cercaria but on the other hand cercaria itself bore their way into the body of the fish. Here they parasitize the fish, their new host.

The cercariae have a long tail with long dorsal and short ventral fin. The surface of the body which is covered by the cuticle possesses fine spine. They have seven pairs of penetration glands, 14 cytogenous glands, eye spots and brownish pigment. The secretion of the penetration glands helps the cercaria to enter into the muscles of fish. Now the cercariae create a protective cyst and form metacercariae. This protective covering of the cyst protects the growing parasites in the stomach of the human host.

The metacercarical cysts are oval in shape have thin walls and measure 137 × 115 μm in size. They lose their eye spots and possess sac like excretory bladder filled up with coarse refractile granules (Fig. 3.17).

A number of species of carp family like Cyprinidae act as second intermediate hosts. According to Hsü the metacercariae encyst in the flesh of fish are occasionally under the scales or in the gills (Chandler and Read 1961).

When the fish with encysted metacercariae are consumed by man, they are infected. The acid resistant cyst wall remains intact in the acidic gastric secretion of human beings and passes safely into the small intestine. On reaching the small intestine unharmed metacercariae search the free passage towards liver, the final residence. Actually *C. sinensis* penetrates the intestinal wall and via hepatic portal system enter into the liver. *Clonorchis* within the liver feed on bile and grows to sexual maturity to produce and release eggs after 3 weeks.

According to Wykoff and Lepes (1957) migration to the liver by metacercariae is not performed only through bile duct. They showed that if the bile duct of the experimental animal is tied off, some worms still find their way into the liver.

3.4.3.6 Symptoms

The disease caused by the *Clonorchis sinensis* produces pathological manifestations from inflammation and intermittent obstruction of the bile duct. In acute condition the patient expresses abdominal pain, diarrhoea, nausea and vomiting. In chronic condition the patients feel fatigue, abdominal discomfort, anorexia, weight loss, diarrhoea and jaundice. In later stage cirrhosis and ultimately cholangiocarcinoma may progress.

The adult while residing within the bile duct and liver consumes voraciously the bile produced in the liver, which inhibits the fat digestion in the human hosts. The obstruction of the bile duct may also occur due to the presence of the parasite or large number of eggs resulting cholangitis.

Eighty cases of CSR (Central Serous Retinopathy) are observed in Hong Kong by Dr. Jah Chiao-nen Cheng, Dr. Yin-Ping Wang both are M.D.

Faust and Khaw show how *Clonorchis* infection thrives among many human beings. In the mulberry growing areas near Canton the latrines are situated over ponds where culture of fish is done. Faeces fall directly into the ponds an arrangement called integrated fish farming. In this process the main principle is that waste product of one species is used as food for the other species. The night soil as fertilizer will provide nitrogen in the ponds and as a result zooplankton and phytoplankton will develop in large quantity. The planktons are natural food for the fishes.

The suitable freshwater snails are present in such ponds and feed on the faecal material. The cultivated fishes are infected by the cercariae coming out from the snail. The people become infected when they consume raw sliced fish with vegetable salad. Raw fish are not eaten all the times. The raw sliced fish are kept on the top in a dish on steaming rice to remove raw taste by heat and vapour. It is seen that metacercarial cysts in the flesh of fish remain unaffected if they are used with vinegar or sauces.

The infection can be controlled by preventing the consumption of raw or half cooked fish flesh.

3.4.3.7 Pathogenesis

The adults take residence in the large bile ducts of the host. Here the parasites cause erosion and inflammation of the bile ducts take place. The result is fibrosis and necrosis of the liver cells around the inflammatory tissue. Now from this damaged liver cells worm antigens are also detected. Due to the inflammation in chronic cases perforation of the bile duct may occur. Infiltrating eggs may produce cysts and liver function is interfered. If obstruction of the bile duct takes place due to the presence of the parasite then obstructive jaundice may results. It is seen that the gall bladder stones sometimes formed where entire worm is used as nuclei.

3.4.3.8 Diagnosis

The saline and iodine preparations of faeces are examined under microscope and operculated eggs with a knob like structure at the either end if found, then can be suspected of Clonorchis infection. The eggs can also be detected in bile.

3.4.3.9 Epidemiology

The disease Clonorchiasis is common in the countries where people used to consume raw fish. The disease can be prevented simply by taking boiled and cooked fish. But it is very difficult in those countries to change the food habit. However, the infection is due to ponds where human faeces are used as fertilizer. The people of other countries are also become infected because of consumption of imported fish from the endemic areas.

3.4.4 *Paragonimus westermani*

3.4.4.1 History

The trematode parasite, *Paragonimus westermani* was first discovered from two Royal Bengal tigers in 1878. Then two tigers died in Zoo in Europe. Following this discovery it was found in Human beings in Formosa. The life cycle was described by Kobayashi and Yokagawa.

3.4.4.2 Geographical Distribution

Paragonimus westermani is a lung fluke which parasitises the lungs of man, cats, dogs, tigers, pigs and opossums.

After the discovery of the parasite in human beings they are found in Japan, Philippines and Korea where millions of people are seen infected. They are also found in Africa, Japan, Asia, America and India. In India it is mainly restricted to West Bengal, Assam, Kerala, Tamil Nadu and Maharashtra. First fluke that was described from human beings was named *Paragonimus ringer*. Another lung fluke *P. kellicotti* was found in mink In North America

There is a controversy about how many types of lung flukes are there and at the same time are they really different from each other? However, the lung fluke was first discovered from the Royal Bengal Tiger in India.

The life cycle and morphology of these above mentioned flukes have some differences in their morphology. A species from rat *P. ilontsuensis* was found by Parasitologist Chen in Canton (China) which could not be established experimentally in carnivorous, pigs or monkeys. Another example is *P. ohirai* found in pigs and dogs in Japan have same features and characteristics with *P. westermani* in adult stages but there are some differences in their larval stages.

Chandler and Read are of opinion that *P. westermani* found from man and *P. ringer* found from tigers are same and identical.

The infection of man from *P. westermani* rather lung fluke is endemic in Korea, Japan, Philippines and parts of Indo-China. It is recorded that 40–50% human population is infected with lung flukes. Not only the places already mentioned but infection is also found in New Guinea, Indonesia, India, Belgium, Congo, Cameroon, Nigeria, Ecuador and the United States. Reservoir hosts are Feline group who serve as reservoirs of the infection.

3.4.4.3 Morphology

The adult flukes are reddish brown in colour and about 8–12 mm long and 4–6 mm in diameter (Fig. 3.18). The body is covered by a cuticle with minute simple spines. They live in lungs of human beings. In living condition they are reddish brown in colour. Oral and ventral suckers of the parasite are more or less equal in size. The lobed testes are present at the posterior one-fourth of the body. There are cirrus and cirrus pouch. The genital pore is located below acetabulum. The lobed ovary is placed to the left of the midline of the body. The uterus is coiled and opens into genital atrium along with vas deferens. Vitelline follicles are present on the lateral sides extensively. They are hermaphrodite and oviparous. They are generally found in three or four in a cluster. The hosts form a cyst like pocket around them. The lifespan of adult worm is about 20 years.

The pockets rupture and eggs are escaped into the bronchial tubes which are excreted with sputum. These cysts are like hazel nests. Many of the eggs enter into the tissues and cause tuber like abscesses. In case of *P. westermani* it is found that

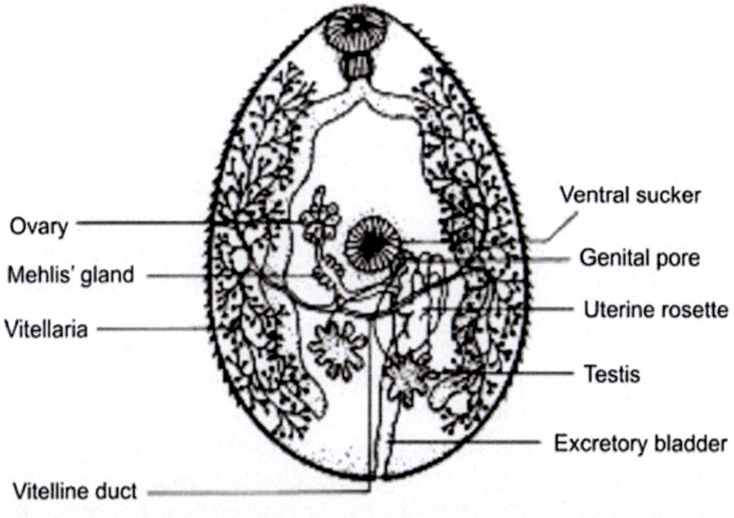

Fig. 3.18 Adult *P. westermani*

some of the worms most of the time miss their destination and reach spleen, liver, brain, urinary system, intestinal wall, eye and other muscles.

The eggs are yellowish brown, 80–188 μm in length and 40–60 μm in diameter. They have an operculum which is set to the rim. The eggs are seen in faeces of the host as they are most of the time swallowed with the mucous coming from lungs in the pharynx. Each egg consists of an unsegmented ovum with yolk cells around the ovum. The eggs when laid are in immature condition.

3.4.4.4 Life Cycle

P. westermani requires three hosts to complete its life cycle. So the production of number of eggs and polyembryoni (this is a phenomenon when from single larva by asexual reproduction a large number of larvae are produced) is an astounding factor by which the parasites overcome the hazards of infecting three hosts.

The two intermediate hosts are snail of Phylum Mollusca and crab or crayfish of Phylum Arthropoda.

The life cycle of this parasite was first established by several Japanese workers. The complete account of the life cycle of *P. westermani* was described by Ameel in the year 1934 whose experimental specimen was American lung fluke in Michigan.

First Intermediate Host

In Asian countries rather oriental region the first intermediate hosts snails are *Semisulcospira libertina*, *Thiara gramifera* and *Semisulcospira amurensis* all these snails of class Gastropoda was once included under the genus Melania.

In China *Gyncera lutea* is the intermediate host of *P. westermani* found in rats.

In Michigan (USA) *Pomatiopsis leopidaria* is the first intermediate host of this lung fluke of mink.

The adult lung flukes usually residing in lungs in pair form a cyst. Usually cross-fertilization takes place. It is observed that some of the lung flukes are triploid and some are tetraploid. The triploid individuals cannot produce sperm. In this case parthenogenesis takes place. Now question arises how tetraploid individuals arise? It is proved that when a triploid oocyte fertilized by a diploid fluke's sperm produce tetraploid. Occasionally the eggs remain trapped in some other surrounding tissue. In that case eggs cannot enter into air passage. They reach the pharynx and subsequently enter into alimentary canal and escape through faeces. The lung flukes usually remain in lungs are found in faeces of the definitive host. The first stage of their development is known as miracidium. Miracidia develop within the eggshell slowly after they leave the body of the host. In moist condition the miracidia develop in 3 weeks.

The developed miracidia now hatch in water. The swarms of ciliated free swimming miracidia search for the suitable snail for further development. If suitable snails are found the miracidia enter into the body of the snails through the mantle where they take water for their respiration. On entering the tissue of the snails, according to some, in the digestive gland the miracidium sheds its cilia and changes into sac like sporocyst, the next stage of their development. The phenomenon of polyembryonae now starts. Sporocysts produce about 12 first generation rediae. The first generation of rediae next produce same number of second generation of rediae. Redia is next stage of development at the same time product of asexual reproduction. The last developmental stage is the cercaria. Twenty or 30 fully developed cercaria are formed from a single redia. We can imagine now the number of cercaria produced from the large number of eggs. The cercariae are 175–240 μm long with a small tail, spiny cuticle and 14 penetration glands, the secretion of which helps the cercaria to enter the second intermediate host, the crab or crayfish. The time taken by a miracidum to transform into cercaria is near about two and half months after their entry into the snails' body. The cercariae do not swim in the water but their movement in the water is creeping or passive movement floating in the direction of stream.

Second Intermediate Host

These cercariae developed within the snail come out and escape into the water and enter into second intermediate host.

Second intermediate hosts are crustacean of Phylum Arthropoda, the crabs or crayfish.

Crabs of genera *Eriocheir* and *Potamen* are second intermediate hosts in Japan. *Eriocheir japonicas* is the usual second intermediate host in which 90% are infected. These crabs are found in rice fields near the sea in small islands and are used very much as food.

In America species of *Canbarus* are acting as second intermediate hosts living in shallow, sluggish streams where large number of crayfish is found infected which are used as food by human beings.

In the body of the second intermediate hosts like crab or crayfish, they are encysted in the muscles and gills and are known as metacercaria.

Encysted metacercaria within the body of the crab or crayfish are more or less spherical and 0.5 mm in diameter. The metacercaria lies straight with the cyst. They are recognized by their large excretory vesicle filled with refractory granules and convoluted intestinal caeca on both sides.

Infection to human beings occurs from consuming crab or crayfish infected with metacercaria without cooking or semi cooking like smoked. In China, Philippines, Japan and Korea the crabs are eaten raw with salt and dipped in wine or vinegar with which the muscles of crabs or crayfish are marinated. In some parts of Japan there is another practice for consumption. The crabs or crayfishes are chopped into small pieces and are mixed in the preparation of other delicate food. In some places people used to drink juice of the crab as medicine. All these practices make the human beings vulnerable to *P. westermani* infection.

Final or Definitive Host

The metacercaria with cysts in this way enter into the G.I. tract of human beings. The cysts dissolve in the duodenum and are freed from the cysts. In the duodenum the free metacercariae bore through the wall of the intestine and escape into the abdominal cavity and roam about through peritoneal cavity. A considerable number of worms then penetrate the diaphragm within 4 days and reach thoracic cavity.

From here the worms go to the lungs penetrating the pleural membrane via pleural cavity in about 15–20 days.

Now *P. westermani* unlike others a considerable number of metacercariae instead of going to their destined address become vagabound and take residence in any important organ like spleen, liver, kidney, brain, etc. where they form cyst.

When they reach bronchioles the host forms a cyst wall around them. The cyst is nothing but fibrous capsule. Within the final host in the cyst the worms mature and produce eggs. The eggs are liberated into bronchiole after rupturing the fibrous wall. The presence of the parasites in bronchiole create irritation and are coughed out with sputum. The sputum when swallowed with mucus the eggs are found in the faeces. The cycle is thus completed.

3.4.4.5 Pathogenicity

The infections are generally not so serious. But the diagnosis often made wrong because the symptoms are like tuberculosis, fever, cough, blood stained sputum, mild anaemia and weakness. Then enlargement of spleen, eosinophilia, pleural effusion, pneumothorax may occur due to wondering worms. Chronic granulomatus reaction may occur in response to the presence of worm and the laid eggs. The symptoms are fever, dysponea, cough with haemoptysis and chest pain. Serious complication may happen if the wondering worms reach brain. The disease is known as Jacksonian epilepsy, sometimes brain tumour develops.

Sometimes worms in the capsule of granuloma tissue if present in the spinal cord the patient becomes totally paralysed. There is a report of worms present in heart cause death of the patient. The lung flukes when in lungs produce chronic coughs, blood in the sputum and breathing difficulties.

3.4.4.6 Epidemiology

The definitive and reservoir hosts of *P. westermani* are usually carnivores, rodents and pigs. Infection in human beings is not so common. But human beings may also be infected if they consume raw or undercooked crabs. Some ethnic people use crushed crab juices as medicine. They are also prone to Paragonimus infection.

3.4.4.7 Diagnosis

The infection is diagnosed by detecting the eggs in sputum or faeces. Concentration technique by centrifugation may help to detect the eggs. Chest X-ray findings often misled with tuberculosis. But in chronic cases 'soap bubble' calcification is seen in X-ray. Serological tests like CFT and IHA are available in the market to detect the infection.

3.4.4.8 Host Defence Mechanism

The tissue migratory stages of *P. westermani* cause the activation of peritoneal macrophages of host eosinophil associated tissue inflammation. There are no changes in the activity of CD4+ cells and CD8+ cells in the infected hosts.

3.5 Schistosomiasis

3.5.1 *Schistosoma haematobium*

3.5.1.1 History

The Trematode parasite *S. haematobium* causes schistosomiasis or bilharziasis in human beings. The parasite was discovered by Theodor Bilharz in the year 1852 and it was proposed that the name of this parasite may be given Schistosoma in the year 1858 by Weinland. It is interesting that in the year 1799–1801 a surgeon was detected with the disease. He is the first European who was with Napoleon's army.

One of the most important human diseases of today is the schistosomiasis. It is caused by a blood fluke known as *Schistosoma haematobium*. There are other species also like *S. mansoni* and *S. japonicum*. All the Schistosomas are digenetic trematode.

3.5.1.2 Geographical Distribution

The disease is prevalent in the various region of Africa, Middle East and some parts of India. In the Ratnagar district of Maharashtra in India the disease is reported.

3.5.1.3 Habitat

The mature adult male and female parasite lives in the venous plexus of urinary bladder of human beings.

3.5.1.4 Morphology

The male worm usually measures about 10–16 mm in length and has a cylindrical appearance having 1 μm in diameter. They are actually flat in the sides of the body posterior to ventral sucker and rolled ventrally to form a groove known as gynecophoric canal in which the cylindrical, longer and slender female projects free at each end, but enclosed in the middle. The female is 15–20 mm long with a diameter of 0.25 mm. The female lives under the safe arms of her spouse at least for 40 years. In most Schistosomas they seen to remain permanently married and monogamous, the uncoupled females remain spinsters. The female worms do not become sexually mature if they are not coupled with the male.

Both male and female worms have two suckers: one ventral and one oral. In male the ventral sucker is larger and quite powerful. The alimentary canal does not have any pharynx. The oesophagus is forked just anterior to ventral sucker and the two arms of the fork unite in the middle of the body to continue as a single tube. In male there are several testes placed just behind the ventral sucker. The male genital opening is located here.

The mouth is at the anterior terminal end of the female and oesophagus is forked and the intestine is straight. The ovary is elongated and situated near the forked oesophagus where the intestinal caeca rejoin. Most of the posterior half of the worm is occupied by the yolk gland. Anterior to the ovary is a straight uterus which contains 1–50 eggs. Some are of opinion that female Schistosoma produce several 100 eggs daily (Karyakarta and Damle 2003).

Unlike most flukes, the Schistosomas do not develop large number of eggs at a time but instead develop them gradually (Chandler and Read 1961) and have only a few in the oviduct at any point of time.

Eggs

The female worm lays eggs in the small venules of vesicle plexus. The female extends its anterior end into the smallest venule and deposit the egg one at a time in a line. Each time an egg is laid the female worm withdraws itself a short distance and lays another egg immediately behind the first. In this way the venules of the plexus are filled up with eggs. The worms in the state of copulation enter into adjacent venule.

The eggs are oval with terminal spine. They measure 110–170 μm in length and 40–70 μm in breadth and are covered with a shell. By the histolytic secretions of the embryo the eggs go out of the vessels and enter into the wall of urinary bladder. The eggs with the embryo now pass through the wall of the urinary bladder and drop down into the cavity of the bladder. Finally they escape with the urine at the end of micturition.

3.5.1.5 Life Cycle

Schistosoma passes its life cycle in two hosts: definitive host, man and intermediate host freshwater snail like *Ferrissia tenuis* in India and *Bulinus truncates* in Africa.

The adult worms reside and perform sexual reproduction in the venous plexus or pelvic venules around the urinary bladder of the host.

The eggs with embryo inside passed along with urine and drain into the water. Eggs hatch and ciliated miracidia larvae come out to swim in the water in search of the intermediate host, the freshwater snail. The ciliated miracidia have photoreceptors that help them to search snails. They on reaching the proper snail host penetrate the soft tissues of the snail and ultimately reach digestive gland of the snail. Now the larvae shed their cilia and undergo developmental changes in 4–8 weeks. The miracidium is transformed into oval sac like sporocyst.

Sporocyst now undergoes asexual reproduction within the liver of the snail, the intermediate host. Such generations of sporocysts are produced from the germ cells of first generation. After about 2 weeks first generation sporocysts produce second generation sporocyst. Several weeks after the infection, when no further multiplication occurs the daughter sporocyst gives rise to the fork tailed cercaria which is infective to the human beings. Thousands of cercariae developing from a single miracidium due to asexual reproduction produce worms of one sex.

The lifespan of miracidium is only 24 h so they must find a snail of suitable species within that short time. Further development of the miracidium takes place in the digestive gland of the snail.

The cercariae are the infective forms. The cercariae are 175–240 μm long and 55–100 μm in breadth. They have a bifurcated tail of 75 μm long called furcocercus. The cercariae come out from the second generation of sporocyst by breaking out the egg cover and escape from the snail into water in a large number at a time. The cercariae alternately swim and take rest in the water for 2 or 3 days. If they fail to reach the final host they perish.

The actively motile furcocercus cercariae penetrate through the skin of human beings using the histolytic secretion containing hyaluronidase of their penetrate glands when they swim or bath in the water contaminated with cercariae. On entering the body of the definitive host they migrate to the liver. In the meantime they cannot off their forked tail and enter into the bloodstream to reach the liver to mature as adult flukes. In order to avoid detection by the immune system inside the host, the adults coat themselves with host antigen so they are not recognized as non-self.

Entering the body of the host after casting off their forked tail they go to the peripheral venules. From here they are carried to the right side of the heart to the pulmonary capillaries, then to the left side of the heart into systemic circulation. The majority is carried to the abdominal aorta and gain access to the mesenteric artery. From here to the capillary bed of the intestine and enter into portal circulation to reach liver. In the liver the immature worms grow into adults. Sexual maturity comes after 3 weeks from the time of entry of the parasite into the body. After gaining sexual maturity they move out of the liver against the current of the bloodstream migrating into the inferior mesenteric vein and finally enter into the pelvic plexus of venules. The sexually mature male and female become coupled and start copulate (Fig. 3.19). The inseminated female then lays fertilized eggs into the venules and reaches urinary bladder. The eggs ultimately pass through urine and the cycle is completed.

3.5.1.6 Osmoregulatory Function

The osmoregulatory system of digenetic trematodes is like protonephridia, a tube closed at one end and open at the other. The flame cells produce current by vigorously beating tuft of cilia. This system serves both as excretory and osmoregulatory in function.

The major nitrogenous waste products like ammonia, urea and other nitrogenous compounds are excreted by the flame cells in coordination with excretory bladder.

3.5.1.7 Pathways Associated with Energy Production

The main source of energy of the parasitic trematodes is the substrate level phosphorylation via glycolysis of the glycogen and glucose. In aerobic condition in the blood vessels glycolysis provides primary energy supply. As Schistosoma cannot synthesize fatty acids, sterols, purines, nine essential amino acids, arganine or tyrosine they exploit fatty acids and cholesterol from host's blood plasma.

The dependence of digenetic trematodes on glycolysis for energy production gives us the opportunity to device effective drugs for treatment.

The miracidia and cercaria of all degenetic trematodes are obligate aerobes, they depend upon oxidative phosphorylation for energy requirement.

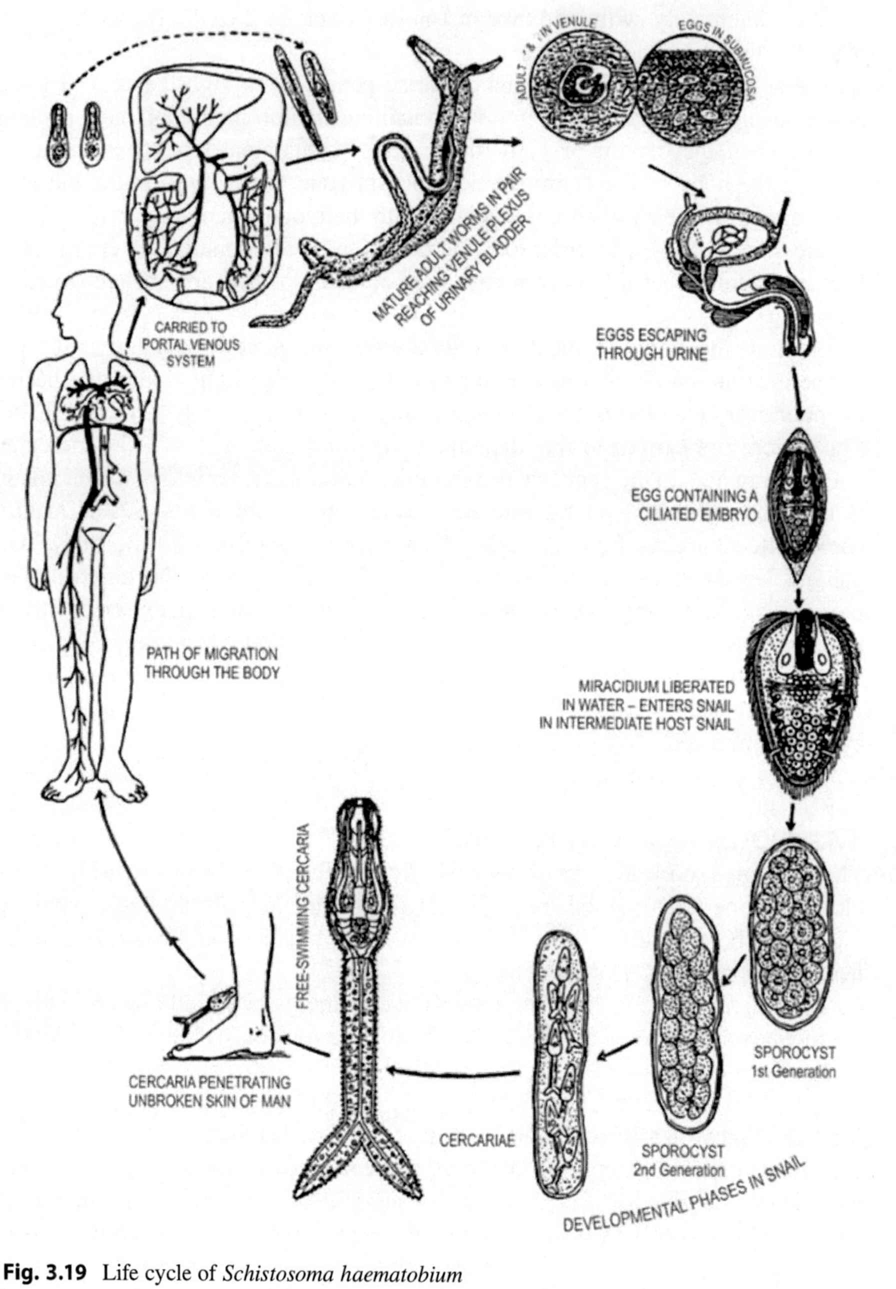

Fig. 3.19 Life cycle of *Schistosoma haematobium*

3.5.1.8 Pathogenicity

Swimmer's Itch

The cercarial dermatitis is also known as Swimmer's itch. This condition of the disease is not life threatening but it has a big impact on the economy of the regions popular with tourists. A number of lakes and seas shore resort in Michigan, Minnesota, New-Jersey, New-England, North Carolina, Wisconsin, Massachusetts, Connecticut in the USA and Canada where large number of tourists are driven away for fear of swimmer's itch. This condition is caused by the cercaria of blood flukes penetrating human skin, the site is swollen and cause itchy rash. As human beings are not usual definitive hosts, the cercarae after penetrating the skin are destroyed by the immune response of the victims. The dead and decaying cercariae release allergic substances and produce local inflammation of the punctured site creating swimmer's itch.

S. haematobium causes urinary schistosomiasis. The penetration of skin by the cercariae causes irritation and skin rash produce within 24 h. The cercarial dermatitis better known as swimmer's itch disappears in 2–3 weeks.

During migration of the cercariae in the blood vessels symptoms like fever, cough, lymphodenopathy, enlargement of liver and spleen are felt by the patients.

Symptoms like haematuria and fibrosis of the bladder are seen after 8–12 weeks of infection. Intense inflammatory response of bladder is felt by the patients. The bladder becomes calcified and there is increased pressure on ureters causing hydronephrosis. Inflammation of the genital organ may contribute to the propagation of HIV. It may associate with carcinoma of bladder. *S. haematobium* most of the time invade female reproductive system of human beings. With the help of blood circulation they may migrate to internal and external female genitalia and eggs of the parasite may be detected in the genitals of the host.

3.5.1.9 Host Defence Mechanism

In the skin penetration stage of cercaria, its surface coat and penetration glands secretion create antibody reaction. During second or early developmental stages, there is an integument change in the migrating Schistosoma along with marked increase in Th1 cell responses and Th2 cell responses induced by eggs of the parasite. The migrating eggs react in response to immunogens released from migrating eggs. The immunogens are glycosylated macromolecules present on the egg surface. The migration of the eggs of the parasite through the tissues of the host produces granulomatous responses which cause pathological changes. The granulomatous response at chronic stage of the disease causes CD4+ cells activity induced by the antigens of the eggs. During this phase Th2 response occurs which causes reduction in size of granulomatous structure and fibrosis increases under the influence of IFN-γ and IL-13 cytokine. The important immune signal in CD4+ lymphocytes is detected by the parasites. In the development of these particular parasites, i.e. Schistosomes hint is necessary from the host's immune system. The egg production of the parasites is stimulated by TNF. The growth of the parasites is roused up by the secretion of IL-7 of the host immune system. Thyroxine is also known to stimulate the growth of the parasite.

3.5.1.10 Epidemiology

The schistosomiasis is caused by *Schistosoma haematobium*, whose life cycle depends upon suitable snail species like Bulinus and others in the water where human beings are used to come for swimming and other purposes. If this water is contaminated with human waste then the disease becomes rampant. To prevent the disease hygienic waste disposal is very much necessary.

The infection spreads within the fisherman who wades in this contaminated water for fishing and the farmers remain longer time in their irrigation water. In some countries where both sexes of human beings work in the agriculture field become highly infected.

3.5.1.11 Diagnosis

Haematuria suggests the diagnosis of *S. haematobium* infection. Haematuria may be microscopic.

Urine examination after concentration technique under microscope will reveal the presence of eggs of *S. haematobium.*

Urine is collected around mid-day centrifuged or allowed to settle for 1 h and then deposit is examined under microscope.

For detection of eggs of the parasite filtration of urine with membrane filter is also important.

Serological test for ectopic Schistosomiasis is performed. Cercarian-Hullen reaction, circum oral precipitation and miracidial immobilization tests detect antibodies against cercaria, egg and miracidium. Other tests like IHA, IFA, ELISA and RIA, for detection of schistosome antigen are used. Antigen detection test kits are also available in the market.

3.6 Systematic of Studied Parasitic Helminthes

3.6.1 *Taenia solium*

Kingdom: Animalia
Subkingdom: Metazoa
Phylum: Platyhelminthes
Class: Cestoda
Order: Cyclophyllidae
Family: Taeniidae
Genus: Taenia
Species: solium

3.6.1.1 Kingdom Animalia

1. Multicellular organisms that are without chlorophyll.
2. They are eukaryotic organisms.
3. Absence of cell wall.
4. Nutrition heterotrophia.

3.6.1.2 Subkingdom Metazoa

1. Multicellular animals.
2. Cells are differentiated into tissues and organs.

3.6.1.3 Phylum Platyhelminths

1. Body bilaterally symmetrical.
2. Absence of coelom or body cavity.
3. No segmentation.
4. Dorsoventrally flattened, body tape like.
5. Triploblastic.

3.6.1.4 Class Cestoda

1. All are parasitic.
2. Completes life cycle on vertebrates.
3. Tape like body covered by cuticle.
4. Anterior portion of the body transforms into scolex.
5. Hooks and suckers are present on the scolex for the attachment to the body of the hosts.

3.6.1.5 Order Cyclophyllidae

1. Parasites of humans and domesticated animals.
2. All have multiple proglottids.
3. Presence of four suckers on their scolex.

3.6.1.6 Family Taeniidae

1. Strobila either large with numerous segments or small with few segments.
2. Scolex with four suckers.
3. Presence of rostellum armed with two rows of hooks.

3.6.1.7 Genus *Taenia*

1. Ribbon like body.
2. Presence of a single axoneme in the sperms.
3. Presence of a creasted body and spiralled cortical microtubules and nucleus in the spermatozoon.

3.6.1.8 Species *solium*

1. All are hermaphrodite.
2. Completes life cycle in human beings as definitive host and pigs as intermediate hosts.

3. A cysticercous develops into an adult worm in the small intestine of human beings.

3.6.2 *Echinococcus granulosus*

Kingdom: Animalia
Subkingdom: Metazoa
Phylum: Platyhelminthes
Class: Cestoda
Order: Cyclophylidae
Family: Taeniidae
Genus: Echinoccus
Species: granulosus
Up to Family same as Taenia.

3.6.2.1 Genus *Echinoccus*

1. The parasites have cysts which contain daughter and granddaughter cysts.
2. Have a number of scolices.

3.6.2.2 Species *granulosus*

1. Possess four segments only.
2. The terminal proglottid is mature and gravid.
3. Rostellum with double rows of hooklets.

3.6.3 *Hymenolepis nana*

Kingdom: Animalia
Subkingdom: Metazoa
Phylum: Platyhelminthes
Class: Cestoda
Order: Cyclophyliidae
Family: Hymenolepididae
Genus: Hymenolepis
Species: nana
Up to Order same as above.

3.6.3.1 Family Hynenolepididae

1. Dwarf tapeworm of man.
2. Does not require intermediate host.

3.6.3.2 Genus *Hymenolepis*

1. Smallest human tapeworms.
2. Larvae develops in the intestinal villi of man.

3.6.4 *Diphyllobothrium latum*

Kingdom: Animalia
Subkingdom: Metazoa
Phylum: Platyhelminthes
Class: Cestoda
Order: Psdeudophyllidae
Family: Diphyllobothriidae
Genus: Diphylobothrium
Species: latum
Up to Class same as above.

3.6.4.1 Order Pseudophyllidae

1. Scolex of the parasite spoon shaped.
2. Rostellum and hooks absent.
3. Onchosphore ciliated.

3.6.4.2 Family Diphyllobothriidae

1. Ova operculated hatch out in water as a coracidium.
2. Head is flattened with two longitudinal suctorial grooves called bothria that are present on each side.

3.6.4.3 Genus *Diphylobothrium*

1. Hookletes are absent.
2. Genital pores situated on the ventral flat surface.

3.6.4.4 Species *latum*

1. Eggs are oval and operculated.
2. Ovary is on the posterior one-third of a proglottid.
3. Testes are numerous and scattered on the upper layer.

3.6.5 *Schistosoma haematobium*

Kingdom: Animalia
Subkingdom: Metazoa
Phylum: Platyhelminthes
Class: Trematoda
Subclass: Digenea
Order: Strigeidida
Family: Schistosomatidae
Genus: Schistosoma
Species: haematobium
Up to Phylum same as above.

3.6.5.1 Class Trematoda

1. Commonly known as flukes.

3.6.5.2 Subclass Digenea

1. To complete life cycle two hosts are required.
2. Larvae develop in snails as intermediate host.
3. Adult stage is found in vertebrate host including man.
4. Development with alternation of generations.

3.6.5.3 Order Strigeidida

1. They are unisexual.
2. The female lives in the gynocopheric canal of male.
3. Pharynx is absent.

3.6.5.4 Family Schistosomatidae

1. Redia stage is absent.
2. Eggs are knobbed.
3. Intestinal caeca fused posteriorly to form a single caecum.

3.6.5.5 Genus *Schistosoma*

1. Found in the lumen of small venules of the urinary bladder.
2. Females are filiform and longer than males.

3.6.5.6 Species *haematobium*

1. Presence of minute spines in the sucker.
2. Sides of male are flattened and curved to form gynocopheric canal.
3. Head of the male has two suckers.
4. Eggs are spindle shaped with terminal spine.

3.6.6 *Fasciola hepatica*

Kingdom: Animalia
  Subkingdom: Metazoa
  Phylum: Platyhelminthes
  Class: Trematoda
  Subclass: Digenea
  Order: Echinostomida
  Family: Fasciolidae
  Genus: Fasciola
  Species: hepatica
  Up to Subclass Digenea same as above.

3.6.6.1 Order Echinostomida

1. Cercaria with simple tail and many cysts producing glands.
2. Life cycle is completed in three hosts.
3. Encystment takes place in invertebrates.
4. Excretory vessels do not open to the exterior.

3.6.6.2 Family Fasciolidae

1. It is a family of trematodes.
2. Resides in the liver, gall bladder and intestine as parasite.
3. In the life cycle the intermediate host is freshwater snail.

3.6.6.3 Genus *Fasciola*

1. Known as liver fluke.
2. The disease caused is called fascioliasis.
3. An important parasite of sheep and cattle.

3.6.6.4 Species *hepatica*

1. It is a flat worm known as liver fluke.
2. It infects liver of various animals and human beings.
3. Infection takes place through plant food contaminated with metacercaria.

3.6.7 *Clonorchis sinensis*

Kingdom: Animalia
Subkingdom: Metazoa
Phylum: Platyhelminthes
Class: Trematoda
Order: Opisthorchiida
Family: Opisthorchiidae
Genus: *Chonorchis*
Species: *sinensis*
Up to Class same as above.

3.6.7.1 Order Opisthorechiida

1. The adult flukes reside in the biliary and pancreatic duct of mammalian host.
2. Definitive host is mammals like cats, dogs and fish eating mammals.
3. Infection by ingestion of fish containing metacercarae of the fluke.
4. They remain attach to the mucosa.
5. Cercariae with two eyespots.

3.6.7.2 Family Opisthorchiidae

1. People are infected after eating raw or undercooked fish infested with metacercariae.
2. Metacercariae excyst in duodenum of human beings.

3.6.7.3 Genus *Clonorchis*

1. Can infect multiple species of snails.
2. Can infect 100 species of freshwater fishes as intermediate hosts.
3. Domestic canids and fields, swine, mustelids and other fish eating mammals are definitive hosts.

3.6.7.4 Species *sinensis*

1. China, Korea, Taiwan and North Vietnam are endemic areas.
2. The name sinensis comes from China.

3.6.8 *Paragonimus westermani*

Kingdom: Animalia
Subkingdom: Metazoa
Phylum: Platyhelminthes
Class: Trematoda
Subclass: Digenea
Order: Plagiorchiida
Family: Paragonimidae
Genus: *Paragonimus*
Species: *westermani*
Up to Subclass same as above.

3.6.8.1 Order Plagiorchiida

1. Secondary excretory pore in the terminal position of cercariae.
2. Body of the adult is covered by cuticle with minute spines.
3. Absence of appendages in redia.
4. Tail of cercaria not bifurcated.

3.6.8.2 Family Paragonimidae

1. Oval shaped body of the adult parasite.
2. Both the oral sucker and acetebulum are round and muscular.

3.6.8.3 Genus *Paragonimus*

1. The excretory bladder extends from the posterior end to pharynx.
2. The testes are lobed located at posterior end.
3. The ovaries are lobed slightly post acetabular.

3.6.8.4 Species *westermani*

1. Parasites remain within the hosts forming a cyst.
2. Pockets rupture and eggs escape into the bronchial tube.
3. The cysts are like hazel nuts.

3.6.9 *Ascaris lumbricoides*

Kingdom: Animalia
Subkingdom: Metazoa
Phylum: Aschelminthes (Nematoda)
Class: Chromadorea
Order: Rhabditida
Family: Ascarididae
Genus: *Ascaris*
Species: *lumbricoides*
Up to Subphylum same as above.

3.6.9.1 Phylum Nematoda

1. Body cylindrical and unsegmented.
2. Body is covered by thick cuticle.
3. Alimentary canal is straight.
4. Absence of any digestive gland.
5. Triploblastic.
6. Presence of pseudocoelom
7. Most of the members are parasites.

3.6.9.2 Class Chromadorea or Phasmida

1. In the tail portion there are a pair of unicellular gland called phasmid.
2. On the lateral lips there are pores or papillae called amphid.

3.6.9.3 Order Ascaridida

1. Amphids poorly developed with small pores on the lips.
2. Presence of phasmids.
3. Excretory system consists of two lateral canals.
4. Caudal and hypodermal glands absent.

3.6.9.4 Family Ascaredidae

1. They are small intestinal roundworms.

3.6.9.5 Genus *Ascaris*

1. Eggs are deposited in faeces or soil.
2. Large intestinal roundworms.
3. Sexes are separate.

3.6.9.6 Species *lumbricoides*

1. Fertilized eggs appear as round or oval.
2. Life cycle simple.
3. No intermediate host.

3.6.10 *Enterobias vermicularis*

Kingdom: Animalia
Subkingdom: Metazoa
Phylum: Nematoda
Class: Chromadorea
Order: Oxyurida
Family: Oxyuridae
Genus: *Enterobias*
Species: *vermicularis*
Up to Class same as above.

3.6.10.1 Order Oxyurida

1. They are called pinworms because females have slender, sharp pointed tails.
2. Definitive hosts include both invertebrate and vertebrates.
3. They parasitize the posterior portion of the G.I. tract of the animals.
4. Possess a posterior pharyngeal bulb.

3.6.10.2 Family Oxyuridae

1. Presence of oesophagus with prominent posterior bulb.
2. Excretory system X-shaped.
3. Males with single penial spicule.
4. Sperm comet shaped.

3.6.10.3 Genus *Enterobius*

1. Eggs flattened on sides.
2. Life cycle direct.
3. In many haplo-diploid condition is seen.

3.6.10.4 Species *vermicularis*

1. Body worm shaped.
2. Buccal cavity tubular shaped.
3. Eggs when laid contain partial developed larva.

3.6.11 *Ancylostoma duodenale*

Kingdom: Animalia
Subkingdom: Metazoa
Phylum: Nematoda
Class: Chromadorea
Order: Oxyurida
Family: Ancylostomatidae
Genus: *Ancylostoma*
Species: *duodenale*

Up to Order same as above.

3.6.11.1 Family Ancylostomatidae

1. They are known as hookworms.
2. The parasite resides in the intestine of the hosts.
3. They remain attached to the mucus membrane of the G.I. tract of the host feeding on blood and tissue fluid.

3.6.11.2 Genus *Ancylostoma*

1. Anterior end of the parasites is curved dorsally.
2. Anterior portion looks like hook.
3. Buccal capsule is large and heavily sclerotized.

3.6.11.3 Species *duodenale*

1. Oesophagus is robust with swollen posterior end.
2. Oesophagus is muscular and acts as a powerful pump.
3. Oesophageal glands are large and remain outside the oesophagus.
4. Copulatory bursa of males consist of two branched lateral lobes and a small dorsal lobe supported by fleshy rays.
5. Spicules are simple and needle like.
6. Presence of a gubernaculum.

3.6.12 *Trichuris trichiura*

Kingdom: Animalia
Subkingdom: Metazoa
Phylum: Nematoda
Class: Chromadorea
Order: Trichocephalida
Family: Trichuridae
Genus: *Trichuris*
Species: *trichura*
Up to Class same as above.

3.6.12.1 Order Trichocephalida

1. Anterior end more slender than posterior.
2. Absence of lips and buccal capsule.
3. Oesophagus is like slender capillary tube.
4. Both sexes are with single gonad.

3.6.12.2 Family Trichuridae

1. Males with one spicule or absent.
2. Eggs with polar plugs.

3.6.12.3 Genus *Trichur*

1. They are called whipworms.
2. Body looks like a whip with a handle.

3.6.12.4 Species *trichura*

1. Absence of lips.
2. Buccal cavity is tiny.
3. Oesophagus is very long occupying about 2/3rd of the body.
4. Presence of stichocytes.
5. Tail of the male is typically coiled.

3.6.13 *Strongyloides stercoralis*

Kingdom: Animalia
Subkingdom: Metazoa
Phylum: Nematoda
Class: Chromadorea
Order: Rhabditida or Phasmidea
Family: Strongyloididae
Genus: *Strongyloides*
Species: *stercoralis*
Up to Order same as Ascaris.

3.6.13.1 Family Strongyloididae

1. Usually long, slender worms.
2. Oesophagus usually swollen posteriorly but no definite bulb like structure.
3. Male with well-developed copulatory bursa supported by sensory rays.
4. Usually females are oviparous.
5. Eggs are thick shelled.

3.6.13.2 Genus *Strongyloides*

1. Excretory system H-shaped tubular structure.
2. First, second and beginning of third stage juveniles are free living or parasitic in invertebrates.
3. In case of female ovijector is complex with well developed sphincter.

3.6.13.3 Species *stercoralis*

1. Females are parthenogenetic.
2. Oesophagus is long and cylindrical.
3. Females are stout and vulva is located more or less at the middle of the body.
4. Males have two penial spicules and a gubernacalum.

3.6.14 *Trichinella spiralis*

Kingdom: Animalia
Subkingdom: Metazoa
Phylum: Nematoda
Class: Enoplea
Order: Spirurida
Family: Trichinellidae
Genus: *Trichinella*
Species: *spiralis*
Up to Phylum same as above.

3.6.14.1 Class Enoplea

1. Anterior end more slander than posterior end.
2. Absence of lips and buccal capsule are much reduced.
3. Oesophagus is a very slender capillary tube.
4. Oesophagus is embedded within one a more rows of glandular cells called stichocytes.
5. Presence of a cillary band.

3.6.14.2 Order Spirurida

1. Males with penial spicule or absent.
2. Parasites are histiotrophic in all organs of the vertebrates.

3.6.14.3 Family Trichinellidae

1. Smallest nematode parasites of human beings.
2. Express most unusual life cycle.

3.6.14.4 Genus *Trichinella*

1. Absence of copulatory spicules in males.
2. Stichocytes are arranged in a row after the short muscular oesophagus.
3. The anus is nearly terminal.

3.6.14.5 Species *spiralis*

1. Presence of a large copulatory pseudobursa on each side of anus.
2. Females are about twice the size of males.
3. Anterior portion is tapered in females.
4. The vulva opens near the middle of the oesophagus.
5. The single uterus is filled up with developing eggs in the posterior portion.
6. Anterior portion of the uterus contains fully developed, hatching juveniles.

3.6.15 *Dracunculus medinensis*

Kingdom: Animalia

Subkingdom: Metazoa
Phylum: Nematoda
Class: Secernentea
Order: Camallanida
Family: Dracunculidae
Genus: *Dracunculus*
Species: *medinensis*
Up to Phylum same as above.

3.6.15.1 Class Secernentea

1. Amphid apertures are pore or slit like.
2. Phasmids are present in posterior region.
3. Excretory system is tubular.
4. Cuticle is striated.
5. Presence of three oesophageal glands.

3.6.15.2 Order Camallonida

1. Males have one testis.
2. Body thread like.
3. Females are larger than males.
4. Absence of lips.
5. Adult females with degenerated bursa.

3.6.15.3 Family Dracunculidae

1. Males have spicules of same size.
2. Absence of bursa in males.
3. Females are oviparous.

3.6.15.4 Genus *Dracunculus*

1. Largest nematode parasitic to man.
2. Mouth is small and triangular.
3. Presence of reduced lips.
4. The oesophagus of the parasite has a large glandular portion.
5. In females the vulva is located near equatorial region.
6. The vulva is degenerated is adult females.

3.6.15.5 Species *medinensis*

1. The gravid uterus has an anterior and posterior branch.
2. The branch of the uterus is filled up with hundreds of thousands of embryos.

3.6.16 *Wuchereria bancrofti*

Kingdom: Animalia
Subkingdom: Metazoa

Phylum: Nematoda
Class: Chromadorea
Order: Spirurida
Family: Onchocercidae
Genus: *Wuchereia*
Species: *bancrofti*
Up to Order same as Trichinella.

3.6.16.1 Family Onchocercidae

1. The parasites live in tissues of mammals and amphibians, reptiles and birds also.
2. The intermediate hosts are arthropods they deposit J_3s on definitive host's skin.
3. The J_3 enters into the body through the bite of the intermediate hosts.
4. The disease caused by the parasite is called elephantiasis of human beings.

3.6.16.2 Genus *Wuchereria*

1. Adult worms are long and slender.
2. The covering cuticle is annulated.
3. The head of the parasites are slightly swollen.

3.6.16.3 Species *bancrofti*

1. Causes bancroftian filariasis.
2. Their mouth is very small.
3. The buccal capsule of the parasites is reduced.
4. Females are oviparous, produce thousands of microfilatriae.
5. Females produce microfiliriae into lymph where they reside.

3.6.17 *Onchocerca volvulus*

Kingdom: Animalia
Subkingdom: Metazoa
Phylum: Nematoda
Class: Chromadorea
Order: Spirurida
Family: Onchocercidae
Genus: *Onchocerca*
Species: *volvulus*
Up to Family same as above.

3.6.17.1 Genus *Onchocerca*

1. Parasites are knotted in pairs.
2. Resides in subcutaneous tissue of the definitive host.
3. The oesophagus is not distinctly divided.

3.6.17.2 Species *volvulus*

1. Males have two penial spicules, spicules are curled ventrally.
2. The annulations pattern of two sexes is different.
3. In the tail end of microfilaria a number of nuclei are there followed by a clear space.
4. Vulva of females open just behind the posterior end of oesophagus.

3.6.18 *Loa loa*

Kingdom: Animalia
Subkingdom: Metazoa
Phylum: Nematoda
Class: Chromadorea
Order: Spirurida
Family: Onchocercidae
Genus: *Loa*
Species: *loa*
Up to Family same as above.

3.6.18.1 Genus *Loa*

1. It is an eye worm that causes the disease loiasis or calabar swellings.
2. The parasite has simple head with no lips.
3. Having a long slender body.

3.6.18.2 Species *loa*

1. Presence of a blunt tail.
2. Copulatory spicules of males are unequal and dissimilar.
3. Vulva of female is located 2.5 mm from the anterior end of the body.

References

Augustine DL (1945) Filariasis. NY State J Med 45:495
Beaver PC et al (1952) Chronic eosinophilia due to visceral larva migrans. Pediatrics 9:7–19
Blacklok DB (1926) The Development of *Onchocerca volvulus* in *Simulium damnosum*. Annals of Tropical Medicine and Parasitology 20(1):1–48
Chandler AC, Read CP (1961) Introduction to parasitology. Wiley, London
Karyakarta RP, Damle AS (2003) Medical parasitology. Books and Allied (P) Ltd, Kolkata
Kouri P, Valdey Dias R (1952) Concepto actual sobre el papel patogeno de *Tricocephalo disper* (*Trichuris trichiura*) sintomatologia gastro intestinale. Rev Kuba Med Trop y Parasitol 8:37–41
Nelson GJ (1990) Onchocerciasis. In: Dawes B (ed) Advances in parasitology, vol 8. Academic Press, New York

Stoll NR (1947) This wormy world. J Parasitol 33:1–18
Strong RP, Sandground JH, Bequert JC, Ochol MM (1934) Onchocercosis. Havard University Press, Boston
Wanson M (1950) Contribution a l'etude de l'onchocercose Africaine humaine. Ann Soc Belge Ode Med Trop 30:667–863
Wykoff DE, Lepes TJ (1957) Studies on *Clonorchis sinensis.* I. Observations on the route of migration in the definitive host. Am J Trop Med Hyg 6:1061–1065

4 Organisms Those Carry Disease Causing Organisms (Vector)

Most of the disease carrying organisms are Arthropods and they are called vectors.

The study of the Arthropods of medical importance is called medical entomology.

The vectors are those who are intermediate hosts of pathogens belonging to three classes:

1. Insecta: The body of the organisms is divided into three parts: head, thorax and abdomen. Thorax is subdivided into three parts: prothorax, mesothorax and metathorax. Each subdivision possesses a pair of appendages or legs, i.e. three pairs of thoracic legs. Presence of a pair of antennae. Presence or absence of one pair or two pairs of wings. The example of wingless insect is bookworm (Lepisma), one pair of wing mosquito, housefly, sandfly, tsetse fly, deer fly, midges and blackfly, while two pairs of wings are cockroach and reduviid bug.
2. Arachnida: The body is divided into two divisions. Cephalothorax and abdomen and have 4 pairs of legs. All have no wings and possess no antenna. Example: Ticks and mites.
3. Crustacea: The body is divided into two parts: Cephalothorax and abdomen. Thoracic appendages are 8 pairs and abdominal appendages are six pairs in number. They live in water. Example: Cyclops.

The vectors are responsible for biological transmission. Biological transmission means the pathogens undergo a part of their life cycle within the vectors and infective forms are developed within them which they transmit to other hosts by injecting the infective forms during their blood meal but mechanical transmitters just carry the pathogens from one place to another (Table 4.1).

P. K. Bandyopadhyay et al., *Biochemical, Immunological and Epidemiological Analysis of Parasitic Diseases*, https://doi.org/10.1007/978-981-16-4384-2_4

Table 4.1 Vectors of different vector Bourne diseases with their name

Vectors	Scientific name	Diseases	Parasites
Mosquito	Anopheles Culex	Malaria Filaria	*Plasmodium* sp. *Wuchereria bancrofti*
Housefly	*Musca domestica*	Amoebiasis, Helminthiasis	Parasitic amoeba, Helminthes
Sandfly	*Phlebotomus papatasii* *P. argentipes*	Kala-azar	*Leishmania donovani*
Tsetse fly	*Glossina palpalis*	Sleeping sickness	*Trypanosoma gambiensae*
Deer fly	*Chrysops* sp.	Surra, African filarial Eye worm	*Trypanosoma evansi*, *Loa loa*
Louse	*Pediculus humanus*	Typhus, Trench fever Relapsing fever	*Rickettsia prowazekii*, *Rochalimaea quintana*, *Borrelia recurrentis*
Reduviid bug	*Triatoma infestans*	Chagas disease	*Trypanosoma cruzi*
Blackfly	*Simulium* sp.	Onchocerciasis	*Onchocerca volvulus*
Rat flea	*Xenopsylla cheopis*	Hymenolepiasis	*Hymenolepis nana*
Hard tick	*Boophilus annulatus*	Babesiosis	*Babesia microti* *Babesia bigemina*
Cyclops		Dracunculosis *Diphyllobothriasis*	*Dracunculus medinensis* *Diphyllobothrium latum*

4.1 Order Diptera

4.1.1 Mosquito

There are two major groups of mosquitos: Anophelini and Culicini. Anophelini group has only one genus Anopheles.

Culicini have 15 genera, the medically important are Culex, Aedes and Mansoni.

The females of the many species of this order are blood sucking insects and often transmit extremely harmful diseases of man and livestock.

Over 3500 species of mosquitoes have been reported from different parts of the globe. Some routinely bite human beings and act as vectors of number of infectious diseases affecting millions of people every year. Others do not routinely bite humans but are vectors of diseases of animals. But many become agents for zoonosis of new diseases when their habitats are disturbed like sudden deforestation.

4.1.1.1 Morphology

Mosquito is an insect so its body consists of head, thorax and abdomen.

Head: It is semi-round and has a pair of compound eyes, a long proboscis, a pair of palpi and a pair of antennae. The proboscis is used for biting. Only the female mosquitos have biting and sucking mouth parts because for the maturation of their eggs in their ovary, they need protein which they collect from the blood. The male proboscis is not adapted for biting (Plate 4.1, Fig. 4.1).

Thorax: Thorax is subdivided into three parts: prothorax, mesothorax and metathorax. Each division possesses a pair of appendages or legs. Thorax bears a pair of wings.

Abdomen: It is narrow, slender and has 10 segments. But we can see only eight segments, other two segments modified to form external genitalia.

4.1.1.2 Life Cycle

The mosquitoes (Plate 4.1, Fig. 4.1) complete their life cycle in four stages and undergo metamorphosis. The stages are egg, larva, pupa and adult, Comparison of the three types of mosquitoes Anopheles, Culex and Aedes are given in the Table 4.2.

Egg: After sexual contact the female mosquitoes are inseminated and eggs are fertilized. The female then lays 200 to 400 eggs on the surface of the water some in stagnant, some lay eggs near the edge of the water, others attach their eggs to aquatic plants. After 2–3 days of laying eggs they hatch into larvae.

Larva: The larvae are restless and feed on microorganisms. The body is slender and elongated and divided into head, thorax and abdomen. The head possesses a pair of compound eyes, a pair of antennae and a pair of feeding brush. Abdomen is provided with respiratory siphons. After 7 to 10 days the larva is metamorphosed into pupa.

Pupa: It is the resting stage and coma shaped with large cephalothorax and abdomen. It does not feed. It floats on the surface of the water. Pupa stage lasts for 1 to 2 days.

Adult: Adults emerge from the pupa by splitting the skin of the thorax and rests on the skin of the thorax of pupa for a while then fly away.

The thorax is specialized for locomotion. Three pairs of thoracic appendages and a pair of wings are attached to the thorax. The Anopheles can fly for up to 4 h at 1–2 km/h speed. They can travel 12 km in a night. Males beat their wings between 450 and 600 times per second.

The abdomen is specialized for food digestion and egg development. The abdomen of a mosquito can contain three times its own weight in blood. The blood serves as a source of protein which is needed for the production of eggs. Prior to blood sucking female mosquitoes inject saliva into the body of the host. The saliva acts as anticoagulant because otherwise the blood may coagulate within the proboscis during sucking. The saliva is the main route by which pathogens are transferred to the host.

The period of development is 7 to 10 days. Lifespan of a mosquito is about 2 weeks (Table 4.2).

4.1.1.3 Saliva

Mosquitoes and other blood sucking arthropods need blood for the maturation of their eggs. So the females are blood sucking. They have mechanisms to effectively

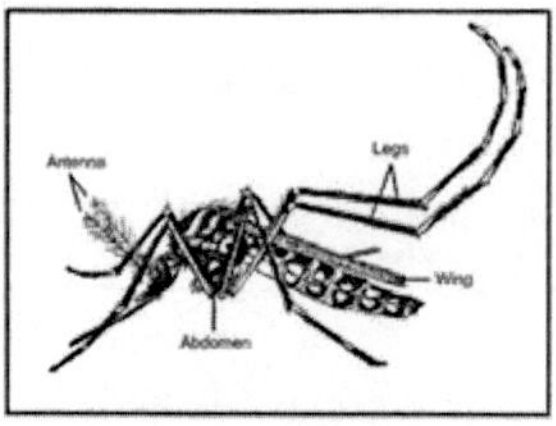

Fig.4.1. Mosquito *(Advanced Parasitology, Das)*

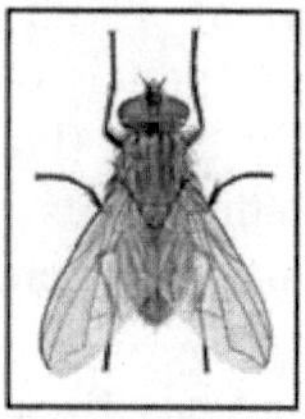

Fig. 4.2. Housefly (From Internet)

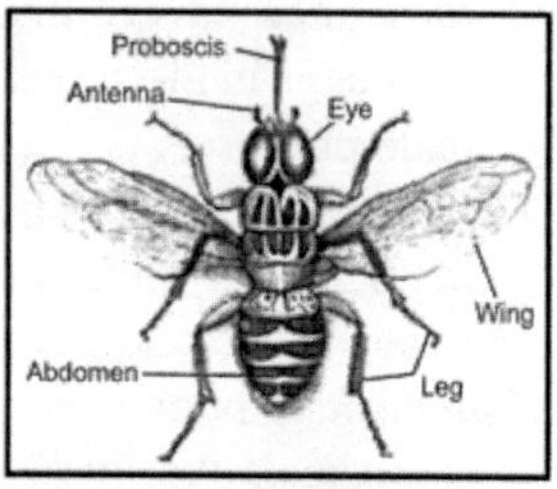

Fig. 4.3. Tsetse fly (*Parasitology*, Chatterjee)

Fig. 4.4. Sandfly *(Advanced Parasitology, Das)*

Fig. 4.5. Deer fly (From Internet)

Fig. 4.6. Black fly (From Internet)

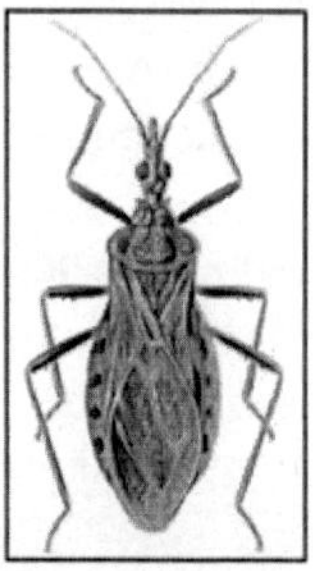

Fig.4.7. *Panstrongylus megistus (From Internet)*

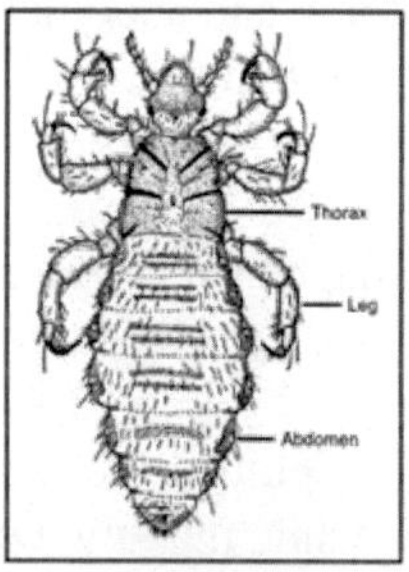

Fig. 4.8. Louse (Roberts *et al.*)

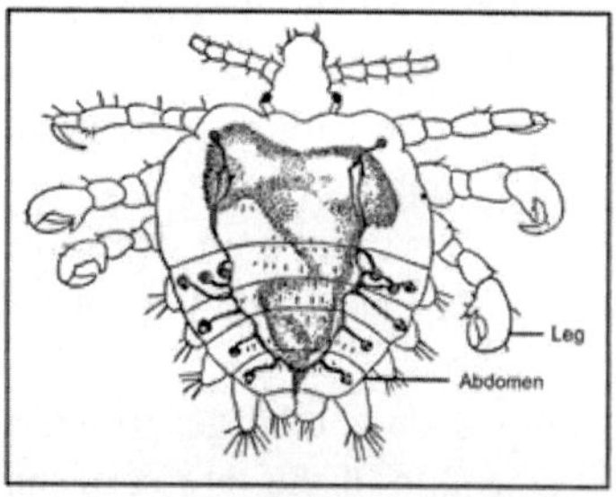

Fig. 4.9. Crab louse (*Parasitology*, Chatterjee)

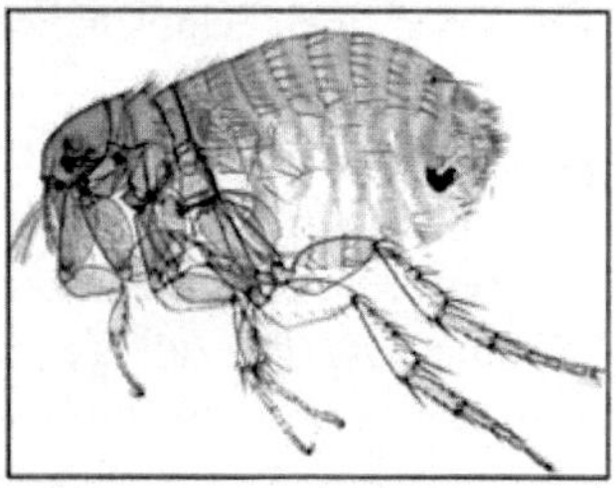

Fig. 4.10. Flea (From Internet)

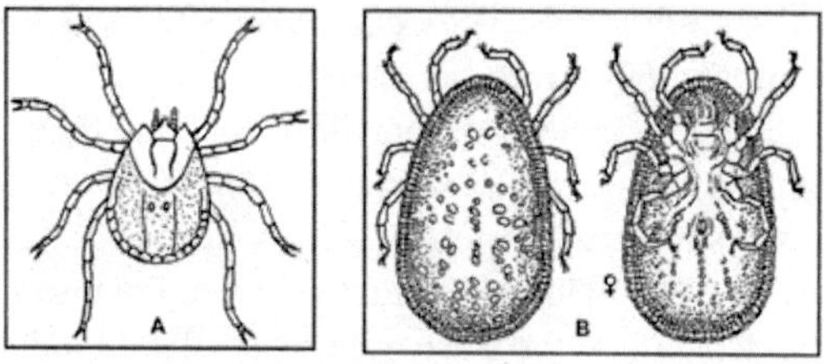

Fig. 4.11. A. Hard tick, B. Soft ticks (From Internet)

Plate 4.1 Figures of all the vectors described in the text

Table 4.2 Difference between the life cycle of Anopheles, Culex and Aedes

Anopheles	Culex	Aedes
Adult:		
(1) The sitting position on the wall forms an angle of 45°, i.e. acute angle (2) Wings have black and white bands (3) Antenna is less bushy (4) Produce sound during flight	(1) The sitting position on the wall is more or less parallel with the wall (2) Wings have no bands (3) Antenna is bushy (4) No sound is produced during flight	(1) It sits more or less parallel to the wall (2) Wings are provided with black and white bands (3) Antenna is bushy (4) Less sound is produced during flight
Eggs:		
(1) Eggs are small cigar shaped and have floats down their sides (2) Eggs are laid in single (3) No. of eggs vary from 200 to 300	(1) Eggs form a raft no air float (2) Eggs are laid in cluster (3) No. of eggs vary from 200 to 800	(1) Eggs form raft but no air float (2) Eggs float separately (3) In large number
Larva:		
(1) It floats, parallel to the surface of the water (2) Respiratory siphon is short	(1) Larva floats obliquely with head hanging down (2) Respiratory siphon is long and tubular	(1) Floats like that of culex (2) Respiratory siphon is tubular
Pupa:		
(1) Colour is green (2) Dorsal respiratory tube small and flat (3) Tail fin large	(1) Colourless (2) Dorsal respiratory tube narrow and elongated (3) Tail fin small	(1) Colourless (2) Same as that of culex (3) Tail fin medium
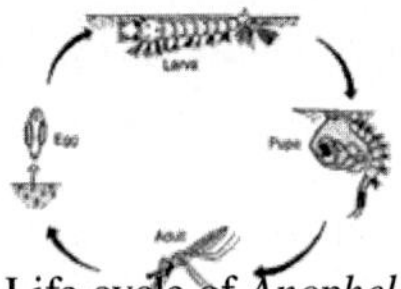 Life cycle of *Anopheles*	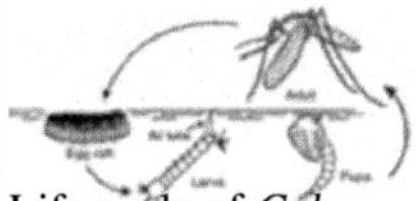 Life cycle of *Culex*	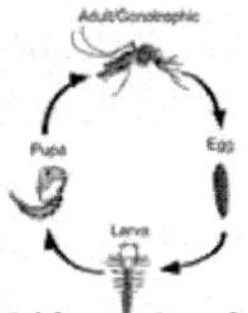 Life cycle of *Aedes*

prevent the haemostasis system of vertebrates from whom they will suck blood. This is done with their saliva, which contains a mixture of secreted proteins. Their saliva negatively affects vascular constriction, blood clotting, platelet aggregation, angiogenesis and immunity and creates inflammation. Saliva of blood sucking arthropod contains at least one anticlotting, one antiplatelet and one vasodilating substance. It also contains enzymes that help in sugar digestion and antimicrobial agents to control bacterial growth in the sugar containing meal. The composition of saliva of mosquito is 20 dominant proteins. Despite the great strides in knowledge of these molecules and their role in blood feeding achieved, scientists still cannot explain the functions of more than half of the molecules found in their saliva. One application is the development of anticlotting drugs, like clotting inhibitors and capillary dilators useful for cardiovascular disease. The mechanism for immune response is due to a factor in saliva that directly suppresses TNF-α release. Experiments by Cross et al.

(1994) demonstrated the inclusion of *Ae. aegypti* mosquito saliva into naïve cultures led to suppression of interleukin IL-2 and IFN-γ production, while the cytokines IL-4 and IL-5 are unaffected by the saliva of mosquitoes. T-cell populations are decidedly susceptible to the suppressive effect that T- and B-cell proliferation was inhibited in a dose dependent manner with concentrations as low as 1/7th of the saliva in a single mosquito. A recent study suggests mosquito saliva can also decrease expression of interferon α/β during early mosquito borne virus infection. Recent research suggests mosquito saliva exacerbates mosquito transmitted viruses infection including the West Nile virus.

4.1.1.4 Disease Carried by Mosquito

Mosquitoes act as vectors and intermediate or definitive hosts for a number of disease causing organisms:

Mosquitoes	Disease transmitted
Anopheles	Malaria, filaria (not in India)
Culex	Bancroftian filarial encephalitis
Aedes	Yellow fever, dengue, Chikungunya

The roles of different mosquitoes in spreading the different parasitic diseases are discussed in the relevant account of particular parasites.

4.1.2 Housefly (*Musca domestica*)

Houseflies are non-biting insects of the suborder Cyclorrhapha and have four species: domestica, vicinia, nebulo, sorbens.

4.1.2.1 Morphology

The adult is 5 to 8 mm long, has a pair of compound eyes, a pair of antennae and a single retractile proboscis. The thorax is provided with 2 to 4 longitudinal stripes, a pair of wings and 3 pairs of legs. The abdomen has light and dark marking. The whole body is covered with hair-like structures. Houseflies have only one pair of wings (Plate 4.1, Fig. 4.2).

The Musca can neither bite nor sting yet they are nuisance to our daily life disturbing us by carrying germs of various diseases which makes us severely ill.

The housefly is diurnal and their presence becomes intolerable to civilized persons.

They simply carry in their legs and wings the germs of the diseases like typhoid, diarrhoea, amoebic and bacillary dysentery, cholera, etc.

They suck the liquid from the waste products, faeces, garbage, decaying organic matters and also lay their eggs there. The drains, garbage, kitchens are the favourite places of their residence, breeding and nursing homes.

They first sit on the waste products and then fly and rest on our food. They are of habit of brushing their hairy legs and in this process they contaminate our food.

4.1.2.2 Life Cycle

The life cycle is completed through four stages: egg, larva, pupa, imago (adult).

Egg: The eggs are white like pearl and 1.2 mm long. They are laid by their mother on decaying organic matter. After 8 to 24 h the eggs hatch and larvae come out. Each female fly can lay about 2000 eggs in her lifetime, i.e. 6 to 8 weeks.

Larva: The larvae are called maggots and they are voracious eater. They moult two times in a week and increase from 3 mm to 10 mm. The larvae after 1 week transformed into pupa. The larvae are pale, whitish, narrow at the mouth and have no legs. At the end of the third instar the maggots crawl to a dry, cool place to pupate.

Pupa: It is the resting stage and lasts for 3 to 6 days. They are colourful, reddish and about 8 mm long. It does not feed. After about a week the adult comes out.

Adult: After about 36 h the female becomes eager for mating. The male mounts on her from behind to inject sperm. Copulation takes a few seconds to a couple of minutes. Usually the female mates once, but sperms are stored in the female body which they use repeatedly. The adult possess all the characters of insect.

4.1.2.3 Transmission of Diseases

Housefly is a very efficient mechanical vector. The germs they carry in their legs and wings are the eggs and cysts of helminthes, amoeba and bacteria and viruses of diarrhoea, trachoma, anthrax and viral hepatitis, dysentery, cholera, poliomyelitis, conjunctivitis.

Houseflies carry the pathogens on their feet and wings from contaminated faeces, sputum and vomit of the patients to our food. They cannot consume solid food. Therefore they vomit on the surface of the food and then suck. The vomit is rich in bacteria. They frequently excrete and excreta contain bacteria, ova and cysts of human intestinal parasites. So the pathogens are carried by their foot, wings, vomit and excreta.

The method is they collect bacteria, spores and eggs of helminth on the hairs those are present on their legs and body surface. The mouth parts and legs of them have sticky pads. They sit on the excreta, vomit, etc. of infected persons to have food and they carry those pathogens to our food while trying to eat our food on sitting upon them and vomit and excrete on our food instantaneously. The bacteria, eggs, spores, etc. do not perish in their intestine.

4.1.3 Tsetse Fly (*Glossina palpalis*)

Tsetse flies are important intermediate hosts or vectors of *Trypanosoma* infections of man and domestic animals.

There are about 20 species of tsetse flies all included under single genus *Glossina*. These flies are restricted to Africa from south of Sahara to the north of Union of South Africa.

4.1.3.1 Morphology

Tsetse flies are elongated in shape and dark brown or yellowish brown in colour. They measure 12 to 14 mm in length. They have piercing and blood sucking mouth parts. They possess a pair of wings and when at rest they fold their wings flat over the back, one on top of the other not spreading them like other flies. Their characteristic movement of flight and buzzing sound makes them recognizable.

The antenna has long feathered bristles. The proboscis consists of a piercing labium with a bulb like base and its tip is adapted for rasping and tearing. The fine hypopharynx contains salivary duct. The elongated maxilla and palpi are closed together to form a cover for the proboscis (Plate 4.1, Fig. 4.3).

The thorax is quadrangular and abdomen is tapered. In the riverine species of *'Glossina palpalis*, vector of *Trypanosoma gambiense*, the hind tarsi are dark in colour. Male tsetse flies have a large oval swelling on the underside of the last abdominal segment called 'hypopygium' which contains the male genitalia.

4.1.3.2 Habits and Habitats

Tsetse flies are diurnal, and *palpalis* species bites on moonlight nights. *G. palpalis* flies 5 or 6 ft. above the ground and bite usually above the waist. They commonly rest on the underside of twigs near the ground. *G. palpalis* are found close to rivers or lakes. The flies have preference for dark colours so Negros are the victims.

Both the sexes of *Glossina* are blood suckers. They have preference for large mammals. They are the most important vectors of *T. gambiensae* and *T. rhodesiense* of man and animals causing African sleeping sickness.

4.1.3.3 Life Cycle

Glossina palpalis is a large dark fly with blackish-brown abdomen and grey thorax with distinct brown coloured markings.

Tsetse flies never lay eggs instead a single larva is produced within the body of the mother and nourished by special milk glands present on the wall of the uterus and remain close to the genital opening of the mother with respiratory spiracles. The larva moults and becomes full grown to pupate before it is born. The larva occupies practically the whole of the swollen abdomen of the mother. When one larva is born the other starts developing and new one is born about 10 to 12 days interval provided food is abundant and temperature is optimum.

In laboratory condition it was found that one fly gave birth to 8 larvae in a span of 13 weeks.

The full grown larva is a third stage larva when born is yellowish white in colour and 8–10 mm in length with a pair of dark knob like projection at the posterior end. Immediately after birth they crawl beneath the soil surface at a depth of 1 or 2 cm in loose soil or under dead leaves and become pupa. The mother selects dry, loose soil in shade and protected spots for deposition of larvae.

Pupa is olive shaped and mahogany colour with blackish knobs still present at the posterior end. The pupa stage remains for 17 days to 3 months depending upon the temperature of the environment. Adults emerge about 70° and below 86 °F.

Adult flies live for only a few months in the wet season and for about 3 weeks under dry and hot conditions. Man and flies both seek wet and water holes. So the tsetse flies find ideal conditions for transmission of the trypanosomiasis. These flies go berserk and create havoc. It is recorded that 4 flies infected 43 people in a hamlet the place being a small water hole.

4.1.4 Sandflies (*Phlebotomus papatasi*)

Sandflies are blood sucking insects classified under order Diptera. These are hairy midges found in nearly all warm and tropical countries.

Sandflies play a great role in transmitting various types of Leishmaniasis, 3-day fever, a filterable virus disease and Oroya fever. The 3-day viral fever is also called papatasi fever.

The transmission of Leishmaniasis was first discovered in 1941 and the role of sandfly was recognized.

Theodor (1948) pointed out that there is no evidence of endemic Leishmaniasis in absence of sandflies.

All the types of Leishmaniasis are not the same and carried by *Phlebotomus papatasi* there are other species of *Phlebotomus* also. Visceral leishmaniasis or kala-azar in India is transmitted by *P. argentipes* where man is the only host.

Cutaneous leishmaniasis or oriental sore is transmitted by *P. papatasi*.

4.1.4.1 Morphology

Sandflies are light or dark brown in colour and smaller than mosquito in size. They measure only 3–5 mm in length. Their body and wings are covered by dense hair. They have a pair of hairy antennae, a pair of palpi and a proboscis. The thorax possesses a pair of hairy upright wings. The three pairs of thoracic legs are long and slender. The abdomen consists of 10 segments. Sandflies in spite of their presence of wings they do not fly but hop about (Plate 4.1, Fig. 4.4).

4.1.4.2 Life Cycle

The female lays eggs in dark, damp soil near poultry and cattle sheds. Eggs hatch in about 7 days. Larvae are hairy maggots which are converted to pupa in 2 weeks.

The pupa stage remains for a week and adults come out. Like mosquitoes the female only sucks blood. Males feed on the plant sap. Female requires blood meal for maturation of ovary third to fourth day. Sandflies bite only during night and during day they remain hidden in cracks and crevices of walls, holes in the trees and in cattle stables.

The female sandfly requires blood meal for the maturation of its eggs and during feeding Leishmaniasis is transmitted horizontally from one person to other. The role of sandflies in spreading Leishmaniasis is discussed in the chapter where *Leishmania donovani* is described.

Sandfly fever: This viral disease resembles in many respect with influenza. The symptoms are sudden fever, headache, pain in the eyes, stiffness of back and neck and rheumatic pain like dengue, leukopenia happens.

The insects become infective after 6 to 7 days of taking blood meal. Since sandflies are very much short lived and suck blood for only one time the transmission is transovarial transmission. Though there is controversy.

Oroya fever: It is also called Carrion's disease. It is an acute febrile disease caused by *Bartonella bacilliformis* occurring in Peru, Chile, Ecuador and Bolivia.

The symptoms of Carrion's disease are high fever, severe anaemia, albuminuria. The disease often becomes fatal. Townsend in 1913 pointed out the role of *Phlebotomus* in transmitting the disease. There is another form of this disease called verruga peruana. It is the cutaneous non-fatal form of disease where eruption of nodules on the skin takes place.

4.1.5 Deer Fly (*Chrysops* sp.)

Chrysops belongs to the family Tabanidae. The family Tabanidae contains horseflies and deer flies. Deer flies are usually smaller than horseflies. Deer flies transmit *Loa loa*, the African eye worm.

4.1.5.1 Morphology

The flies have conspicuous black bands and spots on their wings. Their flight is not so noisy. They bite human beings without being alert. The bite of the deer flies is extremely painful. It seems that in case of timber workers the productivity is hampered because of harassment by these flies.

They are 9 to 10 mm long. The head is triangular having long antennae. Mouth parts of females are used for stabbing and cutting, these are usually short, heavy and powerful. The female sucks blood, the males thrive on plant juices (Plate 4.1, Fig. 4.5). There are over 2500 species spread throughout the globe. But they are mostly found in tropics.

4.1.5.2 Life Cycle

Most of the deer flies breed in damp places, some breed in rotten wood or in tree holes. The female lays several hundred eggs. When deposited the eggs are covered by an adhesive waterproof secretion that binds the eggs together. The eggs hatch within a week.

The larvae are cylindrical in shape and without any appendages or legs. The body consists of 11 segments having a small retracted head. Each segment bears some wart like structures with spines or hairs. The larvae feed voraciously. Most species prey upon soft bodied creatures like earthworms and other insect larvae, sometimes seen to practice cannibalism. Many species thrive on dead organic matter.

The larvae are active and grow rapidly during summer. They pupate on dry soil. The pupa stage lasts only for 1 to 2 days.

Most of the species are diurnal and some often prefer shades. They are very stray flies and may travel more than a kilometre. They like to skid over the surface of the water. Scientists are of opinion that they cause serious damage to the cattle population by disturbing and sucking blood. It is seen that human beings suffer severely from the bites of deer flies.

4.1.5.3 Transmission of Diseases

Some of the deer flies of animals act as vector of *Trypanosoma evansi*, the causative agent of the disease 'surra' in horses, cattle, camels, dogs, etc.

In the Western United States in Utah and Colorado there occurs a disease known as 'tularemia' or deer fly fever. But this disease is caused by a single species of deer fly *Chrysops discalis*.

Anthrax is also transmitted by deer flies. This is a serious dreadful bacterial disease of domestic animals and man.

Certain species of Chrysops are the vectors of African eye filarial worm *Loa loa* in Africa. The microfilariae of *Loa loa* are diurnal. The deer flies suck pool of blood. The filarial larvae develop in the abdomen of the fly and the infective larvae crowd in the proboscis after 10 to 12 days of feeding.

4.1.6 Blackfly (*Simulium* sp.)

The blackflies (*Simulium* sp.) are the most important insect pests and vector of the disease which disturb man and domestic animals. They kill animals and children of human beings when they bite in hordes due to loss of blood and also by stuffing their bronchial tubes resulting in suffocation.

4.1.6.1 Habit and Habitats

Blackflies breed in running water. The genus *Simulium* is found throughout the world but present very much in Africa, Mexico, Central and South America. *Simulium indicum* is the Indian species of *Simulium*.

The nature of the blackflies is diurnal. There are about six genera and are divided according to their venation of the wings (Smart 1945). The species of the genus *Simulium* are important disease carriers. Blackflies are infected during their taking blood meal from infected persons. Their mouth parts are not adapted for deep piercing but their food consists of tissue juices which may contain numerous microfilariae from infected persons.

The first stage larvae of *Onchocerca* sp. move from the intestine of the fly to the thoracic muscles of them. In the thoracic muscles of the blackfly they moult into the second stage sausage shaped larvae. The second stage larvae finally moult to form third stage filariform larvae. The third stage larvae migrate to the labium of the fly. Then the infective larvae of *Onchocerca* enter into the human beings during their bite of the healthy persons.

4.1.6.2 Morphology

The Blackflies are small, robust, hump-backed organisms having short legs, broad wings and short antenna that comprises 11 segments and without whorls of hairs at the joints. The wings are like horns of the mammals (Plate 4.1, Fig. 4.6). The proboscis in males is poorly developed as they are not bloodsuckers. But the proboscis in females is short, heavy and powerful as they are bloodsuckers. They need blood as food for the development of eggs in them.

The mouth parts of blackflies comprise toothed dagger like mandibles and maxillae. There are a hypopharynx and a labrum-epipharynx.

The colour of the body is black in most of the northern species, so their name is blackflies. It is also found that some species are reddish brown or yellowish in colour. They also may be stripped. The colour of the wings is either greyish or yellowish and has heavy veins near the anterior margin. The wings are scaleless.

The females are voracious bloodsuckers and transmit human filarial worms *Onchocerca volvulus* in Africa and tropical America. The bite of the blackflies is very painful but not instantly and followed by itching and swelling at the site of the bite.

4.1.6.3 Life Cycle

Some of the species of *Simulium* breed in slow flowing rivers some in trickling rivulets. S. *pictipes* breeds at the foot of falls.

The eggs are provided with a viscid coat and are deposited on vegetation, stones, etc. wet or partially submerged. One species S. *arcticum* drop the eggs while flying over water avoiding getting her foot cold. S. *damnosum* is found to deposit the eggs on water licked leaves.

The eggs hatch within a few days. The larvae attached them by means of circles of minute hooklets at the blunt posterior end of the body. The larvae then spin cocoons from threads coming from modified salivary glands. There are a pair of mouth fans by the movement of them microscopic organisms are swept into their mouth. The larvae breath by means of minute gills projected through the anal slit present in the last segment of the abdomen. The larvae in batches are attached to the submerged vegetation.

When ready to spin cocoons to pupate then there remains an opening at the anterior end for extrusion of the minute branching gills those are used as breathing organs.

The adults emerge out within 3 days to a week or more. The adults are short lived and start to lay eggs soon after emergence and copulation.

In warmer countries mainly in tropics the generation time is only about 3–4 weeks.

In the tropics the eggs may hatch in 2 days, the larvae go through six moults and pupate in 7 days. The adults emerge from cocoons in 4 days and start laying eggs in 4 days (Wanson 1950).

The blackflies after biting blood trickle down from the biting site and take several minutes to engorge. Out of 57 species only five have been reported to bite human beings in Africa.

4.2 Order Hemiptera

4.2.1 Reduviid bug (*Rhodnius prolixus, Triatoma infestans, Panstrongylus megistus)*

Members of the family Reduviidae are predators of other insects. The Reduviid bugs are valuable as they consume pest species. It is also reported that *Reduvius personatus* sometimes visit residential house to consume bed bugs.

Some of the species live in human habitation hiding them in day time and attack sleeping human beings at night. They are very active and have rapid movement so if light is being switched on they hide quickly and it is very hard to catch them.

The bites of these reduviid bugs are very painful but usually they seldom bite human beings. If the opportunity comes they become human bitters. They feed exclusively on human blood and bite mainly on the face around eyes and lips. So they are called kissing bugs. Biting of kissing bugs is painful.

One subfamily Triatominae is vector of the Chagas disease in Mexico, Central and South America. There are three medically important bugs: Triatoma, Rhodnius and Panstrongylus bugs (Plate 4.1, Fig. 4.7).

4.2.1.1 Morphology

They are 25 to 35 mm in length. Adults have two pairs of wings.

4.2.1.2 Life Cycle

The eggs are white in colour and oval in shape when first laid but the eggs turn yellowish or pinkish later. The total number of eggs laid by a female is generally about 100 to 300. The eggs require 2–3 weeks to hatch. The nymphs hatch out are wingless. They moult 5 times to become adult. The time taken for development from egg to adult is about a year in some species may be 2 years.

At least 36 of the hundred or more species of Triatominae are capable of transmitting Chagas disease, the causative agent of the disease is *Trypanosoma cruzi*. *Triatoma infestans* plays the leading role. These bugs when suck blood have a tendency to defecate at the site. If the infected bug bites then along with the faeces a large number of parasites escape. As a result of irritation of bite of the insect, the host rubs the site and parasites enter within the eye as they bite on the lid of the eye. This is the method by which transmission of *T. cruzi* occurs. Infection of bugs comes from infected mammalian host.

There is another interesting phenomenon. These bugs practice cannibalism. Young bugs sometimes suck blood from the other bugs which have sucked infected blood and the body becomes distended. In this way also the infection transmits from bug to bug. Transovarial transmission does not take place.

4.3 Ectoparasite

4.3.1 Order Anoplura

4.3.1.1 Louse (*Pediculus humanus*)

Louse is under the order Anoplura. The human infesting lice belong to two genera: Pediculus and Phthirus.

Pediculus: **Head and body lice**.

Lice are ectoparasite of human beings, they are prevalent among poorer classes and mostly among children who do not care to clean their body. The head lice (*Pediculus humanus corporis*) and body lice (*Pediculus humanus pules*) are found in human beings. Infestation with lice is known as pediculosis, when the lice bite red papules that exude and itch are developed. The scratching produces dermatitis with secondary infection. There is another louse called crab lice (*Phthirus pubis*) found in the pubic hairs of both sexes of human beings.

Morphology

The head bears eyes and five jointed antennae in the body louse and is larger than head lice. The thorax is almost square and has three pairs of legs with claws at their end. Wings are absent. The thorax has a single pair of spiracles located between the first and second pairs of legs. The abdomen is composed of seven segments of which the first six have spiracles. Pleural plates are well developed. Females are larger than the males and are measured 2.4 to 3.3 mm long. Head lice are 2.4 to 3.8 mm long. Males and females can be distinguished by the indented posterior end of the abdomen and by the slender anterior legs in females. In males the anus and sex openings are dorsal in position (Plate 4.1, Fig. 4.8).

Habit and Habitat

The head lice reside in the hair of the head, sometimes they move to the other hairy parts of the body. They occur in all races of human beings irrespective of sex, religion and colour. Regular cleaning of the head, combing and brushing of the hair helps the population of lice remains at low level.

The body louse mainly lives in the cloth instead of hair of its host. They have evolved from the body of the host but as nudism went out of style they shifted their residence to clothing and most probably they evolved from the head louse. It is seen that in a hairy human hundreds of body lice are found in the body but after removal of the clothing not a single is found in the body. They are also found in the underwear where they lay eggs.

Life Cycle

The eggs called ‘nites’ are oval in shape, white in colour having a little lid at the broad end provided with air cells pierced by pores by which air can enter into the eggs.

The eggs of hair lice are 1 mm in length and are glued to the hairs with the help of a cementing secretion. The number of eggs laid by a female is about 100.

Body louse lays slightly larger eggs and remains attached to the fibres of the clothing particularly along the creases and stitches. Egg laying temperature is more than 80 °F. At 80–85 °F the eggs hatch in about 8 to 10 days.

There is a peculiar phenomenon of hatching larva. The young lice suck air into the body and expel it from the anus until a cushion of compressed air is formed which is sufficient to open the lid of the eggs. The newly hatched lice are not larvae but nymphs as they look alike the adult but miniature in form only difference is they have three segmented antennae. They start eating immediately after birth. They moult first after 3 days, second in 5 or 6 days and third, the last one in 8 or 9 days and transform into adult.

The female starts laying eggs 3 to 4 days after gaining maturity. The average lifespan is 35 to 40 days for both sexes.

4.3.1.2 Crab Louse (*Phthirus pubis*)

It has some differences in their morphology from hair or body lice. It has a very broad and short body with long legs bearing claws, looking like crab. So they are called crab lice. The first pair of legs is smaller than the other two pairs. The thorax is very broad with all segments fused, abdomen is compact and shortened. Its first three segments fused into one and have three pairs of spiracles. The last four segments bear wart like processes on the sides, the last pair is comparatively larger. They are greyish in colour. The females are 1.5 to 2 mm in length. The males are smaller in size. The favourite residing place is pubic hair, armpit or in beards, eyebrows and eyelashes. These lice remain restricted to white people.

The females lay eggs 25 or more in number and glue them in the course hairs (Plate 4.1, Fig. 4.9). The eggs hatch in 6 to 7 days and they become adult in 2 to 3 weeks. The adults suck blood for hours at a time.

Transmission of Diseases

Body louse is the vector of epidemic typhus, trench fever and relapsing fever. Crab louse is not reported to be a vector except causing allergy and irritation.

Typhus: The disease is caused by *Rickettsia prowazekii* causing epidemic typhus. The rickettsia are sucked by the lice during blood meal from an infected person. They multiply in the stomach epithelium of the lice. The infected cells of the stomach epithelium distend and burst in 10 days and kill the louse. However before the death of the louse the excreta of infected louse contain a large number of rickettsia. They usually defaecate while biting and due to irritation human beings scratch the site. The disease causing organisms get access into the body of the new host. Human beings can also be infected from inhaling the dried louse faeces.

Trench fever: Another type of rickettsia *Rochalimaea quintana* causes trench fever. The disease was first reported during World War I it caused more sickness than any other disease within the soldiers.

The infection is transmitted like typhus by louse faeces. The symptoms of the disease are headache, body ache, high fever, presence of albumin in the urine and skin rash. The lice are not killed by the rickettsia so the louse remains infective for the whole life after receiving the infection.

Relapsing fever: The disease is caused by a spirochete named *Borrelia recurrentis.* The lice get the infection from the blood meal of infected person. The spirochetes die in the alimentary canal of the louse but a few survive and reach into the body cavity by penetrating the stomach wall of the louse. Here within the body cavity they multiply and become abundant in number.

Transmission is performed by crushing the lice during scratching and through the wound caused by bite and scratching the organisms enter into the human body.

4.3.2 Order Siphonaptera

4.3.2.1 Flea (*Xenopsylla cheopis*)

The oriental rat flea is also known as tropical rat flea. These are primary vector of Bubonic plague and murine typhus.

Fleas are small, bilaterally compressed, wingless insects. They are blood sucking ectoparasites of mammals and birds.

Xenopsylla cheopis with two other species *X. astia* and *X. brasiliensis* are of medical importance because they can transmit bubonic plague and endemic typhus in human beings. *X. cheopis* also plays a role in the transmission of *Hymenolepis nana* and *H. diminuta*.

Morphology

The length of the body of flea is about 2.5 mm. The body is composed of three parts: head, thorax and abdomen. The head and the thorax possess rows of bristles. The abdomen consists of eight segments. They can jump 3 to 4 in. in height. Their body is covered by backward pointing spines (Plate 4.1, Fig. 4.10).

The mouth parts are used for squirting saliva and for sucking blood from the host. Fleas detect host by the CO_2 they exhaled like man and animals and jump to the source of the food.

Fleas are wingless so cannot fly but can jump long distance with the help of their powerful legs.

Life Cycle

Like all other insects fleas also have four stages in their life cycle: egg, larva, pupa and adult.

The female lays very small eggs white in colour mainly on ground. The eggs that fall into crevices remain there for 10 days to hatch. They hatch into larva which is like their parents in form but miniature in shape just 2 mm long. It has a small body and powerful mouth parts. At this stage they consumed dead skin cells, flea droppings and smaller organisms lying on the ground around them. Then the larva forms a silken cocoon around itself to form a pupa. The pupa stage may stay up to 1 week to 6 months depending upon the climatic condition. Within the cocoon the pupa metamorphoses into adult flea. The adult now starts sucking blood and then mate at opportune moment. A female flea mate once and lays eggs every day up to 50 per day.

Transmission of Diseases

The species acts as vector of plagues, the pathogen is *Yersinia pestis*. *Rickettsia typhi* is also transmitted by the *X. cheopis*. *Hymenolepis nana* and *H. diminuta* tapeworms are also transmitted by *X. cheopis*. The pathogens can be transmitted from one generation of flea to next transovarially.

4.3.2.2 Tick

Ticks are ectoparasites of animals and feed on vertebrate blood. There are two types of ticks which are parasites of human beings and other mammals transmit diseases. These two types are hard tick and soft tick. Hard ticks are called Ixodid ticks belonging to family Ixodidae and soft ticks are called Argasid belonging to Argasidae.

The entire body of a tick is composed of a capitulum and the body proper. The capitulum is moveable and bears rostrum enclosing toothed chelicerae and toothed hypostome. A pair of pedipalpi originates from antero-ventral margin of the capitulum. The legs are 4 pairs with two claws and pad.

The basic features of two types of ticks are basically similar but there are some differences. The differences are given in Table 4.3:

Life History

Almost all ticks are parasitic to human beings and animals. Some ticks attack birds or small mammals in larval or nymphal stage but adult always attacks large mammals. This habit is important from the point of disease transmission. Rodents and birds are the reservoir of rickettsiae and viruses. Many ticks recognize their feeding host by smell up to a distance of at least 50 ft.

Ixodid female ticks in adult condition take a single enormous blood meal after which they drop from the host, lay eggs at once from several hundred up to 18,000. The protein of the sucked blood helps developing the eggs within their body and it takes several days. The eggs need 2 to 3 weeks to develop. The eggs hatch during spring and are called 'seed ticks' and have six legs (Plate 4.1, Fig. 4.11a).

The 'seed ticks' of Ixodidae wait for the suitable host and for that take a strategic position to attack. In the search of suitable host they may have to wait patiently for 1 year or more.

When fortunately they come across any suitable host after a long period of wait they feed for only a few days take full blood meal, the body is distended with blood and then drop on the ground. Now they hide themselves and rest for a week or more. Ultimately shed their skin and come out as eight legged but sexually immature known as nymphs. The nymphs climb up on vegetation and again have to wait for a period of patient delay. After second feed they again drop off to the ground to digest the meal, moult and transform into adult and sexually mature. In this condition they again require host for blood meal if successful the female searches for male partner,

Table 4.3 Morphological differences between hard and soft ticks

Characteristics	Hard tick	Soft tick
1. Body covering	Body is covered with hard cuticle but without tubercles and granulation	Body is covered with leathery cuticle with tubercles and granulation
2. Capitulum	Terminal end is visible from dorsal view	Subterminal end cannot be seen in dorsal view
3. Pedipalpi	Rigid but not leg like	Leg like
4. Scutum	Scutum is present on the back of the male entirely but only anterior portion in females	Scutum is absent
5. Marking on scutum	Scutum have silvery markings called ornates	Absence of any markings
6. Festoon	Scutum is extended beyond the body in female called Festoon	Festoon is absent in male
7. Mouth parts	Mouth parts are anterior in position and are directed forward	Mouth parts are ventral in position
8. Legs	The coxa are armed with spurs The tarsi are armed with ventral spurs	The coxa are unarmed Absence of ventral spurs
9. Pulvii	Pulvii always present	Pulvii are absent
10. Warts	Always present	May not be present
11. Sexual dimorphism	Well defined	Not well defined
12. Parasites	Parasites on mammals	Parasites on birds and bats
Example	*Boophilus* sp., *Dermacenter* sp., *Ixodes* sp.	*Argas* sp., *Otobius* sp., *Ornithodoros* sp.

copulates and starts final feed which distends them out of all proportion. Some Ixodid female ticks start feeding before mating. The males usually die shortly after copulation. It is seen that most ticks spend more time off their hosts than on.

According to the number of times the ticks risk their life by leaving their host to moult on the ground and then seeking another host, they are called one host tick, two hosts or three hosts ticks. The disadvantage of finding new hosts is compensated by the extraordinary longevity and tolerance of fasting. Larvae may live for a year or more without food and adult for 5 years or so without food (Bequart 1945).

The soft ticks or Argasidae are quite different in their habit and habitat from hard ticks. They inhabit the houses of their hosts instead on the body of the hosts and are not dropped on the ground to face hazardous condition of the outside world.

The female lays eggs in batches and in hundreds and young are reared in the houses of the hosts (Plate 4.1, Fig. 4.11b).

The larvae of some soft ticks moult and become nymphs after a few hours of hatching. The nymphs feed once or repeatedly between moults.

During feeding some ticks excrete fluid from a pair of coxal glands opening which covers the ventral surface of the tick, contaminate and help the disease transmission through bite and exudates. The lifespan of the organism may be 15 to 21 years including several years of starvation.

Lifestyle of the ticks depends largely on chemical control. It is proved that pheromones secreted by the ticks influence aggregation, attachment, reproduction, mate recognition and courtship. It is also seen that these pheromones or chemicals secrete from the anus, coxal glands and female genital pore. The reproductive behaviour of ticks is complex and strange and mainly under chemical control.

Mammals mostly are resistant to hard ticks. But cattle and other species of ticks require vaccination as a practical means of controlling economic losses due to acarine infections.

As there are different types of pathogens transmitted by ticks and also consequently the different types of antigens vaccination against ticks are the best method for control of tick borne diseases. Now it is proved that when a tick sucks blood from an immunized animal, antibodies present in the blood meal attack the gut lining, so the populations of vector reduce.

Injury from Bites of the Ticks

When the tick bites it is very painful and the forcible removal cause tearing of the capitulum. This may cause infection and as a result inflammation or ulcers at the site of the wound takes place.

Tick Paralysis

The bite of the tick may also result in paralysis. The paralysed effect is produced by the ticks those who are rapidly engorging females if they bite on the back of the neck or at the base of the skull. The cause of the paralysis is still unknown.

Some are of opinion that bite may cause deep piercing and comes in contact with a nerve or nerve ending. The paralysis starts in the legs of the human host. It slowly affects the arms and ultimately the thorax and throat. It is found that if heart and lungs are not affected recovery may be within 1 to 6 or 8 days after the removal of the feeding tick. Most human cases are in children and most in girls because of their long hairs which conceal the biting ticks.

Transmission of Diseases

The main diseases that can affect human beings by the bite of the ticks are: (1) Spirochetes of relapsing fever, (2) Rickettsiae of spotted fever, (3) Babesiosis, (4) Tularemia, (5) Anaplasma and (6) Filterable viruses of several diseases.

1. *Spirochetes of relapsing fever:*

   Many species of soft ticks transmit Spirochetes of the genus Borrelia, the causative agent of relapsing fever of birds and mammals.

   The ticks those are infected with the Spirochetes, the germs after entering into the digestive tract soon leave the alimentary canal and invade haemocoel and tissue where they live for years. The spirochetes are also transovarially transmitted in vertical transmission. The infection to vertebrate host takes place by bite or from coxal gland fluid of ticks.
2. *Rickettsias of spotted fever:*

Both hard and soft ticks can transmit epidemic and endemic typhus. A type of Rickettsia, *R. pavlovskyi* causing severe kidney damage is found in Russia.

In Australia tick typhus is caused by *R. australis* and the symptoms are rash on the skin including face, palms and soles, headache, body ache and fever.

Tick typhus: The disease is transmitted by the bite of ticks. It is reported that 2 h of attachment are necessary to successfully transmit the germs. The germs invade all the tissues of the tick and also transmission between ticks may take place transovarially.

3. *Babesiosis:*

   The disease is caused by *Babesia* sp. The vector of the disease is ticks of genus *Boophilus*. In the infected tick the parasites after fertilization in the hindgut enter into the reproductive organ of the tick. The germs become enclosed with the eggs and undergo multiplication there. The germs spread to all the tissues of the developing tick embryo.

4. *Tularemia:*

   Tularemia is also known as Rabbit fever. It is the disease of Rabbits and Rodents and man is also affected sometimes.

   The symptoms are local ulcer at the site of bite of the ticks, enlarged and painful lymph glands near the site of bite, fever, body aches, etc.

   The disease is caused by bacillus, *Pasteurella tularensis*.

5. *Anaplasmosis:*

   This is a fatal disease. The symptoms are jaundice, severe destruction of blood corpuscles in cattle and other animals. In stained preparation of blood film of the patient under microscope dot like structures are seen in the blood corpuscles, called Anaplasma.

   These are minute round, deep staining dots about 1 μ in diameter found in RBCs.

   The disease is transmitted by 17 species of ticks belonging to different genera.

6. *Q. fever:*

   This is a rickettsial disease caused by *Coxiella burnetii* and was first reported in Australia in 1937. The disease was transmitted among bandicoots by hard ticks *Haemaphysalis humerosa*. Infection comes from bandicoots to cattle and occasionally human beings by *Ixodes holocyclus*. Small mammals and birds are thought to be reservoirs of the disease.

   Q. fever is not primarily dependent upon ticks for transmission. It is usually transmitted among Bovids through placenta or by mother's milk and to man by raw milk, raw eggs, dust of cattle sheds. No direct transmission from man to man takes place.

   The symptoms of Q fever are fever, chest pain and influenza like cough.

Pathogenesis

Anaemia: Ticks may suck blood from the host which may lead to anaemia of the host.

Skin ailments: Tick bite may cause inflammation, ulceration, swelling and itching at the site of bite. If the tick is forcibly removed from the skin pieces of mouth parts may remain in the wound and that causes secondary infection by bacteria.

Tick paralysis: When tick bites near the base of the skull of human beings or animals, then paralysis may cause to the host. But it is seen that if the tick is removed carefully, then condition reversed. It happens from the toxic secretions.

Otoacariasis: If ticks enter into the ear canal by any way then severe infection may cause which is known as otoacariasis.

Infection: Ticks sometimes transmit diseases by transmitting virus, bacteria, rickettsias, spirochetes, etc.

4.4 Molluscs Intermediate Hosts of Digeneans

Digenea is a subclass of Class Trematoda under the Phylum Platyhelminthes. The members of the subclass Digenea include orders like: Paramphistotomiformes, Echinostomatiformes, Hemiuriformes, Strizeiformes, Ophisthorchiformes and Plagiorchiformes.

Digeneans are parasitic flatworms having a syncytial integument. They have two suckers for attachment with the host: one ventral and one dorsal. The habitat of parasitic adults is the lumen of alimentary canal of the final host. But they may remain in the different organs and systems of all classes of vertebrates.

Digeneans are about 6000 species discovered still today. But only 12 of the 6000 species described infect human beings. The diseases caused by them affect nearly about 200 million people throughout the world.

The human infective digeneans are grouped into two: Schistosomes and non-Schistosomes.

The Schistosomes are *S. haematobium, S. mansoni, S. japonicum, S. intercalatum*.

Among non-Schistosomes seven (7) major species infect human beings.

The normal habitat of all of them is the alimentary canal of human beings.

To complete the life cycle of all these digeneans they require molluscs, so molluscs are the intermediate hosts of these parasitic organisms.

It is seen that families of the pulmonata, the Class Gastropoda, the Planorbidae and Lymnaeidae are acting as the intermediate hosts of Class Trematoda. But these intermediate hosts also like others of the Phylum Arthropoda could not be called as vector. Because these intermediate hosts of Class Gastropoda, Phylum Mollusca only release the infective forms of the parasites passively in the water they live. They do not inoculate infective forms into human beings by biting or any other means. They only release the infective forms of the digenean parasites, cercaria a stage in the life cycle of these parasites which transform into metacercaria by secreting cyst around them and remain on the leaf of the aquatic vegetables and wait for the final host to be consumed along with the leaves.

Table 4.4 List of the intermediate hosts of the digeneans

Parasites	Intermediate hosts	Geographical distribution
Schistosoma		
S. haematobium	*Bulinus truncatus*	Middle East
S. mansoni	*B. rolfsi*	West Africa
S. japonicum	*Biomphalaria pfeifferi*	Africa
S. intercalatum	*B. glabrata*	West India, Brazil
	Oncomelania hupensis	China
	Bulinus sp.	Middle East
Non Schistosoma		
Fasciola hepatica	*Lymnaea trancatula*	Europe, Asia, Africa
Fasciolopsis buski	*Segmentina hemisphaerula*	Far east
Clonorchis sinensis	*Bulinus* sp.	Far east
Heterophyes heterophyes	*Pirenella conica*	Egypt, Far east
Metagonimus yokogawai	*Semisulcospira libertine*	Far east
Gastrodiscoides hominis	*Helicorbis* sp.	Far east
Paragonimus westermani	*Semisculospira cancellata*	Southeast Asia, Japan

Snails are very important to them because of the factor that they complete their life cycle within these molluscs. A part of the life cycle of these parasites is performed within the body of the snail (Table 4.4).

The intermediate hosts are freshwater snails and can also be amphibious (*Lymnaea* or *Onchomelania*). The normal habitat of these snails is lakes, ponds, slow flowing streams and rivers.

4.4.1 Life Cycle of Snails

Snails are usually monoecious but reproduction takes place by mainly cross fertilization. Eggs of these snails are laid in masses in water. Only exception is *Oncomelania* sp., they are dioecious and lay eggs on soil out of water. The mass of eggs are up to 1 cm in size and may have 30 eggs. Eggs hatch at the temperature from 24° to 30 ° C in 1 or 2 weeks after laying. They mature within 3 to 6 months.

The snails lay eggs at the beginning of rainy season and they have a heavy reproductive potential. It is found that the stimulus for egg laying is either change in the atmospheric temperature, supply of nutrients and dilution of chemical factors in solution and change in the size of the habitat. All these factors are available during rainy season.

It is also found that snails for their reproduction require usually 22° to 24 °C and presence of suspended solids in the habitat and habitat should be calcium rich which turns the habitat alkaline. Other factors those are required, water conductivity (presence of salts), light intensity, pH, geology of the area which provides different ions in the habitat. The reproduction of snails also depends upon range of atmospheric temperatures, too hot or too cold will tell upon the reproductive potential. Natural food supply depends upon presence of phytoplankton which sometimes support the organisms in absence of higher aquatic plants. However, it is also found that the snail–plant interaction is an important factor for their sustenance.

But still today no method is available by which it can be linked about the susceptibility of different trematodes to different species of snails.

The period of aestivation, i.e. hibernation during summer is quite different in different species of snails. Some aestivate during dry season, some hibernate in cold water. This type of snails has operculum in their shell to close the shell during unpleasant weather. For example *Oncomelania* hibernates in cold weather. *Lymnea* aestivate during hot and dry weather. It is found that trematodes do not develop during aestivation of the snails. Development of *Fasciola* stops if the atmospheric temperature is below 10 °C. The life cycle of the trematodes within snails starts when free swimming miracidia penetrates within the body of the snails. The particular chemical attractants are responsible for attraction to particular species of snail produced by the snails themselves.

The development of trematodes within the snails creates change in the normal physiology of the snails.

4.5 Vector Identifications

4.5.1 Mosquito (*Culex pipiens*)

They are efficient vectors of Bancroftian filarial, encephalitis, etc.

(a) Mosquitoes are vectors of some diseases of human beings.
(b) The colour of the body of insect is brown.
(c) Females of the species lay eggs in rafts in any small receptacle of stagnant water.
(d) Nocturnal in behaviour.
(e) Females have round tips in their abdomen.
(f) The length of the palp is half that of proboscis.
(g) Absence of thoracic spiracular bristles.
(h) Presence of long, slender, air tube having hair tufts in the larvae.

4.5.2 Mosquito (*Anopheles* sp.)

They are the vectors of vivax Malaria and cerebral Malaria.

(a) Female Anopheles lay eggs in water in thousands.
(b) Presence of long proboscis, a pair of palpi and a pair of antennae.
(c) Proboscis is used for biting in females.
(d) Only females have biting and sucking mouth parts.
(e) Abdomen is long, narrow and slender.
(f) Presence of a pair of wings in thorax.

4.5.3 *Musca domestica* (Housefly)

The housefly is an efficient mechanical vector. Their body construction is favourable for carrying bacteria. They are mechanical vector of a number of diseases like trachoma, anthrax, viral hepatitis, dysentery, cholera, poliomyelitis, conjunctivitis and a number of helminthic diseases.

(a) They measure 5 to 8 mm in size.
(b) Presence of four longitudinal stripes which are dark in colour on the top of the thorax.
(c) Presence of a pair of eyes, a pair of antennae and a single retractile proboscis.
(d) Abdomen has light and dark markings.
(e) The whole body is covered with hair-like structure.
(f) Presence of only one pair of wings.

4.5.4 *Glossina palpalis* (Tsetse fly)

They are primary vectors of Rhodesian sleeping sickness, *Trypanosoma gambiense.*

(a) They are 7.5 mm to 14.0 mm long.
(b) The colour of the body is brownish grey.
(c) When they take rest their wings cross like scissors.
(d) The length of the palpi is equal to the length of proboscis. The palpi protrude from the head.
(e) There is a large oval swelling on the underside of the last abdominal segment.

4.5.5 *Phlebotomus papatasi* (Sandfly)

These are hairy midges which transmit leishmaniasis, 3-day fever, oroya fever, etc.

(a) They are 3 to 5 mm in length.
(b) The body and wings are covered by dense hair.
(c) Presence of a pair of hairy antennae, a pair of palpi and a proboscis.
(d) Thoracic appendages (6) are long and slender.
(e) They do not fly but hop.

4.5.6 *Chrysops* sp. (Deer fly)

This deer fly transmits the causative agent of surra disease, the *Trypanosoma evansi.* Anthrax is also transmitted by deer flies. The bite of this fly is extremely painful and causes blood loss of the man and livestock.

(a) They are large measuring from 6 mm to 25 mm in length and have brown-spotted wings.
(b) Males have no mandibles.
(c) The fascicles are composed of six piercing organ, mandibles are two in number, flattened and blade like having tooth like serrations. There are two narrow maxillae, a median hypopharynx and presence of a median labrum-epipharynx.

4.5.7 *Triatoma infestans* (Reduviid bugs)

These bugs are vector of Chagas' disease, the causative agent is *Trypanosoma cruzi*.

(a) They measure up to 34 mm in length.
(b) Presence of narrow head, eyes are large and situated far back on the sides of the head.
(c) Behind the eyes there are two oceli.
(d) Antenna is four segmented.
(e) Labial tube is three segmented and it rests into a groove between the fore legs.

4.5.8 *Pediculus humanus humanus* (Head louse)

This is an ectoparasite living in the head of human beings. But these body lice may serve as hosts for typhus causing rickettsia, and vector of Trench fever and relapsing fever.

(a) Females are large than males.
(b) Females are 2.4 to 3.3 mm long.
(c) In males the anus and sex openings are dorsal in position.
(d) Presence of a single pair of thoracic spiracle situated between the first and second pair of legs.
(e) Abdomen is seven segmented, first six have spiracles.
(f) Pleural plates are well developed.

4.5.9 *Pthirus pubis* (crab louse)

It is an ectoparasite but does not transmit disease. But the bites cause intense pruritus. They live in Pubic hairs and may cause allergy and irritation.

(a) They are greyish in colour.
(b) They measure 1.5 to 2 mm long.
(c) Presence of grasping tarsi on the two large pairs of legs.
(d) First pair of legs is smaller than other two pairs.
(e) The thorax is very broad with all segments fused.

(f) Thorax is broad with all segments fused.
(g) The abdomen is compact and short.

4.5.10 *Xenopsylla cheopis* (Flea)

This is a flea and is a vector of the disease plague and murine typhus.

(a) Absence of genal and pronotal ctenidia.
(b) The ocular bristles are from front of the eye.
(c) Presence of dark coloured clearly visible spermatheca in females.

4.5.11 Tick

There are two types of tick those are ectoparasite of human beings. The ticks also transmit diseases like Tick paralysis, spirochetes of relapsing fever, Rickettsia of spotted fever, Babesiosis, tularemia, Anaplasmosis and Q fever.

Two types of ticks are hard and soft tick. Hard ticks are called Ixodid ticks and soft ticks are called Argasid ticks.

4.6 Systematic of Vectors

4.6.1 Diptera

4.6.1.1 *Culex pipiens*

Kingdom Animalia

1. Multicellular organisms, those are without chlorophyll.
2. They are eukaryotic organisms.
3. Absence of cell wall.
4. Nutrition heterotrophic.

Subkingdom Metazoa

1. Multicellular organisms with tissue, organ and system.
2. Cells are differentiated into tissues and organs.

Phylum Arthropoda

1. Body triploblastic, bilaterally symmetrical, segmented and covered by exoskeleton.
2. The exoskeleton is made up of cuticle.
3. Segmented body with a pair of jointed appendages in every segment.
4. Open circulatory system.
5. The excretory and reproductive systems are within true coelom.

Subphylum Myriapoda

1. Presence of a pair of antennae.
2. Respiration by trachea.
3. Excretory system consists of malpighian tubules.
4. The appendages are uniramous.

Class Insecta

1. Body is divided into three parts: head, thorax and abdomen.
2. Thorax is subdivided into three portions: prothorax, mesothorax and metathorax.
3. Each subdivision possesses a pair of appendages (legs), i.e. three pairs of thoracic legs.
4. Presence or absence of one pair or two pairs of wings.

Order Diptera

1. One pair of wings.
2. Hindwings are reduced.
3. A comparatively large and moveable head.
4. Presence of compound eyes.
5. Sucking, piercing mouth parts adapted for liquid diet.

Family Culicidae

1. Presence of a long proboscis.
2. Presence of scales on the wing veins.
3. Females with short maxillary palps.
4. Presence of small hairs on the antennae.

Genus Culex

1. It holds its body parallel to the resting surface.
2. Proboscis is bent downwards.

Species pipiens

1. Sucks vertebrate blood.
2. Do not hibernate during winter.

4.6.1.2 *Anopheles stephensi*

Kingdom	Animalia
Subkingdom	Metazoa
Phylum	Arthropoda
Subphylum	Myriapoda
Class	Insecta
Order	Diptera
Family	Culicidae
Genus	Anopheles
Species	stephensi

Up to Family same as above.

Genus Anopheles

1. Body slender, dark brown to black in colour.
2. Known as malaria mosquito.
3. Presence of spots on the wings.
4. When at rest inclined at an angle 45° to the surface.

Species stephansi

1. Vector of malaria disease in human beings.
2. Antenna is less bushy.
3. Eggs are laid in single.
4. Eggs are small, cigar shaped and have float on their sides.
5. Larva floats parallel to the surface of water.

4.6.1.3 *Aedes aegypti*

Kingdom	Animalia
Subkingdom	Metazoa
Phylum	Arthropoda
Subphylum	Myriapoda
Class	Insecta
Order	Diptera
Family	Culicidae
Genus	Aedes
Species	aegypti

Up to Family same as above.

Genus Aedes

1. It is small and dark in colour.
2. 4 to 7 mm in length.
3. White markings on the legs.
4. Females are larger than males.

Species aegypti

1. Presence of small palps tipped with silver or white scales.
2. Transmits dengue fever.

4.6.1.4 *Musca domestica*

Kingdom	Animalia
Subkingdom	Metazoa
Phylum	Arthropoda
Subphylum	Myriapoda
Class	Insecta
Order	Diptera
Family	Muscidae

(continued)

Genus	Musca
Species	domestica

Up to Order same as above.

Family Muscidae

1. Antenna 3-segmented.
2. Calypters well developed.
3. Arista usually plumose for the entire length.
4. Generally more than one sternopleural bristle.

Genus Musca

1. Adults are grey to black.
2. Presence of four dark longitudinal lines on the thorax.
3. Slightly hairy bodies.
4. Presence of single pair of membranous wings.

Species domestica

1. Presence of red eyes.
2. Eggs hatch into legless white larvae, known as maggots.
3. Adult flies have short antenna.
4. Abdomen in males consists of 8 and females have 9 segments.
5. Their wings are translucent and fold back straight at rest.

4.6.1.5 Tsetse fly (*Glossina palpalis*)

Kingdom	Animalia
Subkingdom	Metazoa
Phylum	Arthropoda
Subphylum	Myriapoda
Class	Insecta
Order	Diptera
Family	Glossinidae
Genus	Glossina
Species	palpalis

Up to Order same as above.

Family Glossinidae

1. Cuticle made of chitin.
2. Elastic abdomen.
3. Greyish-brown coloured thorax.
4. Two antennae comprised of three segments.
5. Two sensory organs.
6. Two pairs of spiracle called sensory pits in the sides of thorax.
7. Two wings lie one on top of the other over the back of abdomen at rest.

Genus Glossina

1. The rear trailing edge of the ways is not protected by a thickened vein as the front edge is increasingly damaged with age.

Species palpalis

1. Some of the veins in the wings form a cell in the shape of hatchet or machete.
2. In male folded hypopygium at the posterior tip of the abdomen.
3. Hairy plates just in front of it called hectors.
4. Two salivary glands. The glands are transparent small tubes.
5. Labellar teeth penetrate the skin and cut small blood capillaries.
6. In females no such structures instead small hole surrounded by a variable number of small flat chitinous plates.

4.6.1.6 Sandfly (*Phlebotomus papatasi*)

Kingdom	Animalia
Subkingdom	Metazoa
Phylum	Arthropoda
Subphylum	Myriapoda
Class	Insecta
Order	Diptera
Family	Psychodidae
Genus	Phlebotomus
Species	papatasi

Up to Order same as above.

Family Psychodidae

1. The body segments are all subdivided into two or three parts some of which may have dorsal plates.
2. Complete head capsule.
3. Their size is small and narrow wings which have few veins.
4. Wings are spotted.

Genus Phlebotomus

1. The oval lanceolate wings are carried erect on the humped thorax.
2. Small hairy insects of 2–4 mm in length.
3. During day time rest in cool dark places such as corners of houses, cracks and crevices in walls, termite hills, etc.

Species papatasi

1. Yellowish colour with conspicuous black eyes, hairy bodies, wings and legs.
2. Males possess long prominent genital terminalia known as claspers.
3. Females have a pair of anal recti.
4. Adults do not fly well but shows a typical ‘hopping’ flight.

4.6.1.7 Deer Fly (*Chrysops* sp.)

Kingdom	Animalia
Subkingdom	Metazoa
Phylum	Arthropoda
Subphylum	Myriapoda
Class	Insecta
Order	Diptera
Family	Tabanidae
Genus	Chrysops
Species	sp.

Up to Order same as above.

Family Tabanidae

1. Called horseflies or deer flies.
2. Prefer sunlight, avoid dark and shady areas.

Genus Chrysops

1. Large flies with brightly coloured compound eyes.
2. Large wings with dark bands.
3. Blood sucking insects.

Species

1. Compound eyes have red and green reflections.
2. With dark spots.
3. The transparent wings have dark brown patches.
4. Legs are black.

4.6.1.8 BlackFly (*Simulium indicum*)

Phylum	Arthropoda
Subphylum	Myriapoda
Class	Insecta
Order	Diptera
Family	Simuliidae
Genus	Simulium
Species	indicum

Up to Order same as above.

Family Simuliidae

1. Fan like mouth brushes.
2. The prothorax has a ventral median proleg.
3. The female adults of some species are nasty biting insects.
4. Small little flies that can cause a lot of discomfort.

Genus Simulium

1. Small flying insects with two wings.
2. Small dark, stout-bodied and hump-backed.
3. Size from 3 mm to 7 mm.

Species indicum

1. Short antennae having 11 segments and without whorls of hairs at the joints.
2. The wings are like horns of the mammals.
3. The proboscis in males is poorly developed.
4. Proboscis in females short, heavy and powerful.

4.6.2 Hemiptera

4.6.2.1 Kissing Bugs (*Triatoma infestens)*

Kingdom	Animalia
Subkingdom	Metazoa
Phylum	Arthropoda
Subphylum	Myriapoda
Class	Insecta
Order	Hemiptera
Family	Rediviidae
Genus	Triatoma
Species	infestans

Up to Class same as above.

Order Hemiptera

1. Large in size from 1 mm to about 156 mm in length.
2. Sucking mouth.
3. Presence of maxilla, mandible, labium.
4. Presence of channel for sucking.
5. Presence of channel for pumping out saliva or venom in the mouth parts.

Genus Triatoma

1. Called kissing bugs.
2. Blood sucking insects.
3. Vectors of Chagas' disease.

Species Infestants

1. Length between 0.5 and 0.75 in.
2. Dark brown in colour.
3. Lighter margin along the abdomen.
4. The wings are flat.

4.6.3 Ectoparasites

4.6.3.1 Louse (*Pediculus humanas*)

Kingdom	Animalia
Subkingdom	Metazoa
Phylum	Arthropoda
Subphylum	Myriapoda
Class	Insecta
Order	Phthiraptera (Anoplura)
Family	Pediculidae
Genus	Pediculus
Species	humanus

Up to Class same as above.

Order Phthiraptera

1. Wingless insects.
2. They are obligate ectoparasites living externally on warm blooded hosts.
3. They complete their entire life cycle on single host.

Family Pediculidae

1. Blood sucking lice of man.
2. They spend their entire life cycle on the scalp of human beings.

Genus Pediculus

1. Presence of mandibles is called chewing lice.
2. Presence of piercing mouth parts called also sucking lice.

Species humanus

1. They are ectoparasites of human beings mostly in children.

4.6.3.2 Pubic Louse (*Pthirus pubis*)

Kingdom	Animalia
Subkingdom	Metazoa
Phylum	Arthropoda
Subphylum	Myriapoda
Class	Insecta
Order	Pthiraptera
Family	Pthridae
Genus	Pthirus
Species	pubis

Up to Order same as above.

Family Pthridae

1. Having very short and broad body.
2. Presence of long legs having claws.

Genus Pthirus

1. The first pair of legs is smaller than other two pairs.
2. Thorax is broad with all segments fused.

Species pubis

1. Greyish in colour.
2. Resides mainly in pubic hair.
3. Adults suck blood for hours at a time.

4.6.3.3 Flea (*Xynopsylla cheopis*)

Kingdom	Animalia
Subkingdom	Metazoa
Phylum	Arthropoda
Subphylum	Myriapoda
Class	Insecta
Order	Siphonoptera
Family	Pulicidae
Genus	Xynopsylla
Species	cheopis

Up to Class same as above.

Order Siphonoptera

1. Small wingless insects.
2. Body laterally flattened.
3. Blood sucking insects capable of jumping.

Family Publicidae

1. Presence of one row of bristles on each abdominal segment.

Genus Xynopsylla

1. They can jump 3 to 4 inches in height.
2. The mouth parts are used for squirting saliva.

Species cheopis

1. Body is covered by backward pointing spines.
2. The abdomen consists of eight segments.
3. The species act as vector of plagues.

4.6.3.4 Tick

Kingdom	Animalia
Subkingdom	Metazoa
Phylum	Arthropoda
Class	Arachnida
Subclass	Acarina
Super order	Parasitoformes
Order	Ixodida
Family	Ixodidae (hard ticks)
	Argasidae (soft ticks)

Up to Phylum same as above.

Class Arachnida

1. The body is divided into two parts: cephalothorax and abdomen.
2. All are with 4 pairs of legs.
3. They are wingless.
4. Absence of antenna.

Subclass Acarina

1. Most are minute to small.
2. Consists of two tagmata: Prosoma and opisthosoma.
3. Pro and opisthosomas are fused.
4. Presence of capitulum and idiosoma.

Super Order Parasitoformes

1. Organisms are parasites of mammals, birds and reptiles.
2. Absence of segmentation.

Order Ixodida

1. No division of body.
2. Mouth parts are associated with sensory structures.
3. Blood feeding arachnids.
4. Capitulum is moveable bear rostrum.

Family Ixodidae (Hard Tick)

1. The body is covered with hard cuticle without tubercles and granulation.
2. Terminal end of capitulum is visible from dorsal side.
3. Presence of scutum on the back of the male entirely but in females only anterior portion.
4. Parasites on mammals.
5. Boophilus sp., Dermacentor sp.

Family Argasidae (Soft Tick)

1. The body is covered with soft cuticle.
2. Subterminal end of capitulum cannot be seen from dorsal view.
3. Absence of scutum.
4. Parasites of birds and bats.
   Argas sp., *Otobius* sp.

4.7 Vector and Its Importance in Transmission of Parasites

Vector is an organism, often an invertebrate Arthropod that transmits parasites from reservoir or infected host to a new healthy host. Transmission is the passage of a disease from an infected individual or group to a previously uninfected individual or group. The microorganisms or parasites that cause disease may be transmitted from one person to another by so many methods, one of them is vector borne transmission, i.e. carried by insects or other invertebrate organism.

Transmission may be horizontal, i.e. from one individual to another in the same generation by direct or indirect contact, i.e. vectors.

Transmission may be vertical, i.e. from one generation to another from parent to offspring. Generally mother transmits the disease by means of body fluid and sometimes through breast milk.

Parasites or pathogens must have a way to be transmitted from one host to another. Disease causing organisms are generally specialized for a particular method of transmission. In vector borne transmission, it is by the bite of a vector. A vector is an organism that does not cause disease itself but transmits infection by conveying pathogens from one host to another.

There are two types of transmission: mechanical transmission and biological transmission.

4.7.1 Mechanical Transmission

The vectors are passive carriers of the disease causing agents or acting as vehicles of transmission of the diseases. Example of this type of transmission is transmission of typhoid causing bacteria from contaminated excreta of the diseased person which adhere to the legs and wings of the insects and is transferred to the food and drink of a healthy person.

4.7.2 Biological Transmission

The vectors within the body of them the disease causing organisms proliferate and/or undergo cyclical development to convert to infective forms.

Biological transmission is of four types: propagative biological transmission, cyclopropagative biological transmission, cyclodevelopmental biological transmission and transovarial transmission.

In propagative biological transmission, the disease producing organism reproduces within the vector but undergo no further development. Examples of this type of transmission are the plague bacillus in the flea and yellow fever virus in the mosquito.

In cyclopropagative biological transmission, the disease causing organisms not only reproduce within the vector but undergo cyclical change to produce infective forms. Examples of such transmission are Plasmodium in mosquito and Trypanosomes transmitted by tsetse flies.

In cyclodevelopmental biological transmission, the disease producing organism undergoes cyclical changes to produce infective form but reproduce elsewhere. Example filarial worms, Wuchereria bancrofti pass a part of their life cycle in the Culex vector and transform microfilariae to filariform larvae, the infective form which will infect human beings but no reproduction takes place within the vector.

In transovarial transmission some disease causing organisms are transmitted from infected parent to their offspring through ovum. Examples rickettsiae that produce Rocky mountain fever and scrub typhus are transmitted from parent ticks and mites to their offspring via ovary.

Once the disease causing agent infective to human beings reaches that stage there are a number of ways for transmission to human host.

Infective forms reside within the vector either at anterior portion or posterior portion of the insect vector. They are called anterior stationary or posterior stationary parasites.

Some of the anterior stationary parasites in their infective stage go to salivary gland of the vector and victimized healthy person during their blood meal by discharging the parasites through proboscis when they puncture the skin to suck blood. Others do not enter into the salivary gland but replicate in the proventriculus from where they enter into the buccal cavity and infect the human beings by biting. At the biting site, infective forms enter into the body of the host at injured site through labium by saliva of the mouth.

The posterior stationary parasites like Trypanosoma cruzi reside in the faeces of the vector, kissing bugs, when the kissing bugs bite at the face of the human beings, they defaecate at that site. The host due to irritation rubs his or her hand on the site of the bite. The parasites get entry into the body of the host through that punctured site.

In some like *Dracunculus medinensis* larvae are discharged into the water and the larvae are ingested by the crustacean microscopic larva, Cyclops. The Cyclops with the ingested larvae of *D. medinensis* in their haemocoel when they drink water contaminated with infected Cyclops. Here the entire infected vector is consumed by the human host.

The infective forms of parasite are produced in large number within the vector and infect the host in large number so that it becomes difficult for the host to combat them by the defensive mechanism of the host.

The infective forms which are formed within the vector sometimes produce host like molecules on its body surface before attacking the host and when within the host, the host cannot recognize the antigen as non-self.

All these are the mechanisms of pathogenic parasites to attack and infect human beings and vector helps the parasites in their horizontal and vertical transmission to the host.

Common vectors	Parasites
Cow (*Bos* sp.)	*Taenia saginata*
Pig (*Sus scrofa*)	*Taenia solium*
Dog (*Canis* sp.)	*Echinococcus granulosus*
Snail (*Bulinus* sp.)	*Fasciola hepatica*
Reduviid bug (*Triatoma* sp.)	*Trypanosoma cruzi*
Tsetse fly (*Glossina* sp.)	*Trypanosoma brucei, Trypanosoma gambiense*
Sandfly (*Phlebotomus argentipes*)	*Leishmania donovani*
Mosquito (*Anopheles* sp.)	*Plasmodium vivax*
Mosquito (*Culex* sp.)	*Wuchereria bancrofti*
Deer fly (Chrysops)	*Loa loa*
Blackfly (Simulium sp.)	*Onchocerca volvulus*
Water flea (Cyclops)	*Dracunculus medinensis*

4.7.3 The Method of Transmission

The eggs of *T. saginata* may be deposited on the leaves of grasses from there they enter into the GI tract of cow by ingestion. Upon ingestion the oncospheres are liberated from eggs and bore through the intestinal wall and make their way via blood or lymph to the muscles. They remain in muscle as cysticercus. The muscle with cysticercus is called measly beef. The ingestion of this type of beef causes infection in human beings. The transmission is same in case of *T. solium* also where the pigs are vector. The pigs are coprophagus. Dog is the definitive host of *E. granulosus*. The eggs are passed through faecal matters. They are swallowed by intermediate host like sheep, pig, cattle, horse, goat grazing on the field. Man especially children gets infected due to intimate handling of infected dogs.

F. hepatica lives in the liver of sheep, goat, etc. The eggs after liberation go out from the body of the host through faeces. The first stage larva is called miracidium which enters into the snail. Within the digestive gland of the snail miracidium transforms into cercaria. Cercariae leave the snail and encyst on water vegetation. They are finally taken by the host sheep during grazing. Occasionally the metacercariae are taken by the man from water vegetation and develop in man.

T. cruzi, causative agent of Chagas disease passes its life cycle in two hosts, one in man, the final host and vector is Reduviid bug (*Triatoma infestans*).

Man is infected either by faecal matter of the bug being rubbed into wound or exposed mucous membrane. The bug generally bites on the eye lid, when the site is rubbed the infected stage of *T. cruzi* in the faecal matter enters into the conjunctiva. The trypomastigote forms are liberated in the blood. The reduviid bug sucks the infected forms of *T. cruzi* into their stomach during blood meal. Within the bug they transform into amastigote forms. Within the alimentary canal of bug finally they change into trypomastigote forms by asexual reproduction. These forms appear in excreta which ultimately infect human beings.

T. brucei causative agent of African trypanosomiasis in man, having its insect vector *Glossina palpala* called Tsetse fly.

The insect vector during its blood meal taken from infected man consumes trypomastigote form in their alimentary canal. They continue to multiple in the proventriculus from there go to the buccal cavity and from there to the salivary gland via hypopharynx. In the salivary gland, they change into trypomastigote forms which are infective to man. The trypomastigote forms invade the bloodstream of human beings.

L. donovani causative agent of kala-azar has two prominent stages in their life cycle, the amastigote forms occurring in man and promastigote forms in sandfly.

Sandfly, *Phlebotomus argentipes* is the main vector of the disease kala-azar. The parasites are liberated from RE system into blood circulation. From the bloodstream the free amastigote forms are sucked by the sandfly during their blood meal into their alimentary canal. The amastigote forms develop into promastigote forms. These forms multiply in the buccal cavity of sandfly but they do not infect salivary gland. The transmission to man is effected by bite of the fly.

P. vivax the causative agent of Malaria resides in RBC of human beings, an intracellular parasite and it is transmitted by female *Anopheles* mosquito. The macrogamatocyte and microgametocyte forms, result of asexual reproduction enter into the alimentary canal of mosquito during blood meal. These forms develop into sporozoites by sexual reproduction within the alimentary canal of the mosquito. The sporozoites ultimately infect salivary gland of the vector and during bite of healthy person is infected with the infective form, sporozoites. The sporozoites undergo asexual reproduction in liver, the RBC and form macro- and microgamatocytes to be sucked by mosquito.

W. bancrofti, the causative agent of the disease filaria resides in the lymphatic channels of man. Live embryos are discharged into the bloodstream. The embryos are called microfilariae. These microfilariae are the infective forms which are taken by the *Culex* mosquito during their blood meal. Within the mosquito the microfilariae undergo further development after which they become infective to man.

L. loa lives as parasite in the subconjunctival tissue of eye of human beings. The vector of *L. loa* is insect *Chrysops* so. (deer flies). The infection is transmitted by a day biting female *Chrysops*. The microfilariae enter into the blood of human beings by the infected bite of the *Chrysops*. On entering the human host the microfilariae migrate to the eye and develop into adult.

Onchocerca volvulus, the causative agent of human onchocerciasis is transmitted by the blackflies *Simulium damnosum*. Adult female blackflies become infected with the microfilariae while sucking blood from infected human host.

The microfilariae migrate from haemocoel to wing muscle and to the proboscis of blackfly and are deposited on host's skin when the insect feeds. The microfilariae transform into filariform larvae, the infective forms within the insect vector.

Paragonimus westermani lives in the lungs of the host. Eggs come out with sputum and some of the eggs are swallowed with sputum which pass out through faeces of the host. When the eggs reach water, miracidium larva is produced. The miracidium larva penetrates suitable snail host. The cercariae formed within the snail come out and penetrate crab or crayfish.

Human beings acquire infection by consuming crab muscle undercooked. So *P. westermani* needs two vectors to reach their final host, human beings where they remain as lung flukes.

4.8 Conclusion

From above discussion, it is understood that in case of certain parasites, they need living carrier or vector for their transmission to new hosts. The parasites within the vector undergo physical and morphological changes, multiply in number and develop into infective forms to enter into the final host. These infective forms avoid the immune system and other hazards produced by the hosts for their own protection.

So vectors have a tremendous importance in the transmission of certain diseases rather than disease causing organisms.

Some of the important vectors:

Order Diptera	Ectoparasites	Order Hemiptera
1. Mosquito	1. Louse	1. Reduviid bugs
2. Housefly	2. Rat flea	
3. Sandfly	3. Hard tick	
4. Tsetse fly	4. Soft tick	
5. Deer fly		
6. Blackfly		

References

Bequart JC (1945) The ticks or Ixodoides of the northeastern United States and Canada. Entomol Am 25:73–232

Cross ML, Cupp EW, Enriquez FJ (1994) Differential modulation of murine cellular immune response by salivary gland extract of *Aedes aegypti*. Am J Trop Med Hyz 51:690–696

Smart J (1945) Classification of Simuliidae (Diptera). Trans R Entomol Soc Lond 95:463–532

Theodor O (1948) Classification of the old world species of the subfamily Phlebotominae. Bull Entomol Res 39:85–116

Wanson M (1950) Contribution l'etude l'onchocerose Africaine humaine. Ann Soc belga de méd trop 30:667–863

Epidemiology

5

5.1 Transmission Between Hosts

The maintenance and sustenance of any parasite depends upon the safe transfer of them from one host to another. The transfer may be horizontal or vertical. Horizontal transmission is from one individual to other of the same generation by direct or indirect contact, i.e. by vector. Direct transmission is by contact between hosts or by reaching of the infective stage of the parasite to the host either by inhalation, ingestion or penetration of the skin of the final host. Indirect transmission is by biting of vectors like flies, mosquitoes, ticks, etc. those act as intermediate hosts (where the parasites pass through some stages of their development). It may be that a predator or scavenging host may consume the infected vector. The vertical transmission is possible if the infection is transferred by the mother to her offspring who has not yet born through eggs. The examples are rubella virus and human immuno virus (HIV). The vastness and intensity of any infection by parasites depend upon the population of the host and rate of transmission between hosts. The rate of infection depends upon many factors like climatic condition, host parasitic behaviour and densities of host and parasites. The rate of transmission is measured by several methods. This depends upon the type of the parasites and its way of transmission.

5.1.1 Transmission of Parasites by Connection Among Hosts

The direct transmission of diseases of any kind where infection is due to connection among the hosts or by transient infective agents, the rate of transmission is directly related to the recurrence of meeting between uninfected but susceptible and infected hosts.

Here in this case nobody has made any research to count the parasites' number transferred among the hosts. This is not possible because of the very small size of the

P. K. Bandyopadhyay et al., *Biochemical, Immunological and Epidemiological Analysis of Parasitic Diseases*, https://doi.org/10.1007/978-981-16-4384-2_5

parasites and at the same time as soon as the parasites reach the host multiply rapidly. So the research workers try to focus on the rate of the infection of the hosts.

The parameter of infection as it is called per capita infection is the product of two constituents like the average recurrence of connection among hosts, multiplied by the possibility of meeting among receptive and infected hosts which cause the transfer of illness.

If the interruption of time is less a differential equation reproduce that through time there is a change in the number of infected hosts. In case of epidemic the population of the infected hosts represents bell shaped curve. The epidemic spreads more rapidly if the host population is highly dense. It is intelligible that infection rate of parasites is high in dense population while it is less in thin population.

It is assumed that transmission of many micro parasites is directly balanced to the rate of contact between hosts and this is being supported by observation and theory. In some communities of human beings the rate of transfer of some bacterial and viral diseases is not linearly related to the density of population.

The value of transmission rate sometimes varies seasonally because of the influence of different climates on the frequency of contact among hosts and also depends upon the lifespan of the infective forms of the parasites. This occurrence decides the probability of transfer of the parasites during the meeting of hosts.

Seasonal rate of contact is very important in some common viral infections like measles, mumps and chickenpox.

It is seen that a disease may have repeated epidemic or may remain in a stable endemic condition in large communities where there is continuous flow of immigrants through new births by susceptible. In some viral infection like measles the persons recovered from the disease become immune from further attack. But the lasting of the parasites depends upon the inflow of the rate of susceptible hosts. In large communities infection likely to stay more if the birth rate is high in the host population.

Mass vaccination and the high birth rate in the community actually break the chain of infections. The infection rate is also dependant on the pattern of the human behaviour, it is very important 'who mixes with whom' which indicates the transmission of infection in case of direct transmission of the disease.

5.1.2 Transmission by an Infective Agent

The infective forms which live for a long time outside the hosts are likely to be transferred directly or indirectly. The miracidiae and cercariae stages of schistosomes, the filariform larvae of hookworms and the eggs of Ascaris remain viable for a considerable period of time outside the body of the hosts.

Another factor in case of helminthic parasites is the parasitic burden which is related to the severity of infection resulting in severity of the disease and the condition of the diseased hosts.

So it is important to know the acquisition of individual parasite by the host. It is directly related to the recurrence of connection among the infected host and forms of the infection of the parasites.

By mathematical equation the linear relationship is observed in experimental situation. A host experimentally infected by a number of parasites and the presence of linear relationship between the density of infective stage of the parasites indicate that the net rate of transmission is explicitly proportional to the density of infective stages and host population density.

Expected lifetime of the infective stage of the parasite will influence the number of parasites taken up by the host during an interval of time.

Rate of infection has no relation with the number of parasites taken up within the host population as very high infection rate will not influence the number of parasites acquired and accumulated within the host population.

It is observed that the life expectancy of the miracidia of Digeneans is short but that does not matter with the infection of the molluscs. But if contact between parasite and host is low even the parasites will accumulate within the host if the lifetime of the infective stage is long. Example: the infection of *Ascaris* eggs in human beings.

If the population density of hosts increases, then density of total number of parasites also rises during a fixed interval of time.

There is a relationship between the mean number of parasites introduced in the host and existence of the infection in the population of the exposed hosts.

If there is a highly heterogenous host population with regard to the prone to infection the infection do not increase rapidly. There is an increase in density of the infective stages of the parasites. In this case it is observed that some of the parasites will establish in a few hosts, while a portion of the host population remains uninfected.

5.1.3 Transmission by Ingestion

The feeding behaviour of host influences the entry of parasites in the hosts and the rate of infection of the parasite.

The actively preying host on infected fish (fish consuming digeneans cercariae), consuming food like vegetables contaminated with eggs of Ascaris or consumption of vector which is infected with Diphyllobothrium will lead to the infection due to predator–prey association present among final host and vector.

Here in this case the relationship is not linear among the number of parasites accumulated and density of infective forms or the density of infected vectors.

The net rate of infection is related with the feeding rate of hosts and the relation is not linear with density of food or prey.

This is controlled by the effect of satisfaction for food and time taken to prey and consume the food. The satisfaction for food decreases when they consume enough food and consequently with food parasites are ingested which do not depend upon the density of infection or infected vectors.

Predator always consumes a finite number of preys in a specified time interval because the time is taken to hunt, to capture and to consume the prey by the predator.

5.1.4 Transmission by Biting Arthropod

Sometimes parasites complete their life cycle within two or three hosts. In these cases one of the intermediate hosts is the biting Arthropod like mosquitoes. The transmission of infection is done by the biting Arthropods. Some of the micro- or macro-parasites transfer themselves from one host to another with the help of biting Arthropods. Examples are yellow fever by mosquito, sleeping sickness by tsetse fly and filariasis by culex mosquito.

The intermediate hosts are called, in these cases, vectors. This vector bites a constant number per unit time irrespective of number of available final hosts to feed on. So the transmission of the parasite from infected vector to uninfected human beings and from infected human beings to uninfected vector depends upon the biting rate of the vector. Biting rates indicate probability of infectious biting which leads to infection to human beings. At the same time the susceptible vectors acquire infection from infected human beings.

The existence of many parasitic diseases in their vector population is observed not so high, though rate of infection is high among the population of the vertebrate host.

It is estimated that in the malaria endemic area where the 50% of human population is infected only 1–2% of mosquito vectors is infected. This is also true in case of Bancroftian filariasis and Onchocerciasis.

The number of infected hosts and rate of loss of infected hosts are seen in many living population because of the consequence of an inverse relation.

There is also considerable difference in the lifespan of infection in the vertebrate and invertebrate intermediate host.

Malarial infection depending upon the species of malarial parasites and immunological position of the host may be from a few months to many years while the duration of infection in vector is restricted typically due to short lifetime of mosquitoes.

The transmission of vector-borne diseases is motivated by development of the parasite within the intermediate host as at that time the vector cannot be infectious though infected.

Transient lifetime of intermediate host and long incubation period of the disease affects low existence of infectious vectors even if the disease is existing within human population.

5.2 Control of Parasitic Richness Within the Population of Host

The procedures which regulate the population size of parasite within a host are important to realize the survival and sustainability of parasite within host altogether. It is helpful to scrutinize such phenomena under the following subheadings.

5.2.1 Host Categories

The epidemiological review of many human infections, specifically protozoan, viral and bacterial diseases, depends on different categories of host population. Hosts may be categorized into three distinct varieties like immune, infected or susceptible. Vertebrate hosts show responses against pathogen as an immunological reaction. Triumphant immune reaction will eradicate the parasites or curb the population magnitude of the same to a small volume within the host body. Acquired immunity might help second and subsequent infections to be eradicated without any noticeable symptoms of disease and without the host remaining infectious to additional organisms; therefore, hosts with acquired immunity actually become part of an immune category which is safe from the infection. Briefly, there are two major points pertinent for our concern to control the magnitude of infected host individuals: (1) hosts might be recovered from infection and (2) host individuals which recover may have a level of resistance to subsequent infection.

Host's immunity is obviously a crucial controlling mechanism in the spreading as well as survival of the parasites within the population of host organism. The greater the proportion of immunes within host population (generally considered as level of herd immunity), the lower would be the possibilities for dissemination of infection. Inherently the immunity of host constrains the number of infected hosts. A vast number of infected that become immune in due course cause a reduction in the number of susceptible latter.

A variety of factors regulate the spread and survival of an infection once it is found in a population of host. Primary transmission or proliferation is governed by: (1) number of susceptible, (2) percentage of host comes in contact and (3) the average time span during which an infected individual remains infectious. These points can be represented as follows: if small number of infected enter into a host population that consists of X susceptible, Y infected and Z immune, the preliminary transmission of contamination can be represented by the below mentioned differential equations:

$$dX/dt = -\beta XY$$

$$dY/dt = \beta XY - \gamma Y$$

$$dZ/dt = \gamma Y$$

All infected are lucidly presumed as infectious. The variable β denotes the spreading rate ($dY/dt = \beta XY = \beta(N - Y)Y$), its degree is regulated by the product of two parameters, viz. the rate of contact among hosts and the possibility that a contact causes infection. The variable γ represents the recovery rate where $1/\gamma$ is the mean time span spreading of infection.

The entry of a small number of infected individuals (Y_0) into a population of N susceptible leads to an epidemic (representing the proliferation of the infection) provided that

$$N > \beta/\gamma$$

It is assumed again that after the recovery hosts become immune for lifetime, as, for example, in case of measles. Hence, in a closed community the disease or infection in the first instance is transmitted (causing epidemic) but in due course is extinguished when the number of susceptible individuals is depleted. The infection may survive if immunity becomes dropped to some extent, such that hosts revert to the susceptible category from the immune one.

So, total influx of susceptible into the host population either as a result of natality or colonization is responsible for the perpetuation of the disease within the host population. Thus along with the threshold bulk of susceptible needed to develop an epidemic, long-term maintenance of endemic also depends on the size of population as larger populations tend to generate more susceptible than a smaller group of individuals. As, for example, epidemics of measles are observed in school children of a few hundred susceptible students, but the pathogen becomes endemic on a continuous month–month and year–year basis in urban societies with more than 50 lac individuals. Such type of diseases is the problem of present-day societies as in ancient time the introduction of susceptible people was probably very small that could not lead to an infectious disease.

A number of pathogenic infections of us are present naturally and favour to generate lasting immunity against further infection. These infections only crop up endemically in sizeable communities. In contrast, nearly all protozoan infections are able to sustain in hosts for long period of time and do not favour the development of lasting immunity to successive infection. These parasites generally survive endemically in little host populations with small inflow of susceptible. An occurrence of protozoan diseases is usually more anchored than viral and bacterial infections.

5.2.2 The Parasitic Category

Parasites may be broadly categorized into two varieties: micro- and macro-parasites. Most of the helminthes (macro-parasite), as opposed to the micro-parasites, do not proliferate in the vertebrate hosts but lay eggs or free living larvae that come out of the host for their development. The parasitic load within an individual host is therefore regulated by the rate of new infections in contrast to the rate of the dying of established parasites. Hence, it is an immigration-death mechanism rather the natality–mortality one that rules the proliferation of micro-parasites within their host individuals. So, it is important to keep in mind that helminths tend to be sustainable within the vertebrate hosts and they are not observed to generate secondary immunity to subsequent infection generally.

Vertebrate hosts are capable to develop immunity against helminth infection but the immunity usually acts to curb rather than removal of parasites within the host body. In density-dependent manner the hosts act to respond against helminth invasion. The proportionate decrease in survival and propagation is higher in concentrated subpopulations of pathogens than scattered groups. These mechanisms show negative feedback to restrict parasitic proliferation in individual host organism. Though immune reactions usually generate corresponding impacts, struggling among parasites for limited resources (both for internal and external parasites) might be crucial also. Regulation in density-dependent pattern can also be observed in parts of helminth life history that does not involve the vertebrate host. For instance, the percentage of pathogen-induced death rate of intermediate host is usually harmonious with the parasitic burden. Hence, hosts with high worm load show rapid death than those with low parasitic load. Host death inevitably causes the expiry of the pathogens residing in that host and therefore, such phenomenon leads to density-dependent parasite regulation. Density-dependent restriction also acts on the amount of gaining of parasites, specifically if the parasite gets entrance into the host by ingestion. Now it is known that the volume of parasite load in these situations is governed by host feeding manner to a large extent and hence it does not depend much on the abundance of infective stages or intermediate hosts that are infected. Hindrance on the rate at which hosts ingest foodstuffs aids to control the inflow of pathogens at high amount.

When the life history of any parasite involves two or more hosts it may contain several density-dependent affairs. As, for example, in human schistosomes survival of adult worm and its egg laying capacity are affected in a density-dependent mechanism in the human body and cercarial development in the molluscan host is effectively independent of the amount of miracidia that creep into intermediate host like snail.

Control of parasite density by density-dependent mechanism in the body of hosts takes action to curb the proliferation of the total parasite load within the host population. The endemic level of helminth infection within host population depends on the intensity of these mechanisms and the number that occur in the different phases of a parasite's life history. Such pitch is generally calculated by average worm load per individual host which is a statistic of the frequency distribution. Such arrangements are always highly accumulated in form, where most of the hosts possess few parasites and a small number of hosts bear major parasitic load. Specifically it can be considered that in this small number of hosts the density-dependent impacts are quite critical and therefore they will affect most of the parasitic community.

5.3 Survival Outside the Host Body

The life history of nemathelminth parasite chiefly consists of two stages: the free living and parasitic. The free living is not infective but infection happens when they are transported by an intermediate host. After identifying and attacking the primary host, the parasitic phase initiates. In case of obligate parasites there are some circumstances where existence of a population depends upon the viability of the free living stages of life cycle. It may happen while the host remains unavailable or abiotic factors are not favourable for the continuation of development. At first, the parasites have to live for a satisfactory time so that they can invade a host and then the parasitic organisms have to ensure the viability of offspring when the host is no more nurturing.

The parasites in true sense depend upon another organism for their food and shelter. So without the organism from which they get food and shelter, i.e. the host they cannot survive. The parasites require host to complete their life cycle. It is seen that the parasites sometimes require two or three hosts to complete their life cycle. Out of these hosts the parasites pass their sexual cycle within one type of hosts and asexual cycle within another type of hosts. The type of hosts within which parasites pass their sexual cycle are called definitive hosts and asexual cycle are called intermediate hosts. There are some parasites those require single host to complete their life cycle. Some require double hosts and some three hosts to complete their life cycle.

Whatever they may be the parasites for transmission to another host have to come outside the body of the hosts and live sometime in the outside environment and that is mandatory. The most of the transmissions are horizontal. So they have to leave the primary host and come outside and stay in the outside environment at any stage for some time.

In case of most of the Protozoan parasites the exit form is cyst. In case of Aschelminthes the monogeneans the exit form is fertilized eggs here first stage larvae remains within egg case called embryonated egg. Monogenic parasites like *Ascaris* either eggs or larvae with egg cases come outside the host and wait for the opportunity to transfer to other hosts. In the period between coming out of a host and entering another host body, the parasite is exposed to natural conditions for an uncertain period which might be hazardous for it, and therefore, a protective layer covers its body which helps in its survival in the external environment. For that they require some nutrition also. So within the protective layer there is some provision of food in the form of glycogen or any other nutritious chemicals for the first stage larvae. In case of *Entamoeba*, the parasitic protozoa they come outside with protective layer in the form of cyst. The cyst protects the organism outside from desiccation and the source of food is chromatoid bars and glycogen in the glycogen vacuole.

Others belonging to Aschelminthes have exit form microfilaria but these forms are directly taken by the intermediate host. Then the third stage larva enters human beings directly from intermediate host. So the parasites have no stage to live outside the hosts.

In case of Platyhelminthes the exit form is fertilized eggs those come outside with the faeces of the definitive hosts. Then the eggs survive in the moist faeces for a considerable time. So question of desiccation does not arise. The food for the embryo remains within the embryophore.

In case of Dyphyllobothrium the exit form is fertilized eggs. The eggs contain an unsegmented ovum and yolk cells. The eggs hatch into coracidium. These larvae escape into water and wait for Cyclops to consume. As the eggs contain yolk cells larvae do not have to starve. Question of desiccation does not arise as they remain in water.

Some parasites like flea female lays fertilized eggs on the ground. They hatch into larvae. They consume dead skin cells, flea droppings and smaller organisms lying on the ground around them. Then the larva transforms into pupa. Pupa does not eat anything they remain within cocoon.

The reduviid bugs in young stage practice cannibalism and sometimes suck blood from other bug that has sucked infected blood.

In two hosts type of parasites, i.e. digeneans the exit form is most of the time taken directly by the intermediate hosts. So they need no protection. But in triple hosts parasites they have to come outside but the length of time is very short and most of them remain in water. So question of desiccation does not arise but provision of food remains there.

But in case of ticks they have to pass a period of starvation which may extend up to 1 year in case of larvae and adult up to 5 years. They are adapted in such a way.

In all situations oxygen remains available through the protective layer.

5.4 Population Dynamics

The factors that have governing impact on the dissemination of infection (or parasites) are already considered. The aforesaid factors regulate the potentiality of parasites to combat with biotic or abiotic disturbances and are very much pertinent to interpret the endemic patterns of disease. Now it will be reviewed that how all the various rate-determining mechanisms govern the population functioning of a parasite and its potential to maintain itself within its host.

5.4.1 Transmission Thresholds

A set of the rates of pathogen dissemination are there, natality as well as mortality, below which the parasites cannot infect but above these rates the infection will sustain. This rate or magnitude is recognized as *transmission threshold* and its calculation is extremely crucial to develop and implement disease control programmes.

Host–parasite relationship should be reviewed considering both the micro- and macro-parasites. In case of micro-parasitic infections, the host individuals must be categorized into the following classes: susceptible, infected but not infectious, infectious and immune hence recovered. In comparison, the pathogenic effects on a host by most of the macro-parasitic organisms depend on the expanse of the parasite load within an individual host. Therefore, mathematical models on simple micro-parasites make no discrimination between infection and disease but for macro-parasites an important distinction is there in between infection by one or more parasites and disease that is caused by an immense parasitic load (May 1976).

5.4.2 Micro-Parasites: Direct Transmission

In case of micro-parasites, infection will be sustained when an infected host brings about no less than one new infection during its introduction into a susceptible population. If one infectious host is introduced into a susceptible population (N), the average number of new infections is termed as the basic reproductive rate and is represented by R. The transmission threshold is hence given by the condition $R = 1$.

In case of a directly transmitted viral disease like measles (that stimulates long-lasting immunity in recovered individuals) R is represented as

$$R = \beta N/(b + \gamma)$$

In this case, β is the transmission rate of disease, N represents the population density of susceptible individuals, $1/b$ stands for the longevity of the host and 1 Iy is the average time span in which an infected host remains infectious. R is straightforwardly the biotic potential of the infection (βN), multiplied by the expected longevity of the infectious host [$1/(b + y)$]. The postulation of a fundamental reproductive rate brings about the below mentioned important epidemiological principles.

1. The effective reproductive rate of an infection, R, within a group of N hosts of which only a section q are susceptible, in contrast to its basic reproductive rate R which measures its reproductive potential in a group of host individuals completely susceptible to infection, is

$$R = Rq.$$

2. If the pervasiveness of infection $y(y = Y/N)$ is somewhat persistent over time, the disease remains at equilibrium and

$$R = 1$$

3. Threshold value of the population density of host, NT, required for the elementary transmission of an infection is coupled with the value of R, where in N number of host individuals,

$$R = N/NT$$

4. The biotic potential R is allied to the average age of host individuals A at which they grab pathogen, where

$$R = 1 + L/A$$

Here L is the longevity of the host. This interrelation usually aids to calculate R from epidemiological statistics.

5. If there is accessibility of vaccine, the segment of the population, p, which must be safe at any one moment to wipe out the disease, is simply

$$P > 1 - 1/R$$

As, for example, if 92–94% of children become vaccinated, it would provide 100% effectiveness of the vaccine against the disease measles. The expanse of this figure illustrates why measles and alike viral infections are quite back-breaking to eliminate in particularly Western Europe and the USA.

6. On another scale, the fraction of susceptible, s, left after an epidemic has passed through a small group of host individuals, can be applied to evaluate R, since

$$R = [1/(1 - s)](1n(1/s))$$

For instance, if 30% of children in an academic institute remain susceptible after the ending of an epidemic of chickenpox, R is l. 7. So as to prevent this epidemic, 41% of the pupil would have to be given vaccine against the same ($P > 1 - 1/R$).

7. Amplified levels of vaccination will expedite a hike in the average age at which host faces its primary attack of parasitic infestation. If the percentage of the individuals vaccinated is p, the average age of first attack A is

$$A = L/[R(1 - p) - 1].$$

8. Incubation period of the disease within which an individual is infected but does not act as an infectious will remarkably brings down the biotic potential R if the incubation period is lengthy. The equation should be

$$R = \beta Nf/(b + \gamma)$$

where f is the percentage of infected individuals who sustain and become infectious.

9. The rate of R shapes both the connection between the host's age and the fraction of hosts who have encountered infection, and the equilibrium frequency of infection y^* within the host population. The interrelation between y^* and R is represented as

$$y^* = [1 - (1/R)][b/(b + \gamma)]$$

10. In case of sexually transmitted diseases like AIDS, the biotic potential should be represented as

$$R = \beta c D$$

Here β is the expectation of dissemination per sexual partner contact, D is the average time span of virulence of an infected individual and c is the average rate of obtaining sexual partner (Cox 1993).

5.4.3 Micro-Parasites: Indirect Transmission

Micro-parasites may also be transmitted indirectly, i.e. by the help of vector organisms. In a study Fitzgibbon and his colleague have considered two characteristic scenarios: (1) a vector-transmitted disease and (2) an environmentally carried disease. In their models horizontal criss-cross transference from infectious persons of one group to susceptible of the other one has not been observed. Rather pathogen transmission occurs either through roundabout criss-cross communication between vectors and susceptible individuals and vice versa in case (1), and by indirect association between susceptible hosts and the contaminated portion of ambience in case (2). In one of the host populations' parasite might be considered as benign whereas it might be lethal to another host population (Fitzgibbon and Langlais 2008).

Hence, the above mentioned concepts can be extended to include indirectly transferred vector-borne ailments likely malaria, yellow fever, etc. The biotic potential of a vector-transmitted disease is

$$R = \frac{\beta^2 N_2/N_1}{(b_1 + \gamma_1)(b_2 + \gamma_2)}.$$

Here β represents the successful rate of human-biting by the vector. This variable is squared as the behaviour of biting is accountable to spread to vector from man and vice versa. N_1 and N_2 denote the abundance of humans and vectors, respectively, $1/b_1$ and $1/b_2$ represent the respective expected lifespan of humans and vectors, whereas $1/y_1$ and $1/y_2$ represent the time scale of infectiousness in vector and man.

In other words, to carry on vector-transmitted pathogens, the ratio of vector to the host human has to exceed a threshold value to keep R greater than unity. This postulation has broadly been used in the planning of management strategies for diseases like malaria, where the target of control is to diminish vector population below this threshold level. The victory or defeat of such strategies still relies on the degree of perfection attained in evaluating the number of parameters that govern the value of R. There are a number of issues to estimate such parameters and furthermore the physiology of some vector-transmitted diseases of man is not properly interpreted till now.

Though the epidemiology of human malaria has got a fair attention over the last 30 years but still many issues regarding the biology of the disease and its control strategies remain unanswered. The speed with which the occurrence of malarial disease escalates in endemic localities within the junior age groups and its subsequent downturn indicate: (1) that rate of the biotic potential R is quite high in those localities; (2) that time span of the development of acquired immune reaction is relatively short and (3) herd immunity develops slowly as hosts experience many strains of parasites in a given area. The sharp decline in malarial occurrence in the adult individuals highlights the development of a level of herd immunity. However, it depends also on the frequency of mosquito bites to human hosts. As mosquito control projects turns down the prevalence of infection, thus it also lowers the magnitude of herd immunity. Hence, the re-infection of malaria occurs frequently after the withdrawal of control programmes.

5.4.4 Breakpoints in Parasite Transmission

In the population dynamics of different parasitic helminthes additional problems are developed by the difficulties involved in the finding of sexual partner. This is specifically accurate for dioecious species like schistosomes and hookworms. It is also an important factor in case of hermaphroditic species that follows cross-fertilization.

The drop down of the average parasitic load per host individual below critical level leads to irregular cross-fertilization which becomes unable to continue development of adequate number of parasitic transmission stages to maintain the infection. Therefore, such critical worm burden explains a *breakpoint* in parasite spread that is recognizable from the *transmission threshold* illustrated previously.

A number of factors are responsible to determine the specific magnitude of this breakpoint. Specifically the sexual behaviour of parasitic organisms (monogamy or polygamy of parasite, or number of mating to maintain the generation of egg throughout entire lifespan) as well as the frequency distribution of the number of parasite per individual host is very important factors in this respect.

If R is less than 1, the infection does not sustain. Above this level, nonetheless, three equilibrium factors prevail: endemic disease and parasite elimination are two factors which are balanced, whereas the breakpoint is considered as the third one that is fluctuating. If the population load of helminth parasite drops down below this point the worm will be eliminated, whereas if the population load of parasite exceeds this point, then the pathogen will be able to cause endemic disease. Hence, the notion of parasite breakpoint is very important to control the parasitic load (say, by chemotherapy) in the host population.

5.5 Climatic Factors

The transmission of several infectious diseases is somewhat regulated by several climatic factors like optimum temperature, rainfall as well as relative humidity. Prevalence of infectious diseases is thus affected by climate change. Environmental features like temperature and precipitation show temporal variation in most of the habitats that may stimulate cyclic variations in the incidence and magnitude of infectious diseases. Population load of host or parasite does not regulate the effect of climate on them, respectively. Hence, climatic factors are considered as *density-independent* factors.

Climatic parameters affect the epidemiology of parasites in the following ways:

(a) *Behaviour of Host:* Cyclic fluctuations in the prevalence of diseases are generated by the alterations of host's habits influenced by the climatic changes. Agricultural practices like plantation and reaping of crops at separate times of the year, or social practices like the recess periods and session breaks of academic institutes may result in such alterations. Farming practices are crucial for the perpetuation of infections by helminthes such as ascariasis and schistosomiasis, whereas the functioning of students regulates the epidemiology of several directly transferred diseases like chickenpox, etc.

(b) *Abundance of intermediate host:* Temporal fluctuations in the spreading of malaria and schistosomiasis, for example, are greatly the consequence of fluctuations in the prevalence of the intermediate hosts or vectors for the above mentioned diseases.

(c) *Longevity of infective stage:* Climate has a prime impact on the continuation of parasite transference stages like eggs and larval stages of helminth parasites and the protozoan cysts also. Temperature, for instance, is a prime regulatory factor of the miracidia and cercariae of schistosome flukes and the L_3 infective larvae of hookworms. Again the longevity of the eggs of *Ascaris* and larval stages of hookworms are also distinctly affected by the humidity of ground.

(d) *Infectivity:* Temperature regulates the job of schistosome miracidia and therefore affects their capacity to pierce the molluscan host. Moreover, this climatic parameter also influences the amount of the generation of cercariae by snails that are infected. Climate has a regulatory impact on the infectiousness of arthropod vectors also. For instance, the ideal atmospheric temperature for the transference of *Dirofilaria immitis* (worm causing filaria) from dog to mosquito vector *(Aedes trivittatus)* is around 23 °C and it is observed that mosquito bite reduces at higher or lower temperatures than this.

(e) *Parasite development:* Temperature has a determining role in the rate of parasitic development either in the external habitat or within ectotherm intermediate hosts like mosquitoes. Below 17–20 °C temperature the ova and larvae of *Necator* (New World hook worm) cease their development and death occurs in due course. *Ancylostoma* shows ability to develop at slightly lower temperatures than that of *Necator* and is therefore observed in some temperate areas of the globe. Temperature shows a noticeable impact on the period of incubation of *Plasmodium* in the anopheline vector as well as of schistosome parasites within their snail host.

In a recent systemic review work, articles have scrutinized by disease group like food- and waterborne diseases, vector-borne diseases and airborne diseases. Substantial evidence has been observed in this review for a connection of food- and waterborne diseases with climatic factors. Atmospheric temperature and relative humidity are likely to be prime climatic parameters to scrutinize further for bacterial as well as viral diseases which are airborne in nature (Hedlund et al. 2014).

References

Cox FEG (1993) Modern parasitology. Blackwell, London

Fitzgibbon WE, Langlais M (2008) Simple models for the transmission of microparasites between host populations living on noncoincident spatial domains. In: Magal P, Ruan S (eds) Structured population models in biology and epidemiology. Lecture notes in mathematics, vol 1936. Springer, Berlin, Heidelberg. https://doi.org/10.1007/978-3-540-78273-5_3

Hedlund C, Blomstedt Y, Schumann B (2014) Association of climatic factors with infectious diseases in the Arctic and subarctic region—a systematic review. Glob Health Action 7:24161. https://doi.org/10.3402/gha.v7.24161

May RM (1976) Simple mathematical models with very complicated dynamics. Nature 261:459–467

Biochemistry 6

6.1 Energy Metabolism

This chapter basically deals with the major biochemical processes that occur in parasitic protozoa and helminthes. All parasites require maintaining a reservoir of energy for production of the macromolecules, habitual activity, growth as well as for reproduction. There are major needs on energy production mechanisms because the parasitic organisms generally show rapid multiplication or growth. Another major energetic cost is to protect them from the host immune response. The term 'energy metabolism' refers to such biochemical mechanisms that lead to the generation of ATP and additional energy-preserving substances which are in turn utilized in several energy-dependent processes. Metabolism of energy is specifically crucial because the survival of parasitic organism in successive host relies on its aptness to survive in unfavourable ambience.

- *Storage of energy*

Some types of carbohydrates are generally the macromolecules that are found as energy source. In *Entamoeba,* etc. glycogen is responsible for 10–30% of body weight but very tiny amount of carbohydrate is stored in trypanosomes or *Plasmodium*. This happens generally because of a constant level of ample glucose in their host's blood stream. A different kind of metabolic pathway takes place within insect host that gets the benefit of the amino acid supply. In the *Crithidia fasciculata,* a flagellate, polyphosphate is present which is also of scientific interest. It might act as a source of energy, but is thought also to be necessary for regulating metabolic pathways.

Glycogen reserve, which is a distinct feature of parasitic helminth, usually contributes almost 10% of dry weight in some tapeworms. In general, in vitro culture reveals glycogen depletion in starved parasitic helminth, which gives distinct indication that glycogen serves as energy source. Soluble disaccharide trehalose is found

P. K. Bandyopadhyay et al., *Biochemical, Immunological and Epidemiological Analysis of Parasitic Diseases*, https://doi.org/10.1007/978-981-16-4384-2_6

to be reserved in the tissues of few parasites of the said group. *Moniliformis dubius* of the phylum Acanthocephala and nematode *Trichinella spiralis* are some of the examples of such storage of trehalose.

Glucose is also found in all parasitic organisms. Usually it is not deposited, but active absorption process from the outside is extensive in protozoan and helminth parasites. Glucose intake seems to be a rate limiting step of metabolic process in case of the parasite *Entamoeba*. Cestodes and nematodes also show active absorption of glucose from outside the body whereas present findings have revealed that uptake of this monosaccharide may vary in trematodes and it is dependent on the environmental niche of such parasitic organisms. As there is no dearth of glucose in its preference site in the mammalian (host) hepatic tissue, the uptake of glucose is passive in the parasite *Fasciola hepatica*. Conversely the property of active transport for the absorption of glucose is thought to be associated with the obtainability of this monosaccharide in case of some species of the fish fluke *Proterometra* directly. In case of ectoparasitic organisms active transport system is quite developed generally as there is a chance of a scarcity of glucose. Glucose remains as a metabolic pool inside the cells of all of these parasitic organisms where the glucose is absorbed from outside or is derived by breaking down reserve polysaccharides. Then this glucose is oxidized for prompt energy production or remains as a reserved compound.

Lipids are not present as energy storage in adult protozoan and helminth parasites. Therefore except some helminth larvae most of the parasites do not possess enzymes needed for the oxidation of lipids.

- *Regulating Energy Metabolism*

The concept of 'adenylate energy charge' helps us to understand the basics of metabolic regulation. It is illustrated by the following equation:

$$\mathrm{ATP} + 1/2\mathrm{ADP}/$$

$$\mathrm{ATP} + \mathrm{ADP} + \mathrm{AMP}$$

The value of the energy charge of adenylate is around 0.8–0.9 in most of the mammalian tissues; in pathogenic organisms it ranges from 0.6 to 0.9, the lower amounts are generally observed in case of anaerobic pathways. High values indicate that the parasites are in satisfactory physiological state and are possibly engaged in anabolic processes. So, several anabolic enzymes become switched on, while switching off is observed in catabolic enzymes which take place high concentrations of ATP and low amounts of ADP. The reverse is observed when adenylate energy charge level becomes low.

Redox situation, that can practically be explained as the ratio of concentration of NAD(P) to that of NAD(P)H within the cells, is an additional helpful marker. High value of NAD(P) favours catabolism whereas high NAD(P)H favours anabolic ones. NAD/NADH ratios in *Ascaris* sarcoplasm suggest that high level of redox is maintained in muscle cell. It is achieved by coupling NADH oxidation to malic dehydrogenase.

Generally metabolic reactions in parasitic organisms are controlled in the similar manner as in other living beings, except in few cases. Their enzymes exhibit allosteric stimulation and in general alike enzymes are engaged in regulatory mechanisms. The metabolic pathways are subject to positive or negative feedback regulation. In different organisms, distinct features of alike enzymes are magnified or arrested for optimizing the metabolic reactions which help in the parasitic adaptations in different environments.

6.1.1 Energy Metabolism in Protozoan Parasites

Parasitic Protozoans are actually a heterogeneous group and a number of these parasites (like parasites of disease malaria and the trypanosomes) adapt in more than single host in their life. Because of the variations of hosts from mammals to insects, it is not astonishing to investigate that metabolic pathway of any pathogenic organism within different hosts is fairly different and metabolic pathways can switch over their modes when these parasites infect other host. Hence, many divergent pathways involving energy metabolism are shown by protozoan parasites.

6.1.1.1 Metabolism in Intestinal Amoebae

The metabolism of *Entamoeba* sp. is of special interest as it lacks both mitochondria and hydrogenosomes. Like most other intestinal parasites, *E. histolytica* utilizes carbohydrate as its main energy source. Glucose is taken up largely by a carrier-mediated, saturable, equilibrative system in the plasma membrane, but some is also taken up by a pinocytotic process although at physiological glucose concentrations (up to 5 mM) the latter process contributes less than 1% of the total glucose. Because the parasite lacks lactate dehydrogenase, lactate is not the end-product. Pyruvate is converted into ethanol and CO_2, which are the main products of anaerobic metabolism. Although trophozoite contains abundant glycogen, the pathways of its synthesis have not been determined. Several key enzymes concerned with the glycogen synthesis have been found, but the chief enzyme concerned, glycogen synthetase, is lacking.

Since *E. histolytica* lacks mitochondria it is not astonishing that it also lacks a functional Krebs cycle. A unique feature of the glycolysis is that general reactions beyond phosphoenol pyruvate (PEP) are catalysed by a PP-dependent enzyme, pyruvate phosphate di-kinase:

$$\text{PEP} + \text{AMP} + \text{PPi} - \text{Pyruvate} + \text{ATP} + \text{Pi}$$

Even so, the reverse reaction predominates and produces pyruvate with a net yield of 1 mol ATP (Fig. 6.1). An alternative pathway for the formation of pyruvate from PEP has been postulated involving a unique enzyme, phosphoenol pyruvate carboxyl transphosphorylase. The final fate of the pyruvate depends on the prevailing

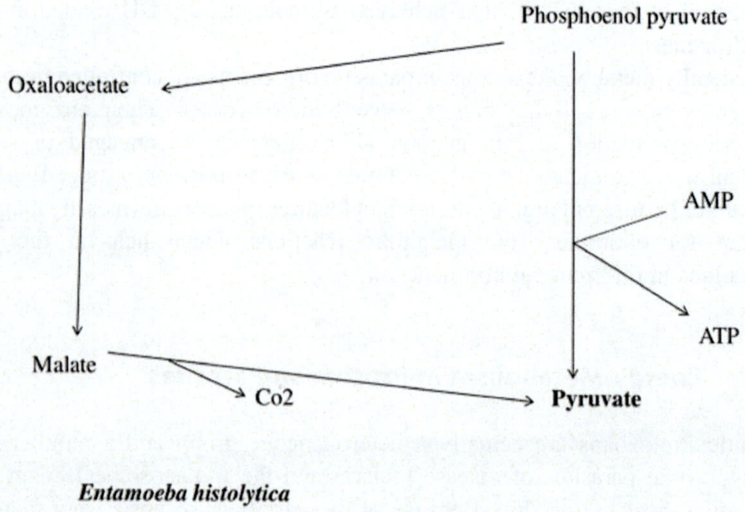

Fig. 6.1 Pathways of pyruvate formation from PEP in *E. histolytica*

rate of anaerobiosis. Under aerobic states, both acetate and ethanol are produced along with CO_2, whereas under anaerobic conditions only ethanol and CO_2 are produced.

Electron transport in *Entamoeba* is very poorly understood. Although no cytochromes or haemin have been found, there is clear evidence of a Fe-S protein, ferredoxin being implicated and it has been isolated from *E. histolytica*. The mechanism of reoxidation of ferredoxin remains unrevealed (Müller 1988).

Under Oxidative Stress

E. histolytica may reside in the intestinal lumen as non-virulent or it may induce the damage of tissues when they are virulent. The virulent trophozoites are able to adhere to the hepatic endothelial layer and then they move to the parenchymatous tissue via sinusoidal endothelial cells of liver. These trophozoites cause inflammatory reaction and ultimately form amoebic liver abscess. Although the host organism utilizes an immense inflammatory reaction against this parasite, the parasitic organism is able to sustain even at this habitat. Parasitic survival and also its virulence are determined by the ability of such parasite to confront NO and ROS. Hamster has been used as a model (ALAH) to find the mechanism how these parasites invade and survive within the host liver. Trophozoites are then collected from host liver and this virulent phase has been utilized for further research. Again, if trophozoites are steadily cultured in laboratory where no other contaminating organisms are present, their virulence is lost. Recently, it has been reported that virulent pathogens collected from ALAH and exposed to explants of human colon are able to alter their gene expression rapidly by escalating the genetic expressions associated with glycosylated residue and carbohydrate metabolic pathways. Such kind of modification in different niche conditions (liver and colon) revealed that alteration of the amoebic mRNA is crucial for parasitic adaptation to survive and continue their life cycle. Again, during oxidative stress, the nonvirulent trophozoites show a decreased antioxidant capacity, and, significantly, glycolytic flux is decreased in a reverse way when these stages are exposed to oxygen which indicates an adaptive relationship between survival mechanisms and virulence under the immune response of host (Pineda and Perdomo 2017).

It is known that *Entamoeba histolytica* is an intestinal parasite which is responsible for haemorrhagic diarrhoea and extraintestinal inflammation in millions of residents. The genome sequencing of this parasite has already been performed and has been used to forecast the metabolic capability of that parasite. As almost 56% of *E. histolytica* genes are unexplained, allied studies like proteomics, metabolic profiling, etc. are crucial to disclose novel, or not well known metabolic reactions. Metabolomics targets to understand the physiology and biochemistry of organism by thorough profiling of metabolites (Jeelani and Nozaki 2014).

6.1.1.2 Metabolism in Intestinal Flagellates (Like *Trichomonas* and *Giardia*)

Trichomonas contains membrane bound organelles, popularly known as hydrogenosomes. These organelles constitute a separate niche of energy metabolism (Fig. 6.2). The paracostal bodies and the paraxostylar bodies described by the study under light microscopes have now been considered as hydrogenosomes. *Giardia* does not possess the hydrogenosomes as well as mitochondria but contains lysosome-like organelles that store ferritin and stain positively for acid phosphatase and aryl sulphatase. Thus, all the enzyme activities of *Giardia* occur in the cytosol.

Their metabolic pathway follows fermentation and the main source of energy is carbohydrates. Glucose is phosphorylated by hexokinase or glycogen by glycogen hydrogen phosphorylase and enters the glycolytic pathway and by a number of steps

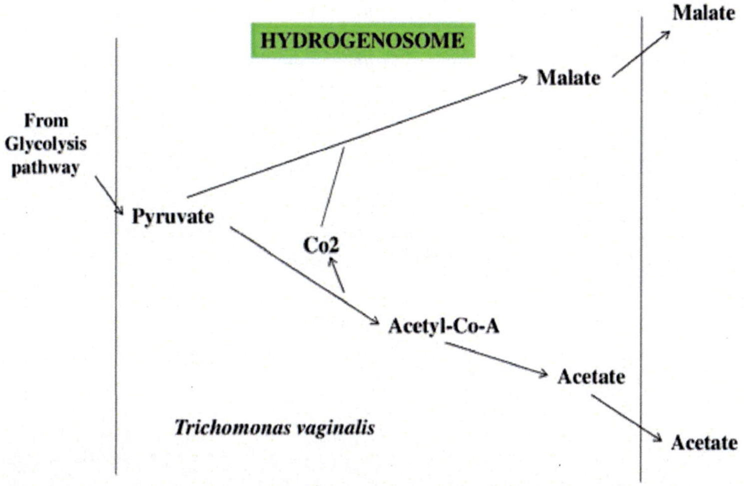

Fig. 6.2 Metabolic pathways within the hydrogenosome of *Trichomonas vaginalis*

it is converted to equimolar amounts of dihydroxyacetone phosphate and glyceraldehydes 3 phosphate. Through the classical Emden-Meyerhof pathway the glyceraldehydes 3 phosphate is metabolized to PEP and ultimately to pyruvate. In *T. vaginalis*, various intermediates of glycolysis produce glycerol, lactate, CO_2 and H_2 as end-products.

Pyruvate can further give rise to lactate in the cytosol by the action of lactate dehydrogenase. In both *T. vaginalis* and *G. lambia* the oxidative decarboxylation of pyruvate, leading to production of acetyl CoA, is triggered by an unusual reversible, iron-sulphur enzyme, pyruvate ferredoxin oxidoreductase and not by the irreversible pyruvate dehydrogenase as in most organisms. Another iron-sulphur protein, ferredoxin is used also. In *Giardia*, the oxidative decarboxylation of pyruvate happens in the cytosol, as hydrogenosomes are absent.

6.1.1.3 Energy Metabolism in Sporozoea (Like *Plasmodium* sp.)

Malaria parasites (*Plasmodium* spp.) are exposed to strikingly non-identical (nutritional) habitats in the complex life cycles of themselves in two different hosts like man and mosquito. A pronounced change in metabolism, from glycolysis in the asexual phases (Fig. 6.3) to tricarboxylic acid (TCA) metabolism in sexual phases, is associated with the conformity with different habitats. In experiments, stable isotope labelling has been used to depict stage-specific alteration in the carbon metabolism of *P. berghei* and find out the functional importance of such kind of alterations for parasitic survival in the human blood as well as in the body of mosquito (sexual stages). It has been revealed that glutamine plays an important

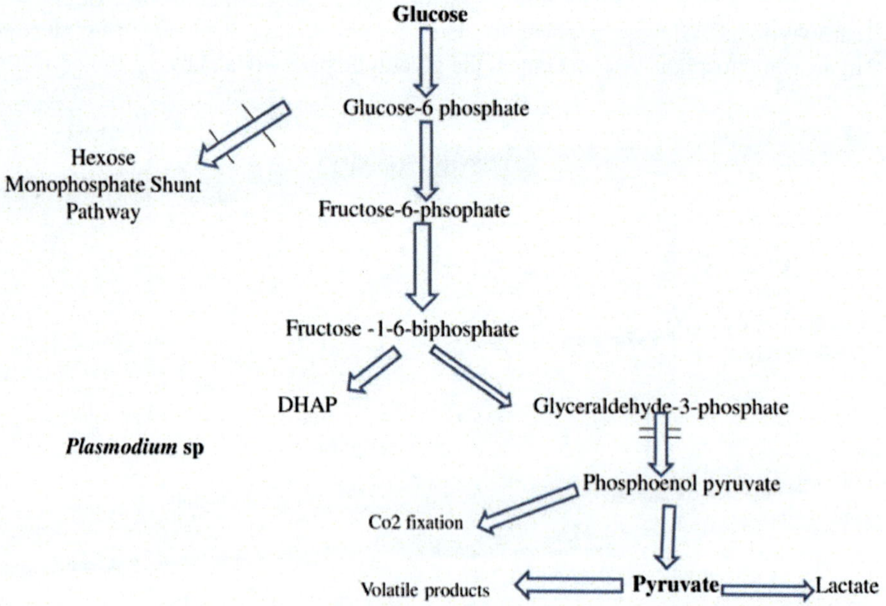

Fig. 6.3 Glycolytic pathway in *Plasmodium* sp.

role in citric acid cycle in both sexual and asexual phases and is also significant in the completion of gametogenesis of male parasite. Glutamine catabolism along with CoA synthesis is indispensable for the transformation of ookinete to the oocyst within the mosquito, i.e. primary host. These findings widen the understanding of *Plasmodium* metabolism and are therefore essential for designing suitable drugs (Srivastava et al. 2016).

Intraerythrocytic stages of *Plasmodium* lack carbohydrate reserves and consequently their primary source of energy is glucose from the blood stream. Although the pO2 in the blood is high, *P. falciparum* does not oxidize glucose completely to CO_2 and H_2O. It has been shown that, in vitro, infected human red cells (10^9) consume 122+− 34 nmol glucose per 24 h compared with 4.6+− 1.5 **m(mue)**mol glucose per 24 h for uninfected cells. The lactate produced was found to have a marked inhibitory effect on growth in vitro. All the enzymes of the Emden-Meyerhof pathway of glycolysis have now been identified (Fig. 6.3) although only a few have been characterized.

The fact, that a deficiency in G6PD in the human erythrocyte (X-linked abnormality) is more common in high malarious regions, indicates that such a deficiency provides some selective advantage in the survival within human host. This hypothesis is also supported by the findings from in vitro experiments in which growth of plasmodia was retarded in G6PD-deficienr cells.

It has also been found that *P. falciparum*-affected RBCs, cultured in vitro, had not liberated $^{14}CO_2$ when incubated along with [1-14C] glutamate even with the glutamate dehydrogenase. It has been concluded that this implied the absence of the activity of alpha ketoglutarate dehydrogenase and the unavailability of an effective TCA cycle.

During the various stages in the life cycle of *Plasmodium* many membrane-containing organelles like mitochondria, food vacuoles, Endoplasmic reticulum, nucleus, pelliculate complex, etc. are developed, the biochemistry and ultrastructure of which have also been reviewed (Sherman 1985). Biogenesis of these membranes involves a pronounced rise in the lipid content of the affected RBC. Amino acids needed for the synthesis of proteins in *Plasmodium* are obtained from three sources; (1) biosynthesis from carbon sources, e.g. by CO_2 fixation; (2) uptake of free amino acids from the plasma or blood cells of the hosts and (3) Proteolysis of host haemoglobin.

Hence, as a whole the above mentioned parasitic organisms depend on the metabolic pathways of glucose in their complete life cycle, while co-used glutamine is the chief operator of citric acid cycle and vital for the generation of the ookinete stage and in the formation of oocyst from ookinete. Both the CoA production and intermediary carbon metabolism are crucial for the transformation of ookinete to oocyst with a speculated role in complex sporogony (Srivastava et al. 2016).

Electron Transport in Protozoan Parasite

The above said 'anaerobic' parasitic protozoans lack mitochondria and haem proteins like catalase or cytochromes. Electron transport in these parasites is fairly distinct from other parasites. Low redox potential iron-sulphur complexes, with the features of ferredoxins remain effective in both anaerobic and aerobic electron transport chain.

The other chief category of protozoans contains mitochondria possessing cytochromes. These parasites follow aerobic respiration but a great variation in electron transport system among different stages in their life history is observed. Most of them show branched electron transport chains, like the helminths. For example, a branched respiratory chain containing two terminal oxidases has been described in *Plasmodium*. Its function in respiratory system is ambiguous as in these organisms lactate is the only end-product from the metabolism of glucose. It is thought that oxygen utilization is linked with pyridine metabolic pathway.

6.1.2 Energy Metabolism in Helminth Parasites

Glucose is utilized by all helminth groups as a respiratory substrate and as a fact tapeworms cultured in the laboratory are capable to intake nearly all the glucose supplied in culture media. These parasites possess active absorption machineries which help in the uptake of glucose. After its absorption, the glucose is either transformed into glycogen for further use or is employed directly. There are many effective alternatives for the metabolism of glucose.

6.1.2.1 Homolactate Fermentation

Perhaps true homolactate fermentation does not withheld in helminths. In homolactate fermentation glycolytic pathway converts glucose entirely into lactic acid. This end-product is excreted and therefore energy production does not depend on oxygen at all. ATP is generated at two steps and NADH is oxidized further by the enzyme lactate dehydrogenase by which the pathway can maintain the balance of redox. However, there are instances that additional energy production pathways also take place in the helminth homolactate fermenters. For instance, low levels of Krebs cycle activity are exhibited by schistosomes in their energy metabolic reactions.

This is also observed in the filarial worm, *Litomosoides carinii.* As a whole experiment with 14C-labelled substances indicates that 2% of utilized carbohydrate might follow total oxidation to generate water and carbon dioxide in normal respiration in vitro. A simple estimation reveals that 98 moles of glucose converted to lactic acid generate 196 moles of ATP whereas 2 moles of glucose converted to water and carbon dioxide by the Krebs cycle produce 72 moles. Therefore, it is logical that at the minimum, 27% of energy in *L. carinii* is produced by the aerobic pathway.

6.1.2.2 Malate Dismutation

The malate dismutation has most extensively been analysed in *Ascaris*, *Fasciola hepatica,* etc. (Fig. 6.4). Oxidation of glucose produces PEP. Then a carbon dioxide fixation step results in the formation of oxaloacetate that in due course is reduced to malate. On the contrary, PEP in mammals is modified to pyruvate that in turn is converted to acetyl CoA and successive metabolism of acetyl CoA occurs by the TCA cycle within mitochondria. Occurrence of such aerobic system allows the

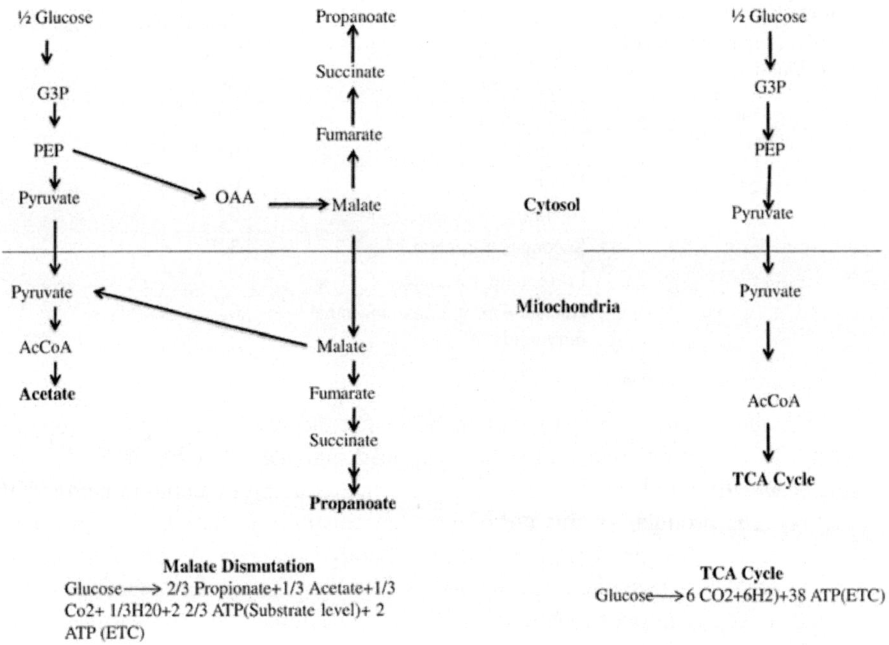

Fig. 6.4 Comparative analysis of malate dismutation and the tricarboxylic acid cycle

production of enormous ATP. Function of PEPCK in class Mammalia is reverse of that observed in parasitic helminth. It provides a diversion of the irreversible pyruvate dehydrogenase step.

Usually the TCA cycle acts at a small scale. Essential enzymes of this pathway are generally lacking or available in low concentration. Citrate synthase, aconitate hydratase and isocitrate dehydrogenase are usually not noticeable in cestodes and trematodes, but three enzymes of the TCA cycle, namely MDH, fumarate reductase and fumarate hydratase are present there. They induce the below mentioned reductive steps, a reversed sequence of portion of the Krebs cycle:

$$\text{Oxaloacetate} \Rightarrow \text{malate} \Rightarrow \text{fumarate} \Rightarrow \text{succinate}$$

MDH fumarate hydratase fumarate reductase

Two other oxidative enzymes, namely the pyruvate dehydrogenase complex and the malic enzyme (ME) are also responsible for the catalysis of the below mentioned reactions in the mitochondria.

$$\text{Phosphoenolpyruvate} + CO_2 + \text{GDP} \Rightarrow \text{oxaloacetate} + \text{malate, GTP} + \text{NAD(P)} \Rightarrow \text{pyruvate} + CO_2 + \text{NAD(P)H.}$$

Table 6.1 ATP production of different patterns of fermentation

Final product	mol ATP formed per mol glucose
CO_2+ water	36
Lactic acid	02
Ethanol	02
Nanine	02
Acetate	02
Succinate + acetate	3.7
Propionate + acetate	5.4
Acetate + propionate and other fatty acids	05

Then the malate comes to mitochondria for the reaction, disproportionation. During this reaction, one molecule is reduced and the other becomes oxidized. This is actually a redox reaction which gives the reducing potential to navigate the reaction. The products of this redox reaction, generally acetate and propionate or succinate, are subsequently removed from the body as excreta. As a final result more amount of ATP is generated than that of its own homolactate fermentation.

The malate disproportionation occurs in two intercellular compartments—the mitochondrion and cytoplasm—both of which must remain in redox equilibrium. Malate and lactate are equivalent in terms of metabolism when these are generated in cytoplasm. The dehydrogenases that help in the production of these components have alike function in regeneration of NAD (Table 6.1).

As each of PK and PEPCK gives rise to the production of ATP, the chain of reactions that generate malate or lactate are energetically comparable. PEPCK's desired cofactor is GDP instead of ADP, and it synthesizes GTP. Therefore, an added reaction is mandatory to convert the GTP into ATP:

Reaction by PK:

$$\text{PEP} + \text{ADP} \Rightarrow \text{pyruvate} + \text{ATP}.$$

Reaction by PEPCK:

$$\text{PEP} + \text{GDP} + \text{Pi} + CO_2 \Rightarrow \text{oxaloacetate} + \text{GTP}.$$

PEPCK and PK stand at a crucial branch point of metabolic pathways in parasitic helminths. Ethanol, acetate, lactate, succinate and evaporatibe fatty acids are the end-products in the metabolism of helminth parasites. Single parasite does not liberate this whole thing as end-products and it remains unrevealed why a specific set of final products has been taken on by one specific helminth. The production of such extremely reduced final-products might be described either in terms of 'energetic advantage' or of redox supremacy. The foremost infers that the specified chain of reactions to a specific final-product or output triggers the storage of energy in the form of GTP or ATP. The next explanation indicates that the reduced cofactors like NAD(P)H, produced in different sectors of the metabolic pathway, are re-oxidized during the end-product formation.

Again, the citric acid cycle is employed in growth and reproduction. It is to be noted that fumarate reductase and succinate dehydrogenase induce the backward reactions. The chief regulatory point of malate disproportionation: phosphoenolpyruvate (PEP) can either produce oxaloacetate (OAA) by means of carboxylation or generate pyruvate by means of dephosphorylation.

Some Specific Examples Are as Follows

Energy Metabolism in Cestodes

Most species investigated have utilized glucose. An attribute of carbohydrate breakdown in cestodes is the range of complex end-products, mainly organic acids, which are generated under both the aerobic and anaerobic situations. In contrast, the pyruvic acids produced during lipolysis are converted into lactic acids in the aerobic organisms like free living metazoans. The lactic acid produced by muscular activity under anaerobiosis is converted to CO_2 and water in the Krebs cycle under aerobic situation. In the absence of sufficient O_2, an oxygen debt is created.

Numerous investigations in cestodes have established that the classical Emden-Meyerhof pathway of glycolysis takes place in cestodes. Many of the enzymes of the glycolytic pathway in cestodes have been purified and/or characterized, chiefly hexokinase, glycogen phosphorylase, phosphofructokinase, pyruvate kinase, lactate dehydrogenase.

Cestodes produce a range of end-products as a result of their energy metabolism. In *Moniezia* sp. the significant outputs of metabolism of the carbohydrate are lactate and succinate where the relative amount of the end-product is related to the presence or the absence of glucose/fumarate and the PO_2. Lactate production is favoured under anaerbiotic condition and this is accompanied by a drop of the intracellular concentration of malate, whereas under aerobic conditions malate concentration increased, pyruvate kinase activity is inhibited and reduction of lactate production is observed. A model to explain these events has also been proposed. These authors mentioned that sections of mitochondria are very heterogenous under TEM study and they postulate the existence of aerobic and anaerobic mitochondria. It is indicated that succinate (or NADH) is oxidized in aerobic mitochondria, which contain a functional electron transport system linked to oxygen. At the same time, in anaerobic mitochondria (lacking cytochrome oxidase), malate is metabolized to succinate via the fumarate reductase system.

The cycle is essentially the metabolic centre at which carbohydrate, fat and protein metabolisms may contact and are able to interchange intermediate compound. Although the importance of Krebs cycle in the energy budget of cestodes in general is difficult to measure, there is evidence that species such as *Schistocephalus solidus,* etc. are capable of catabolizing substantial amount of carbohydrate by an active Krebs cycle. Parallel to this species, specific roles of many of the Krebs cycle enzymes have been studied in *H. diminuta,* etc.

Energy Metabolism in Nematodes

Most of the work on metabolism of nematodes have been carried out on ascarids (especially *Ascaris suum*) and filariids.

Metabolism in *Ascaris*

- Adult: *Ascaris* has an anaerobic pathway—often referred to as the fumarate reductase pathway—which is closely similar to that of some cestodes. This pathway involves CO_2 fixation and accumulates succinate, the major fermentation product or products obtained from this. Glycogen is the main energy reserve in most adult nematodes, being about 20% of the dry mass of the whole *Ascaris* and about 70% of the dry mass of its muscle tissue. During starvation, *Ascaris* consumes 1.3 gm of glycogen/100 gm body mass in 24 h under anaerobic conditions and 0.8 gm under aerobic situation. The energy producing routes in *Ascaris* are shown in figure.
  Muscle cells of *Ascaris* catabolise glucose up to phospoenol pyruvate (PEP) by means of glycolysis. Because of the scanty quantities of pyruvate kinase, the muscle is incapable of synthesizing marked amount of pyruvate in cytoplasm. Instead, CO_2 is fixed into PEP to form oxalaoacetate (OAA), being catalysed by the enzyme PEP carboxykinase. Then OAA becomes reduced and form malate by the help of malate dehydrogenase. *Ascaris* PEP carboxykinase perform in a reverse direction to that of the enzyme found in mammals and it is interesting to observe that a PEP carboxylase purified from the cestode *H. diminuta* shows stronger affinity for PEP than that of mammals.

  The malate formed in the cytoplasm now penetrates the mitochondrial membranes and acts as the mitochondrial component where a dismutation occurs. One mole malate undergoes oxidation to generate pyruvate and CO_2. This reaction is stimulated by the NAD+ linked enzyme to produce NADH which in turn acts to reduce an added mole of malate via fumarate to succinate. The pyruvate as well as succinate generated in the mitochondria help to form acetate and propionate which then serve as precursors for 2-methylbutyrate, tiglate and 2-methylvalerate.
- Eggs: In contrast to the adult worm, oxygen is required for the commencement and continuance of the egg development, processes which generally happen on pasture or in soil, normally highly aerobic ambience. *Ascaris* eggs, which contain relatively more lipid than carbohydrate, have a complete glycolysis and TCA cycle. During the early phases of development both trehalose and glycogen serve as energy sources, whereas in late developmental stages lipid metabolisms become more crucial.
- Larvae-embryonated eggs from infective L2 larvae at about 22 days when eaten by appropriate host undergo a tissue migration, moult to L3 larvae and reach the lungs. L2 larvae metabolize aerobically and have an active Krebs cycle and a cyanide sensitive O_2 uptake. The L3 larvae apparently make the transition from

aerobic to anaerobic metabolism during the third moult as evidenced by the loss of cytochrome oxidase activity.

Since the pathway of Krebs cycle is the terminal oxidative step for the end-products of protein, lipid and carbohydrate metabolisms, as well as playing a major role in synthetic reactions, it is clearly important to determine the extent to which it operates. In some species, under partly aerobic conditions, it plays an important metabolic role. A complete sequence of the enzymes has been detected in the eggs of *Ascaris*, which require oxygen for their embryonation and development, although only a partial reversed cycle operates in the anaerobic adults. The enzymes have also been recorded in different groups of parasites like the transmitting larva of *Ancylostoma tubaeforme*.

Electron Transport in Helminthes

In vitro, all parasitic helminthes uptake a noticeable amount of oxygen; where a part of the oxygen is undoubtedly utilized for synthetic pathways a part of which is employed in respiration. As many adult helminthes show reduced functions of classical electron transport systems, their functional electron transport systems are believed to be anaerobic. Additionally, these ETSs possess several terminal oxidases as they may be branched. As, for example, some helminthes show three branches in ETS: First with cytochrome oxidase, the second with a b-type cytochrome and a third that generates hydrogen peroxide.

By the help of several carriers, electrons are carried either to cytochrome oxidase or to the system of fumarate reductase which is very unique system for parasitic helminth. The end-products of respiratory system are either water or the quite harmful hydrogen peroxide (that is removed generally by peroxidase) or succinate and its derivatives.

Very fascinating anaerobic adaptation of helminthes is that they possess rhodoquinone instead of ubiquinone. The redox potential (E_o) of rhodoquinone is −63 mV, that is markedly lower than ubiquinone (+ 113 mV) and couple of fumarate/succinate (+33 mV). Transport of electron is thus favoured towards the fumarate. Helminthes also contains an enzyme called fumarate reductase, essential in anaerobic glucose metabolism. Fumarate reductase differs from succinate dehydrogenase which induces the reverse pathway in aerobes. It also possesses good quantity of cytochrome b_{558}. Finally, Kohler has mentioned that acyl CoA transferase pathways and acyl CoA-dependent ATP conservation strategy play effective role in parasitic energy metabolic pathway. *Fasciola* and *Ascaris* produce propionate within their mitochondria and this is another instance of the turnaround of a pathway observed in ruminant mammals where propionate synthesized by rumen microorganisms is utilized.

6.2 Metabolism of Lipid

Lipids are classified into three categories, which are as follows: (1) glycerides that are esters of fatty acids and glycerols, (2) phospholipids and (3) fats that are not saponified like waxes, sterols, etc. A major portion of the plasma membranes comprises varieties of lipids.

Occurrence and distribution of different kinds of lipids are described elaborately in different parasitic organisms but the actual metabolic pathways of these chemical compounds are not well documented till now. No evidence reveals that a specific part of host's body possess specific fat profile but some cases reveal that lipid category of parasitic helminth reflects the lipid profile of the host. In living beings generally fatty acids possess even number of C atoms. Most fatty acids are C_{16} or C_{18}. In helminthes, this range of C atom in fatty acids usually depends on the food habit of host and thus it can be altered when the diet of host organism is changed. For instance, sharks possess C_{20} and C_{22} poly-unsaturated fatty acids that are useful for its adaptation in marine temperature as saturated fatty acids show higher melting point than that of unsaturated one and it is interesting to notice that identical composition of fatty acids is observed in tapeworms which are parasitic on shark.

Like all other living creatures, parasitic cells have both internal and external cell membranes. The external is specifically essential as it provides protection from the attack of host and simultaneously regulates the absorption of nutrients from outside. Hence, a number of characteristic pathways or enzymes are traced in few protozoan parasites which are quite dissimilar to that of their hosts. For instance, acetate, which is effective for parasitic lipid production, is specially produced from threonine by the stimulation of two enzymes in *T. brucei*. But this pathway is not observed in their host organism.

Cell membranes of parasite differ in their lipid composition from those of its host. As, for example, membranes of malaria parasite contain more amounts of unesterified fatty acids, phosphatidylinositol, etc. and lesser amounts of cholesterol, sphingomyelin, etc. than those of its host. It is also supposed that parasites also possess divergent processes for providing membrane constituents. It is also observed that unique and developed external membranes are a characteristic feature of some parasites. For instance, adult *S. mansoni* contains a fat-rich external membrane that is formed a few days later the penetration into the host and this composition is modified constantly as it has important function in the host–parasite interaction.

Neither protozoan parasite nor helminth is capable to generate long-chain or non-esterified fatty acids from simple substances. Most of these organisms assimilate acetate into these fatty acids or carbon from glycerol or glucose into the lipid moiety. The above mentioned pathogenic organisms generally have no such enzymes that are needed for fatty acid's beta-oxidation and/or a functional Krebs cycle. Free living larvae of a number of helminth parasites show exceptional feature as they are capable to oxidize conserved lipids in their energy production.

6.3 Metabolism of Amino Acids

20 amino acids and less as well as varied number of purines and pyrimidines are generally obtained in free living organisms. Parasites show the same range of molecules. Generally to get nitrogenous compounds, parasites have to rely on their host organism as parasites have either lost or not developed the capability to generate most of these compounds. Amino acids are produced in all parasites. As, for example, investigation reveals that Trypanosomes are able to synthesize serine, alanine, glutamic acids, etc. The *Plasmodium* produces aspartic acid, alanine, glutamic acids, etc. from a variety of precursor molecules. To generate alanine and glutamate, respectively, some helminthes integrate ammonia straight into 2-oxoglutarate and pyruvate.

Amino acids become easily metabolized by trans-amination. The two major pathways catalysed by aminotransferases are:

$$2-\text{oxoglutarate}+\text{amino acid}\Rightarrow\text{glutamate}+\text{oxoacid, pyruvate}+\text{amino acid}\Rightarrow\text{oxoacid}+\text{alanine.}$$

They are reversible in nature and allocated in almost all parasites. Breakdown of amino acids is catalysed by L-amino acid oxidase:

$$\text{L-amino acid}+H_2O+O_2\Rightarrow 2-\text{oxoacid}+NH_3+H_2O_2$$

Then ammonia is removed from the body by their excretory system and the oxoacid might be utilized in respiratory pathway.

6.4 Biochemistry of Parasites in Relation to Antiparasitic Drugs

Observations, specifically with bactericidal drugs, have helped to identify 5 pivotal biological processes through which microbes become drug-resistant: (1) metabolism of antibiotic to an inert state, (2) alteration in membrane permeability for which the drug cannot be absorbed or is swiftly discharged from the microbial cell, (3) generation of another metabolic cascade so that the metabolic lesion can be eluded, (4) change in target site so that the drug becomes unable to bind to proper target and (5) increase in the concentration of the target enzyme.

6.4.1 Drug Resistance in Malaria Parasite

Very recent experiment to find the causative factors of multidrug resistance in some varieties of the *P. falciparum* which are responsible for the disease malaria in man indicated that in few of these, such resistance is developed due to the occurrence and/or amplification of one of following two genes, pfmdrl or pfmdr2, present, respectively, on chromosome 5 and chromosome 14.

Again resistance of *Plasmodium* against chloroquine is dependent on the ability of pumping out the drug by an ATP driven transporter as one of the adaptive mechanisms of the parasite. For being functional, chloroquine is to be stored within the body of microbe but such parasites are observed to pump out the drug by amplifying the amount of efflux transporter protein. In vitro studies reveal that pyrimethamine resistant strains of *Plasmodium* possess either magnification or alteration of dihydrofolate reductase. In the last-mentioned situation, *K*m generally rises, along with the inhibition constant (*K*i). *Plasmodium falciparum* become resistant to pyrimethamine and proguanil by means of point mutations at the site of dihydrofolate reductase (DHFR)/thymidylate synthetase (TS) gene.

6.4.2 Drug Resistance in Helminth

Studies indicate that resistance to the anthelmintic benzimidazole carbamates (e.g. Oxfendazole) and livamisole has been associated with a modified tubulin and a changed acetyl choline receptor, respectively. In the nematode *C. elegans*, concurrent mutation of three genes that encode α-type subunits of glutamate-gated chloride channel can develop high-level of resistance against ivermectin (Dent et al. 2000). It indicates that both point mutation and transport alteration may develop this drug resistance in these helminthes.

6.4.3 Drug Resistance in *Leishmania*

Another procedure to become drug resistant is the elevated gene expression of the specific proteins under drug pressure. It is known that folate is a crucial cofactor in microbial metabolic chain and dihydrofolate reductase–thymidine synthetase plays a vital role in the metabolism of folate. A number of functional antimicrobial drugs like methotrexate are powerful inhibitors of such enzyme system. Resistance in *Leishmania* against methotrexate is developed by the formation of this enzyme system in excess amount. Again *Leishmania* cells selected for resistance to trivalent antimony or trivalent arsenicals had elevated amount of trypanothione (TSH), which is the major reduced thiol of *Leishmania* cells and contains a N^1, N^8 bisglutathione spermidine conjugate. In these ways parasites become adapted to tolerate antiparasitic drugs and show the resistance status by their molecular and biochemical alterations. These studies might enable us to control several parasitic diseases in better way.

References

Dent JA, Smith MM, Vassilatis DK, Avery L (2000) The genetics of ivermectin resistance in *Caenorhabditis elegans*. Proc Natl Acad Sci U S A 97(6):2674–2679

Jeelani G, Nozaki T (2014) Metabolomic analysis of *Entamoeba*: applications and implications. Curr Opin Microbiol 20:118–124

Müller M (1988) Energy metabolism of protozoa without mitochondria. Annu Rev Microbiol 42:465–488

Pineda E, Perdomo D (2017) *Entamoeba histolytica* under oxidative stress: what countermeasure mechanisms are in place? Cell 6(4):44

Sherman IW (1985) Membrane structure and function of malaria parasites and the infected erythrocyte. Parasitology 91:609–645

Srivastava A, Philip N, Hughes KR et al (2016) Stage-specific changes in *Plasmodium* metabolism required for differentiation and adaptation to different host and vector environments. PLoS Pathog 12(12):e1006094

Molecular Biology and Molecular Genetics 7

7.1 Introduction

The equipped machineries of molecular biology play a crucial role in arena of molecular parasitology. Parasite proteins are the subjects of great interest in this field since genetic manipulation provides a convenient avenue to design antiparasitic drug and vaccine. Hence, parasitic genes for its antigens can be modified and manifested within the bacteria in order to generate vaccine which is recombinant in nature. The very chapter targets to provide a basic idea on the major angles of genetic exchange that command research on parasitology over few decades.

7.2 Genetic Exchange

Sexually reproducing organisms possess a characteristic feature of genetic recombination and from the practical viewpoint it is a paramount in the spread of drug resistance among the parasites.

Mendelian inheritance propagates genes vertically within lineages (VGT), whereas horizontal gene transfer transfers genetic components between or among unicellular or multicellular individuals (Fig. 7.1). Horizontal gene transfer (HGT) is considered as a driving force in prokaryotic evolution. In case of eukaryotic organisms HGT is quite less frequent and its functional implications in eukaryotes are also poorly recognized.

This genetic exchange has been very prominently observed in *Plasmodium* where resistant genes against chloroquine and pyrimethamine are propagated and transferred to the hybrid individuals that ease the spread of drug resistance among these parasites. In asexual organisms exchange of genes is quite problematic to observe but presently there are instances that it happens in case of *Ttrypanosoma* and probably in some other protozoan organisms that reproduce asexually.

P. K. Bandyopadhyay et al., *Biochemical, Immunological and Epidemiological Analysis of Parasitic Diseases*, https://doi.org/10.1007/978-981-16-4384-2_7

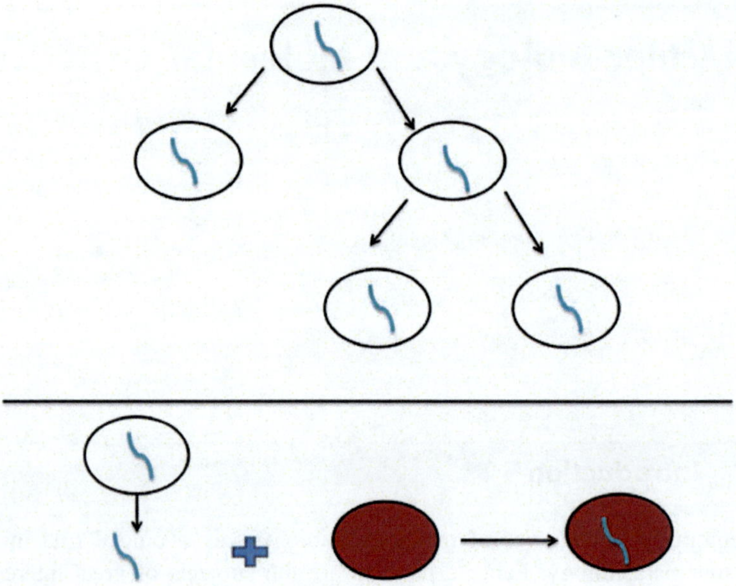

Fig. 7.1 Vertical and Horizontal gene transfer

7.2.1 In *Plasmodium*

In the research on the disease malaria five species affecting man (*Plasmodium falciparum*, *P. vivax*, *P. malariae*, two non-recombining sympatric varieties of *P. ovale* and *P. knowlesi*) and a number of parasites causing malaria in rodents are broadly employed. *P. knowlesi* and *P. falciparum* might be cultured in the laboratory within RBCs of monkey and man, respectively. Studies on genetic mapping and transfection in other human malaria species are problematic as persistent culture of these parasites is not feasible so far. Parasites causing malaria in rodents can be inherited in laboratory rodents but are not culturable in vitro. In this case drug selection is possible also. In addition, the complete life cycle can be carried on in experimental conditions aiding genetic interchanges.

Plasmodium species show sexual dimorphism which is keenly associated with the cycle of disease transmission (gametocyte is the infective stage). These parasites possess two hosts, one of which is mosquito vector. In its life cycle cells that reproduce asexually divert to produce micro (male) and/or macro (female) gametocytes by gametocytogenesis that is keenly knotted with its transmission from secondary to primary host. Each and every merozoites developed from only one schizont are devoted to asexual/sexual process. Additionally, each merozoite from a committed schizont produces either micro or entirely macro gametocytes. Gametocytogenesis is the emergence of these male or female gametocytes from infected erythrocytes.

Plasmodium falciparum shows an array of population structures from 'clonal' or asexual to 'sexual' that widely reveals the potency of localized transmission. For example, African variety of *Plasmodium* with excessive intensity of transmission has more opportunity to outcross and reveal reduced levels of linkage disequilibrium where frequency of association of different alleles is lower than the expected value, whereas Southeast Asian variety with reduced transmission intensity exhibits the reverse situation (Weedall and Hall 2014).

The genome size of *Plasmodium* is 23 Mb which accommodates a 6-kb mt genome and a plastid genome of 35-kb. Compared to most eukaryotic organisms, genomes of both *P. yoelii* and *P. falciparum* show biasness for A + T content in their nucleotide framework. In *P. falciparum*, A T richness is also prominent where A + T content altogether is almost 80% (Carlton et al. 2002; Gardner et al. 2002). Such partiality infiltrates in each and every chromosome hence intergenic portions often show A+ T content higher than 90%. AT richness indicates genetic flexibility and one of the causative factors of this is genetic recombination through genetic exchange. Subtelomeric sites accommodate sizeable multigene families like the *rifin* superfamily (that involves the allied rifin and stevor families) and *var* genes of *P. falciparum*, the *vir* genes of *P. vivax*, etc. These genes undergo genetic variation through gene exchange and are considered to take an essential part to avoid host immune responses. Rate of genetic recombination in *P. falciparum* is quite high (1 cM = 17 kb) that is almost 50-times higher than that of man and 15-times more than that of *Drosophila* (Anderson et al. 2011).

Plasmodium parasite population structure shows dramatic variation in different sites, with a range that is alike to the populations of inbred plant. This has crucial statistical implications regarding linkage disequilibrium mapping. African variety of this pathogen shows a property of high infection rates (like three bites per night per individual person), whereas the varieties of South American and Southeast Asian parasites show much decreased transmission rate (generally <one bite per year per person). As outcome maximum malaria in Africa possess multiple genotypes. Non-random association of alleles or linkage disequilibrium [LD] is run down by genetic interchange. Populations that show high levels of transmission have a tendency to possess greater rate of outbreeding than that of parasites of reduced transmitting regions. LD is swiftly worn down by means of genetic exchange in high infection areas and it leads to the development of least linkage disequilibrium (Fig. 7.2).

Both the association and linkage mapping target to locate marker loci closely positioned to loci underlying drug resistance. The chief variance depends on how many times the genomes are reorganized by genetic exchange. Parental generation and progeny are separated in linkage mapping by one round of genetic exchange in artificial reproduction. Therefore, the dimensions of haplotype blocks kept back from the parental generation are fairly large and comparatively sparse marker placing (5–10 cM, 75–150 kb) is necessary for locating genome sites possessing trait loci. By contrast, genome association research utilizes parasitic specimens from wild populations which remain isolated by an unspecified number of generations and genetic exchange phenomena. As an outcome, their genomes reveal quite higher rates of restructuring, and hence dense marker placing is needed for locating association in between the trait loci and the marker loci.

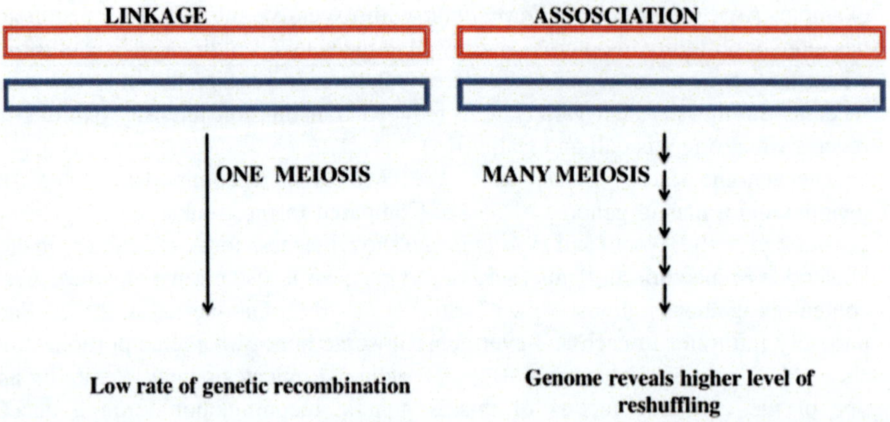

Fig. 7.2 Linkage and association analysis

Horizontal transfer of DNA has a minor role in eukaryotes but this transfer in between varieties of life forms is a paramount strength framing the prokaryotic genomes. The databases of the genome of malaria parasite have been split into tiny parts for comparing to that of man in addition to other malaria parasitic genomes. Such computational technique has showed that the genome of *P. vivax* is twined with multiple DNA fragments that have been possibly obtained through horizontal gene transfer from man. Contamination becomes a vital affair in these observations; furthermore, whether the recognized homologies are observed as a chance factor is to be decided. Such reservations become assisted by the fact that such recognized homologous sequences have been observed to be chiefly within short sequences. Re-sequencing of target regions applying specific isolates of *P. vivax* DNA revealed deletions which are not present in the genome of human. Furthermore, the recognized fragments have been rich in mRNA coding sequences and genes that have been considered as functionally crucial in *P. vivax*, including NO synthase 1 (nervous) adaptor and IL-1 family, indicating an effective contribution. These outcomes are necessary for dual purposes. Primarily, a horizontal gene transfer from man to other eukaryotic fauna is exhibited for the first moment. This throws halo on the co-adaptation, evolution and immune evading mechanism of parasite. Successively, the observed DNA is rich in IL-1 family that is investigated to be important for reducing parasitaemia. This might help us to know in a better way how the *P. vivax* and the host immunity interact with each other.

7.2.2 In *Trypanosoma* sp.

Genetic exchange in Trypanosomes also enables the spread of genes for essential features like drug resistance and/or virulence and may expedite the evolution of new parasitic strains. Genomics of such early evolved eukaryotic organisms will also reveal detailed knowledge regarding the sexual evolution as well as meiotic division

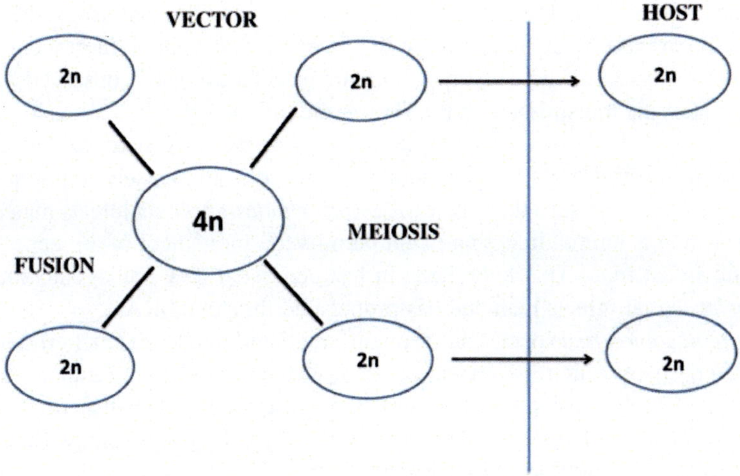

Fig. 7.3 A model of genetic exchange in trypanosomes

in those organisms. The evidence for genetic recombination in trypanosomes arises from experiments on isozymes and genomic DNA where the orientations found are suitable for the exchange of genetic data. Actually, researches regarding cloned lines of *T. brucei,* which generate hybrid generation, reveal clear-cut affirmation of the phenomenon of recombination in the next-generation when transmitted by tsetse fly (Fig. 7.3).

The occurrence of genetic recombination through genetic interchange in *Trypanosoma* is of substantial importance to combat with this disease as this recombination provides the pathogen limitless probabilities to develop different antigenic varieties by which novel ways of drug resistance can be generated. It was considered for a long period that by means of binary fission asexual reproduction of trypanosomes takes place. Occurrence of some sort of genetic recombination in *T. brucei* had been initially published during 1986 (Jenni et al. 1986). Hybrid trypanosomes have been developed by a co-transmission of an amalgam of 2 parental varieties with the help of the vector tsetse fly. Genetic recombination has also been revealed in the South American trypanosome, *T. cruzi* that infects man and this exchange takes place within mammalian host body.

There is insufficient knowledge regarding the procedure of genetic interchange in *Trypanosoma. brucei* till date. This exchange happens in the insect vector and this is quite problematic to notice as rearing of these tsetse flies in lab is not an easy task. A complicated life cycle of trypanosomes is observed in the insect and it follows a twisted pathway which initiates in the middle portion of digestive tract and finishes in the salivary glands in which the contagious stage is generated. In the life cycle of trypanosome genetic interchange seems to be an infrequent and facultative phenomenon. Genetically altered trypanosomes that are tagged by fluorescent dye are utilized to follow the frequency of the presence of hybrid pathogens inside the insect vector. Once green and red coloured parasites are crossed, few of hybrid individuals become yellow in colour and accordingly are differentiated from the parental generation at ease. It has been shown that hybrids with yellow fluorescent tag have only been located in the salivary glands of insect vector and observed at around 13 days after the transmission in fly. Hence, the salivary glands are considered as the area of genetic interchange of *Trypanosoma. brucei*. As genetic exchange allows transfer of harmful traits between strains, hence it is immensely important for the pathogen. Cross between the species *T. brucei rhodesiense* that infects man and *T. b. brucei* which is animal infecting strain permits the movement of the gene *SRA* that helps to infect man. This hybridization has generated new pathogenic strain *T. b. rhodesiense* that infects man and has propagated the gene SRA.

Trypanosoma cruzi species has been differentiated into three chief groups, Z1, Z2 and Z3, each of which possesses the same isozymes. Within Z2 there is distinct variation. An effective correlation has been described in between the clusters of strains resolved by isoenzymes and which found by the RAPD study. Study on the polymorphisms in rRNA genes, in small exon gene, etc. specifies the occurrence of minimum two chief genetic lineages of *T. cruzi*. Lineage 1 seems to be similar with Z2 and lineage 2 with the group Z1. Lineage 2 or Z1 is related with Didelphis, while lineage 1 might be related with primate group as their host. Deviations from H-W equilibrium as well as linkage disequilibrium suggest that *T. cruzi* propagates mainly by clonal or asexual mechanisms. Both genotypic and phenotypic characterizations have changed our perception on the variations and pathological roles of *T. cruzi* entirely. Documentation reveals that *T. cruzi* can show genetic exchange in wild populations, and presently a number of preliminary studies suggest that such genetic recombination might be acquired in laboratory too. It is known that asexual propagation (clonal) takes dominating role in the life cycles of this parasite but genetic recombination in wild may produce new strains of *T. cruzi* showing changed characteristics, such as hostility as parasite and its power to resist available drugs along with capability to propagate in the human hosts (Stothard 1999). Lateral or horizontal gene transfer has aided trypanosomatid ancestor to acquire exotic-genes by which new physiological processes and structural organizations can be evolved in case of the parasite *Trypanosoma*. A number of virus, bacteria and some other organisms have acted as genetic donors in the above mentioned case.

7.2.3 In *Schistosoma* sp.

Schistosoma mansoni is a blood fluke that affects almost 90 million persons. This disease is termed as Schistosomiasis that is native in countries in which sanitation is poor and people have no access to clean drinking water. Schistosomiasis is still a major health concern as it affects approximately 240 million individuals among which more than 90% inhabit in sub-Saharan Africa (WHO 2013). The chief causative agent is *S. mansoni* which results in intestinal form of schistosomiasis. Schistosomiasis is one of the very few helminth infections of man that is controlled in the laboratory as the total life cycle of this pathogen can be visualized within the laboratory. A number of literatures suggest that genetic variations are important for the generation of several pathophysiological traits like parasitic virulence, specificity regarding hosts, mode of transmission as well as power of drug resistance.

The positive correlation between physical size of chromosomes and the genetic map suggests that average rates of genetic exchange are analogous among the chromosomes of *S. mansoni*. Again, it is revealed that the rate of recombination is 1.27 times higher in females than that of male counterpart.

In case of Schistosomes snail act as an intermediate host whereas vertebrate acts as definitive one. Genetic study reveals that these parasites are normal diploids and sexually dimorphic.

Though inter-specific cross in most species is obstructed by reproductive blockade, it happens in case of schistosomes naturally. In addition, crosses among species can be observed in the experimental condition: a majestic study during the 1980 and 1990s described interspecific crosses among the *Schistosoma haematobium* group (*S. mattheei, S. guineensis,* etc.) were performed up to 4 t generation at least and the progeny remained fertile. Population genomic analysis in natural populations aids to prioritize candidate loci from linkage mapping observations. Genome sequence from miracidium larvae or from eggs collected from faecal matter or urine samples can easily be acquired than from this adult parasite.

In non-replicative stages (eggs) which are to be victoriously discharged in the nature, mutations [10^{-4}/generation for microsatellite locus (Valentim et al. 2009)] may only be developed and these have to compete with wild-type individuals. This kind of bottleneck might restrict the development of drug resistance in schistosomes. Rather, helminth parasites those are present in a large population show high level of genetic diversity which has already been noticed in wild helminthes.

Compared to Sub-Saharan zones there is a marked difference in the intensity of disease generated by *S. haematobium* in North-African regions. As a whole, there is a definite hybridization of such pathogen with other *Schistosoma* species in certain portions of Africa. This firmly indicates that *S. haematobium* possess at least a pair of phylogenetic clusters with varied virulence. The goal of the experiment has been the study of probable variance among *S. haematobium* by amplifying genomic DNA of specific individuals simultaneously. Study of genetic variability or diversity in the wild population of *S. haematobium* has been performed by PCR from specific areas of Africa like Egypt, South Africa, etc. where there are separate environmental conditions and chances of natural hybridization. An average to elevated degree of genetic variation has been noticed among three natural isolates. Greater number of bands had been shared by the specimens collected from South Africa and Zimbabwe with similarity index = 0.721 than the bands shared by the specimens from Egypt, revealing that minimum a pair of phylogenetic categories of *S. haematobium* are present in different geographic parts of Africa. Reports on the probable genetic variance in *S. haematobium* may clarify a number of obscure aspects of parasitology-like hostility, ability to bypass host's immunity and resistance.

There is molecular evidence that diversity can be produced within sporocyst generations, supporting the hypothesis of mitotic recombination during the asexual phase of schistosomes.

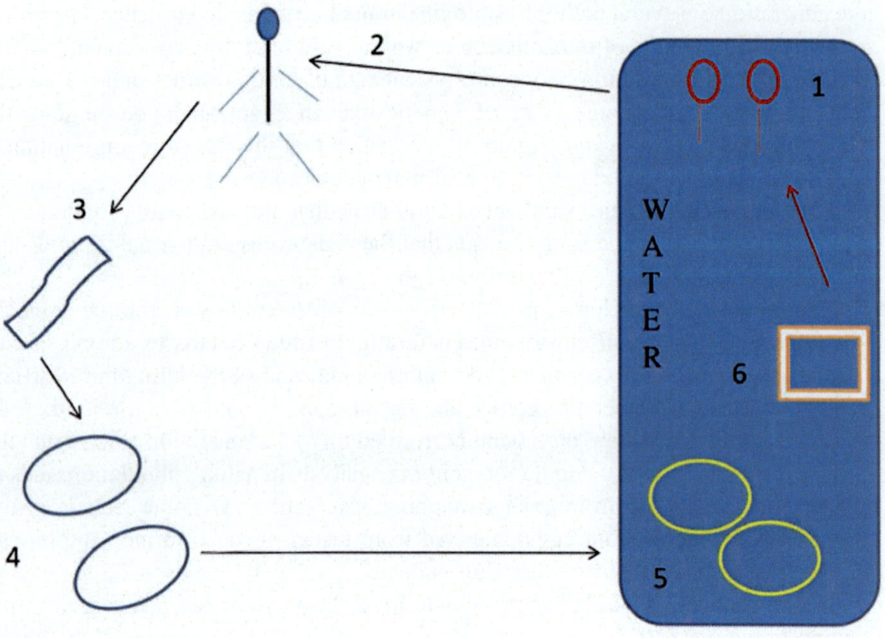

Fig. 7.4 Word diagram of life cycle of schistosomes highlighting clonal genetic recombination

In the life history of *S. mansoni*, the miracidium larvae infect snail, *Biomphalaria* that serves as a secondary host and generates cercariae. The definitive host is human which becomes infected when cercariae pierce the integument. Recombination is usually a rare case in mitosis and if occurs then in lower frequency than during meiotic divisions but still it is evident that genetic variance in the population of this parasite is developed by the genetic exchange during mitosis (mitotic recombination) (Fig. 7.4) in the asexual phase of life cycle. It is also demonstrated that the genetic diversity among asexual cercariae is developed by the mitotic exchange phenomenon that occurs within the intermediate host during the time of sporocystogenesis. Generally mitotic cell division generates only one genotype but mitotic recombination is believed to generate genetic variance as multiple genotypes remain present there (Bayne and Grevelding 2003).

Therefore it can be concluded that both the meiotic (during sexual reproduction) and mitotic (clonal) recombination are crucial factors in the generation of genetic variability among the parasites which help in their adaptation against host's immune reaction as well as to combat with the antiparasitic drugs. Hence, this molecular genetics of parasites will be a basis of wise approach to design drug and to develop vaccines that should also exploit the techniques of molecular graphics and molecular biology as a whole.

References

Anderson T, Nkhoma S, Ecker A et al (2011) How can we identify parasite genes that underlie antimalarial drug resistance? Pharmacogenomics J 12(1):59–85

Bayne CJ, Grevelding CG (2003) Cloning of *Schistosoma mansoni* sporocysts in vitro and detection of genetic heterogeneity among individuals within clones. J Parasitol 89 (5):1056–1060

Carlton JM, Angiuoli SV, Suh BB et al (2002) Genome sequence and comparative analysis of the model rodent malaria parasite *Plasmodium yoelii*. Nature 419:512–519

Gardner MJ, Hall N, Fung E et al (2002) Genome sequence of the human malaria parasite *Plasmodium falciparum*. Nature 419:498–511

Jenni L, Marti S, Schweizer J et al (1986) Hybrid formation between African trypanosomes during cyclical transmission. Nature 322:173–175

Stothard JR (1999) Genetic diversity and genetic exchange in Trypanosoma cruzi: dual drug-resistant "progeny" from episomal transformants. Suppl 1:189–193

Valentim C, Loverde P, Anderson T et al (2009) Efficient genotyping of *Schistosoma mansoni* miracidia following whole genome amplification. Mol Biochem Parasitol 166:81–84

Weedall GD, Hall N (2014) Sexual reproduction and genetic exchange in parasitic protists. Parasitology. Accessed 22 Dec 2014

WHO (2013) Schistosomiasis: progress report 2001–2011, strategic plan 2012–2020

Physiology

8

8.1 Host Recognition

Diverse strategies and specificity in host recognition indicate that finding of suitable host and its recognition are prime determining factors in the parasitic evolution. Efficiency of transmission is habitually dependent on the generation of invasive forms in mass. In fact very much specificity in host-recognizing mechanism is indicated by most of these invasive stages. The recognition of host is generally species specific and it occurs sequentially by which they react to numerous signals from host as well as entire ecosystem. The different host-finding mechanisms generally indicate adaptability to specific environmental situations for transmitting. Again it is to be known in which way the worms search the pathways for their very specific micro-niche in the host body after invasion. Present investigations though indicate that the circulating parasitic stages go along with chemical gradients of integument and blood components, but how these parasites navigate a long pathway within their host is not clear specifically (Haas 2003).

8.1.1 Mechanism for Detecting Host

Active locomotion and movement towards the intermediate and/or final host is accomplished by a variety of helminth larvae like the larvae of monogenea and digenea. Each of these various larval parasites except coracidium explore and fix to or pierce the succeeding host during their life history.

1. Monogenea: In external flukes, chemotaxis is an important method to locate host. Experiment on *Entobdella soleae*, a monogenea, shows that oncomiracidium exhibits a marked preference for the sole fish (*Solea solea*) that is mediated by chemical signals from the fish integument.

P. K. Bandyopadhyay et al., *Biochemical, Immunological and Epidemiological Analysis of Parasitic Diseases*, https://doi.org/10.1007/978-981-16-4384-2_8

2. Digenea: 2 classes of larvae remain entangled with host's active movement—miracidium come out from the egg as free swimming ciliated larva in aquatic habitat and then it attempts to find and perforate a favourable mollusc host. Cercaria is removed from mollucan host and may migrate through their aquatic habitat to find and fix in the succeeding host (e.g. schistosomes) or it can stay as cyst on flora awaiting for consumption (e.g. *Fasciola hepatica*).
   Experiments could not explain the exact mechanism of the nature of host-finding of these above mentioned larvae. The sense organs of many miracidia like surface papillae or eye spot might assist in the orientation of larva in its domain so that they can come in close vicinity to molluscan host. Duly, miracidia respond to environmental cues like light, warmth, gravitational force, current of water flow and changes in the partial pressure of Co_2 (PCo_2) in various manners. Host recognition of host by miracidial larvae is somehow controversial but a number of studies also support that miracidia are attracted to chemical signals that come out from host, possibly in its mucous or faecal matter. Additional observations lead to hypothesize that miracidial parasite detects snails by an arbitrary, trial method. A cercaria of *S. mansoni* is negatively affected by turbulence of aquatic habitat and it remains incompetent to invade even when it remains physically unaltered. Data reveal firmly that an approach to reduce the transference of free swimming larval parasites is to bring a surge in water velocity if feasible.
3. Cestoda: The coracidium is a free swimming larva of pseudophyllidean tapeworm. Here, transmission is characterized as a passive phenomenon as the hatched larvae are consumed by copepods.
4. Nematoda: In many plant and animal parasite nematodes chemotaxis is a predominant process of host-findings. The L3 larva is infective stage that consists of well-developed sensory parts that also assist in the detection of their hosts.

8.1.2 Entry Mechanism

Active penetration through the epidermal covering of host assists a number of parasites to enter their hosts. Secretion from the glands at the apical part of the digenean miracidium larva helps to penetrate the larvae to their suitable snail host. Both the lytic enzymes and lubricants are present in such discharges. Some miracidia remove their epidermal ciliary plates (like *Fasciola hepatica*) when they invade the snail hosts whereas cilia are retained in other larvae. Surface lipids of host commence the perforation to mammalian integument by the schistosome larvae. During the penetration cercariae remove their tail to develop schistosomulum stage and in this phenomenon few fatty acids along with integumentary complex lipids take important stimulatory role. Infective larval stage of hookworms and similar nematodes also pierce host integument actively with the help of lyase class of enzymes.

8.1.3 Circadian Rhythms and its Relation to Parasite Transmission

It is known to us that a circadian rhythm is a natural, internal process that controls the sleep-wake cycle and oscillates roughly every 24 h. This daily cycle might regulate the parasitic transmission to new host. These are of following types:

8.1.3.1 Periodic Cell Division

Within the red blood cell, cell division of parasite occurs every 24, 48 and 72 h in *Plasmodium knowlesi*, *P. vivax* and *P. malariae*, respectively This periodic division causes the development of gametocytes within blood and these seem to be infective stages for the anopheline mosquito. Gametocyte maturation coincides with the blood sucking of the vector mosquito. Daily rhythms of such type are probably associated with the body warmth of the homoeothermic hosts as in laboratory hypothermic monkeys cause derangement in this circadian cycle of the development of disease malaria.

8.1.3.2 Daily Release of Parasites

In order to lay eggs, mammalian pinworms migrate from the rectum during sleep time to the perianal site. Even without faecal contamination, eggs of parasite show hand to mouth contamination specifically during sleep time and this is caused by the movement of female parasitic worm which is led by the decrease of temperature in the rectum. Coinciding with the occurrence of the suitable successive host, the digenean larvae are discharged into aquatic habitat from the molluscan host.. Such episode has greatly been investigated in schistosomes and each species shows specific features in the timing of cercarial release. Most of the schistosomes show a single surge of shedding each day that is circadian in nature. As, for example, *S. mansoni* and few strains of *S. japonicum,* etc. liberate cercariae in day time, whereas *S. rodhaini* and different strains of *S. japonicum* liberate the same at night. Environmental light and temperature cycles are the most crucial factors among those which can shape the periodicity of larval release from molluscan host. Thermal periodicity shows a less significant role than the periodicity of light. Human schistosomes have a tendency to release their larvae during day time whereas rodent schistosome, *S. rodhaini* sheds the same at night and all of these occur according to the presence of potent host fauna.

8.1.3.3 Circadian Migration

Few parasites reside in internal host tissue and are transferred by the non-chordate vectors which bite superficially, hence these parasites are exposed to difficulties to reach peripheral tissues when the vectors come and feed. It can be noticed that some parasites are efficiently adapted to aid their transmission in these situations. In this regard, the microfilarial larvae of nematodes that cause filaria are well-researched examples. The adult worm of *Wuchereria bancrofti* resides in the lymphatic system whereas the adult of *Onchocerca volvulus* stays in thick lumps in the integument and sub-dermal parts and transmission of their larval stages occur, respectively, by mosquitoes or simuliid blackflies that feed on blood at periphery. As adaptive feature, a number of filarial nemathelminth have evolved diurnal rhythms in their larval migration. These migratory behaviours are of different classes and are associated with the feeding behaviours of the insect concerned with the parasitic transfer. These are as follows: (1) microfilareae in peripheral blood of host at night only, missing in day (e.g. *W. bancrofti*), (2) microfilareae more predominant in periphery by day only, not at night (e.g. *Loa loa*), (3) microfilareae more in number at the same site during evening (e.g. the heart worm, *Dirofilaria immitis*) and (4) microfilareae staying in peripheral blood for the whole day but are dominant at the afternoon (e.g. A variety of *W. bancrofti*).

Diurnal migration of the larvae is triggered by the difference in partial oxygen pressure (PO_2) between oxygenated and less-oxygenated blood. The larvae of *W. bancrofti* gather in the pulmonary blood when PO_2 is greater than 55 mmHg while these larvae move to the superficial blood when PO_2 drops to 47 mmHg or less than that.

8.2 Foundation and Continuance Within the Body of Host

The parasitic organism is to establish in a physiologically favourable micro-niche after invading into appropriate host for their growth either to the sexually matured stage or to a transitional stage.

8.2.1 Hatching and Excystation

A number of parasitic organisms launch into their hosts in encysted condition or within egg membranes. Along with food these organisms reach the host and then within the gastrointestinal tract these become stimulated and released from their cysts before their further growth and maturation. It should be kept in mind that all parasites do not enter to their host in encapsulated state; for example, many larval helminth stay in the cells of a transitional host without being encysted.

- Protozoa: Stimulation and excystation of protozoan cyst of a few species are investigated in vitro. If the parasite invades homoeotherm host, then the optimum experimental conditions include mainly increased temperature, neutral power of hydrogen and different PO_2. Parasitic activation in encysted condition might be unmistakable from excystation as the first one requires increased PCO_2 while the next depends upon the activity of proteases. In the different Coccidia, trypsin and bile salts assist in the breakdown of a specific small portion of the cyst wall during sporocyst excystation.
- Digenea: Most of digeneans egg hatch in aquatic medium under favourable conditions of different ecological factors like light, temperature and salinity. Eggs of some additional digeneans are consumed by molluscan hosts and then hatching takes place in the digestive tract of this host organism. Specific wave-length of the light may have important role in the hatching of the operculate eggs

of *Fasciola hepatica*. Activity of miracidium is stimulated by light by which miracidium changes its ability of penetrance in the viscous layer just beneath the operculum. Actually, operculum may be forced off by the hydration that allows the pathogen to get rid of. As eggs of schistosome are without an operculum, the larval parasites are liberated by the rupture of their egg shell. The chief physiological controller in the hatching of their eggs is osmotic pressure, as, for example, a speedy reduction in osmotic pressure is observed to trigger hatching of the egg when the egg is released into fresh water. The species which are infective for mammals and birds and in which cyst wall is papery show the initiation of excystation by increase of surrounding temperature singly.

- Cestoda: Getting proper stimulus the eggs of many cestodes hatch in the nature. Conversely in the Cyclophyllidea, eggs hatch in the digestive tract of their hosts. Although cyclophylladean egg has papery external capsule, the oncosphere larva remains present within a defensive embryophore. In taeniid tapeworms hexacanth larva brings about disruption of the oncospheral membrane firstly and rupture of the external capsule is mediated by the host proteases finally. In cyclophyllideans which are non-taeniid, hatching is chiefly a mechanical phenomenon that takes place by the activity of host mouth parts on egg shell.
- Nematoda: A number of nematodes are observed to hatch their eggs in nature after which either infective larvae are liberated or the larvae mature into the infective phase (e.g. *Ancylostoma,* etc.). Receiving the appropriate environmental impulses like water, alterations in oxygen level or temperature, the covered larva releases lytic enzymes that aid in the escalated water influx into egg. Then hatching seems to be stimulated by an escalation in the turgor pressure inside the egg. In case of other nematodes that are parasitic in nature, eggs hatch only when they are consumed by the suitable host (e.g. *Ascaris,* etc.), eggs of *Ascaris* hatch in vitro at 37degree centigrade in culture media showing elevated PCO2, bi-carbonate of neutral pH, etc. Larvae inside the egg are stimulated to form special kind of fluid that comprises enzymes digesting the chitinous coats of egg shell for hatching.

8.2.2 The Action of Bile Salts in Establishment of Parasite

Bile is a complex fluid that is released into the upper part of small intestine of vertebrates through the bile duct. It consists of bile salts that are composed of salts of 4 various types of acids and the breakdown product of the RBCs also. These salts affect parasites in different ways that are as follows: (1) Change in membrane permeability; (2) stimulation of encysted stages; (3) lysis of surface layers of pathogens; (4) alliance with host digestive enzymes and (5) metabolism.

Echinococcus granulosus, a hydatid organism, has been investigated to reveal how bile salts take role in determining specificity towards host by the parasites. Protoscoleces of larvae after their removal from cyst respond in numerous manners to bile in vitro as, for example, bile with abundant deoxycholic acid can lyse the protoscoleces whereas bile from carnivorous animals, the wild host for

E. granulosus, is poor in this specific acid. Experimental cannulation of the bile ducts of rats affected with *Hymenolepis diminuta* leads to a decrease in dimensions and productivity of the tapeworm. Excystation in lot of parasites like protozoans (e.g. *Eimeria*), digeneans (e.g. *Fasciola*), cestodes (e.g. *Taenia pisiformis*) and acanthocephalans (e.g. *Moniliformis*, *Polymorphus*) is stimulated and initiated by hepatic bile. Bile salts inhibit the anaerobic metabolic reactions in *Hymenolepis diminuta* and *Oochoristica symmetrica*. The hindrance is more at pH 7.0 than the basic solution. Washing the parasitic worms for a short period of time can reverse the hindrance.

8.2.3 Hypobiosis

The developmental state, in which fauna stay in an inactive condition, is considered as hypobiosis. Arrested development can be observed in free living fauna that reside in dry and/or cold locations and therefore hypobiosis is an adaptive mechanism to cope up with climatic adversities. In the case of parasites, hypobiosis occurs habitually; immature larval stages encased within cysts or egg or even at liberty in tissues of host are hypobiotic phases by which the parasite ceases its maturation and awaits an appropriate stimulant to restart its development to adult stage. When some of the nematode parasites of cattle are considered in commercial sense, hypobiosis might be recognized as a significant complication.

ALD (arrested larval development) is observed in cases of several parasitic diseases of cattle by the infection with trichostrongyles along with *Dictyocaulus viviparous* (lungworm), *Haemonchus contortus,* etc. (as a whole considered as the causative agent of gastroenteritis). Ship or cows are affected by these pathogens as they consume the L_3 larval stage at the time of grazing. The larval stages come into the alimentary canal and within 21 days they mature into adult stage either in the lungs, abomasums of stomach or intestine in different species. Though in such parasites ALD stimulating factors are not marked sharply but, immunity of host individual, seasonal factor and parasite population size may trigger the ALD. At the onset of successive grazing session arrested larvae can continue their development to become adult. Artificially ALD has been triggered by acclimatizing third instar larvae in an artificial chamber while most of these 'conditioned' L3s remain hypobiotic in cattle host. In addition to this, these larvae in hypobiosis might show drug resistance and a lot of anthelmintic medicines are not useful against these parasites. Therefore, hypobiosis of trichostrongyles should be considered as a sophisticated adaptative mechanism to environmental extremities. It is now investigated that when most of the larvae are released all at once, as takes place after hypobiotic stage, fluid efflux and diarrhoea are generally observed. Hypobiosis is followed by the disorder termed as type II ostertagiosis and treatment of cattle in this hypobiotic situation is crucial by the use of appropriate medicines to combat with this disorder.

A number of other resting phases are exhibited by nematodes which are adaptive features for specific environmental stress components. Anabiosis is an utmost state of quiescence that may continue for substantial period of times: Few nemathelminth preserved in arid situation for even more than 20 years are able to mature normally. The parasite has no observable metabolic reaction and senescence is also halted in anaerobic situation. Specific anabiotic responses in nematode parasites occur outside of the host body and this adaptive feature is induced by desiccation, low temperature, osmotic stress and low Po2.

8.2.4 Invading Tissues

Most of the parasites invade particular cells of the host body in which they inhabit either for the time being or for a prolonged period. Some examples of such tissue-invading pathogens incorporate those of malarias and/or babesias (within RBCs), leishmanias (in macrophages), helminthes (invading mucosa of the GI tract), trypanosomes (in nervous system) and larval digeneans, nematodes, etc. (in body musculature).

- Cell invasion in malaria: Attention has been paid to basically two discrete characteristics of the cytology of malarial pathogens: recognition of red blood cells and attachment of the infected RBCs to the endothelium of the blood vessels of host. Invasion of the RBCs by malaria parasite occurs in distinct and sequential steps that involves:
  (1) Cell identification by merozoite; (2) inclination of pathogens to the RBC surface; (3) development of a junction at site of contact between the merozoite and the RBC surface; (4) stimulation of invagination of membranes of RBC by merozoite's secretion and (5) arrival of a pathogens with the help of a considerable invagination of plasmalemma of RBC by the formation of parasitophorous vacuoles. Recognition of the RBC by merozoite stage is dependent on unique receptors on the surface and differs in accordance with the age of cell, antigens of blood group and specificity of host. Recent investigations reveal that surface glycophorins are major determinants of this invasion process. RBC surface glycophorins are glycoproteins rich in sialic acid and are of four subgroups (a, b, c and d). Function of these glycophorins as receptor for merozoites of *P. falciparum* is well accepted at present. Cells that lack the glycophorin a or the glycophorin d are resistant to invasion of merozoite to a remarkable extent and by the removal of persistent glycophorin molecules such ability of resistance might be increased.

Though specific function of glycophorins of RBC as receptors is not identified properly but the evidence reveals that they assist in the initiation of red cell invasion and in this process there is participation of some other factors. Once identification of the cells is done, the invasion carries on with the development of a junction site of almost 10 nm size that contains delicate fibrils expanded from the apical projection of the merozoite to thick membrane of erythrocyte. It is observed that invasion-resistant cell does not develop this site of attachment but if they are treated with

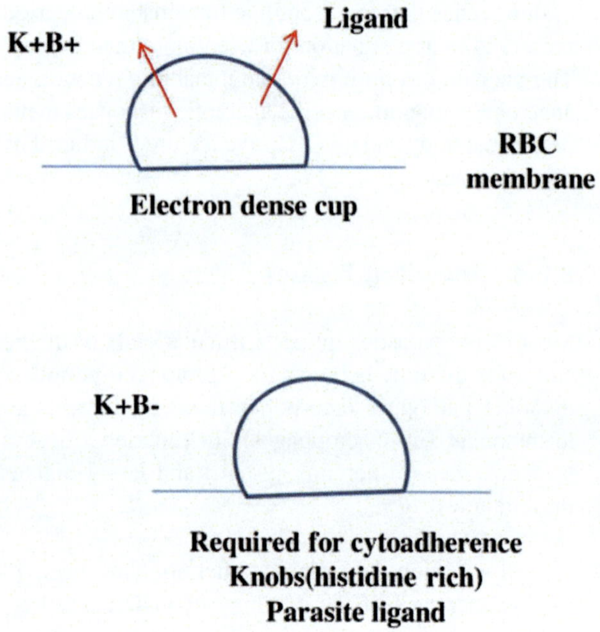

Fig. 8.1 Figure depicting knobs on the RBC surface induced by *Plasmodium falciparum* infection

trypsin then the cells exhibit the process of invasion by merozoites of *P. knowlesi* and a distinctive host cell–pathogen connection is produced. The organelles of the apical part of merozoite like rhoptries and micronemes begin the proper invasion by release of a secretion which contains a histidine-rich protein. The RBC membrane invaginates in progressive way at the time of invasion to surround the merozoite and the site of attachment moves in order to maintain its position at the apex of the developing parasitophorous vesicle. Mobility of the site of attachment may depend on the fluidity of membrane, since treating merozoites by cytochalasin B slows down invasion though a junction is developed but in this case junctional movement does not occur. After fulfilment of general invasion, the attachment site seals up at back of the merozoites which at that time lie totally surrounded in the parasitophorous vacuoles. Erythrocyte membrane forms this vacuole by the molecular reorientation.

The knob at surface is the effective part of cellular adherence of *P. falciparum*-infected RBC and it contains a cup like structure of membrane and allied projection of the RBC membrane (Fig. 8.1).

Unique protein of Mr. 80,000 and rich in histidine and proline is present in those knobs. This protein is not observed in knob less strain of the parasites. Its function to mediate cellular adherence is unrecognized but as immunoglobulins have been shown to obstruct this in a strain-specific manner, it indicates that there are some kind of involvements of added molecules in this cytoadherence.

8.2.5 Cell Invasion in Case of Other Pathogens-

As different parasitic protozoa also inhabit within the cell, therefore they have to invade the cells of host body. Such parasites may also inhabit within parasitophorous vesicle and reach by invagination of the cell membrane instead of active penetration. This kind of invasion is seen in the Coccidia (e.g. *Eimeria* invading cells of digestive tract and *Toxoplasma* in macrophages). Such parasitophorous vacuole might be for the time being (e.g. *Babesia*), so regress soon after accomplishment of this entry. *Eimeria, Leishmania,* etc. show invasion in the host tissues with the help of their own phagocytotic machinery. Way of invasion can be modified by the drugs like colchicines or cytochalasin B that inhibit phagocytosis. After invading host cells *Leishmania* stay within a parasitophorous vesicle that is produced from the surface of host. Attention has been given on the sustainance of such pathogens inside the host cell that ingests and kills invaders as it has an applied consequence. *Trypanosoma cruzi*, the responsible parasite of Chagas disease within the America, shows obligatory parasitism. In cell culture trypomastigotes spread in fibroblasts by means of an interaction in between the parasites and membrane of the host cells. Lectin like protein is produced by parasite and this protein is involved in cellular adherence assisted by proteolysis; penetration is aided by a tunicamycin-sensitive glycoprotein. The role of host cell in these events is the production of glycoproteins which participate actively in adherence and penetration also. On the contrary, reaching of *T. cruzi* to macrophages occurs with the help of phagocytosis and trypomastigotes can sustain themselves inside the macrophage as they are able to escape from the lysosome and can grow within cytosol.

In few occasions such invasion by parasitic protozoa is a prime pathophysiological phenomenon. In *E. histolytica*, released substances like 'amoebapore' (a pore forming protein), proteolytic enzymes, etc. aid in the spreading of this pathogen into the intestinal cells of host and acute intestinal as well as hepatic disorders are observed in this stage of amoebiasis.

8.3 Increased Parasite Survival by Manipulating Host

In some situations, parasite modifies host activity by which though transmission is not assisted yet survival of parasite is aided. It is investigated that a number of parasites get profit from physically altering their host in such a way so that resources usually utilized in the host's reproduction then might be invested in the growth and maturation of that parasitic organism. Host's behavioural castration might be induced by parasites. Then the affected hosts show reduction in their mating behaviour and even less interest in sexual activity. Such actions may enhance the chance of host's survival because of the decreased chance of predation related with the sexual behaviour. Such pattern has been noticed in the amphipod affected by acanthocephala. Likewise, biting rate of mosquito vector can be decreased by *Plasmodium* oocytes. It is known that biting is a chief life threat for a mosquito, therefore decreased biting habit escalates the chance of the survival of parasite–

vector pair prior to transmission. Few parasites are exposed to even more hazardous interplay with the hosts. As they sting on the host, the parasitoid wasp has to overcome the protective mechanism of the host's body. A number of polished plan of actions have developed from the initial meet to manipulate the host. The wasp *Ampulex compressa* pricks the host cockroach at the outset in thoracic region, leading to the prothoracic legs quite paralysed. This phenomenon facilitates the next sting via the neck into a distinct portion of central nervous system (CNS). Cockroach is induced to stay in an enduring sluggish condition because of this injected poison and hence become target for the consumption by developing larva of wasp in its habitat. In braconid wasp an amazing behavioural alteration has recently be revealed. Caterpillar larvae are utilized as a defender for the pupal stage of wasp from the strike of predator. Numerous instances of parasitic manipulation studied to date indicate that the parasite hampers the neural and hormonal activities of their hosts (Cézilly et al. 2010).

8.4 Reproductive Biology

Most of the parasites possess complex life history that reproduce sexually in one host whereas asexual proliferation occurs in another.

8.4.1 Asexual Reproductions

Budding or splitting asexually is common in a number of Protozoan parasites, each and every Digenean parasites and in few Cestodes. In the Protozoa asexual mechanism mainly occurs by means binary or multiple splitting, Schizogony, endodyogeny or single budding, whereas Helminth parasites are increased immensely by budding which is considered as polyembryoni that occurs internally. In the life history of Digenean parasites asexual cycle happens within the molluscan host solely. Generally two prominent asexual cycles take place in this intermediate snail host, mother's sporocyte to daughter sporocytes or rediae from sporocysts, the ultimate fate of which is the emergence of numerous cercariae from only one miracidium entering a molluscan host. All of these cercariae are identical genetically.

8.4.2 Sexual Reproduction

Gametes are quite distinct in their morphology like microgametes in male and macrogametes in female and sexual reproduction may occur alternatively with asexual one that take place in separate host. In *Plasmodium*, the number of merozoites is increased by asexual schizogony in the blood of host and gametocytes (sexual stage) are developed also which enters mosquitoes where fusion of gametes and sexual growth happens. For a long time malariologists are quite concerned about the inducer that regulates the development of gametocytes; present school of thought favours the line of thought that trophozoites are induced towards sexual proliferation by ecological components related to the degradation or lysis of host cell. In this case gametocytogenesis might be considered as one escape machinery from unfavourable situations employing genomic variation developed by random amalgamation of gametes in the anopheline vector.

Trypanosomes show their sexual phase either in mammalian host or within vector tsetse fly. In favour of the second argument, large trypanosome forms, which are able to liberate a good numbers of newly formed trypanosomes, are collected from mid gut cells of vector flies; these large trypanosome forms are seemingly the result of the union of two individuals and this exhibits a probable strategy of genetic exchange. Almost all tapeworms show hermaphroditism and each proglottid (segment) contains a full complement of female and male reproductive structures. Both Self- and cross-fertilization happen here. In cestodes the terminal segments become the oldest and, at gravid condition, contain ripped eggs which are released independently or along with the removed proglottid. Eggs of the tapeworm are surrounded by capsule containing different constituent coats.

The majority of parasitic nematodes follow sexual reproduction and these are sexually dimorphic. Hermaphroditism or parthenogenesis is shown in little number of species but no somatic asexual mechanism is noticed. The male nemathelminth possesses only one testis along with some add-on reproductive structures like copulatory bursa or spicular expansions which facilitate copulation. The female nematodes possess single or double sets of gonads; spermatozoa are deposited in a seminal receptacle and fully developed oocytes are fertilized by these spermatozoa. Formation of egg shell is started off with the help of fertilization and continues during the maturation of egg. Hatching of eggs, in few parasitic groups like the filarial worm, occurs in uterus and accordingly egg shell is reduced both in its chemical complexity and magnitude morphologically.

8.4.3 Reproductive Synchrony

Reproductive processes in few parasitic species seem to synchronize with host sexual and breeding cycles; such interaction aids in the release of infective parasites into the nature at the same time when with vulnerable juvenile hosts are liberated. Prominent instances are the flagellated protozoan parasites of arthropods and amphibian hosts. The liberation of opalinid gametes from the digestive tract of amphibian host is triggered by its sex hormones. The Hypermastigina that inhibit the gut of arthropod hosts might be induced by moulting hormones of host for sexual reproduction. In this case, synchronization of sexual mechanism with ecdysis of the insect host warrant reinfection and in this incident it is an obligatory phenomenon as the parasite releases cellulases essential in host's digestion.

8.5 Neurophysiology of Helminth Parasites

Parasite's neurophysiology is of applied importance. As, for example, Nematodes possess different sensory receptors along with a finely tuned nervous coordination in order to react against various chemical as well as physical inducers. The different types of sense organs of nemathelminth identified till now are discussed here in details. It is tried here to assess the probable functional importance of different receptors. A fundamental structural organization is found in most of these sense organs. This organization involves three cell types: (1) a neural cell that terminates in the shape of one or added highly reformed cilia; (2) a non-neural cell that possesses a kin relationship with nerve cell and (3) another non-neural cell that contains pack of fibres and covers different cells. A probable feedback process is presently recommended in Nemathelminth, where the sensory cilia regulate the activity of glandular cells following environmental cue. Experiments of the bacillary band receptors of *Capillaria hepatica,* etc. reveal that such organization might be a general feature in nematodes (Mclaren 1976).

8.5.1 Nervous System

This system of platyhelminth shows a cerebral complex comprising commissures along with ganglia and this complex is situated at the apical part of the body, anterior to pharyngeal portion. From cerebral ganglia nerves enter different tissues and these nerves are symmetrical bilaterally. The key variation from the general neural arrangement which is seen in the platyhelminthic parasites is related to the innervations in several organs like suckers in the Digenea, scolex in the Cestoda, etc. which are involved in the parasitic attachment. The nervous system in Monogenea is quite similar to that of Turbellaria which is free living in nature. Few members of this group possess quite primitive arrangement where the cerebral ganglia are positioned dorsally and anteriorly to the pharyngeal part (such as Tatraonchidae), but, mostly brain is made up of paired ganglia which are connected by a single commissure that is present very near to the pharyngeal region. In some species two commissures encircling the pharynx are found for the above mentioned function. There are three primary neural trunks running posteriorly from the brain and three to four pairs supply the head region anteriorly.

In Digenea nervous system shows little variation from that of monogenean parasites, as, for example, there is an extra and substantial neural supply to suckers situated ventrally at apical part of the body. Again two cerebral ganglia, connected by commissures, bring about a pair of lateral neural trunks that is stretched throughout the strobila. The parasitic platyhelminthes show a nerve net that contains unmyelinated neurons. This net can be studied under light microscope. Though little is known about the ultrastructure of the nervous system of Platyhelminth but it is not wise to consider this as a degenerated organization. Free living and parasitic nematodes consist of quite identical pattern of neural orientation that contains a nerve ring of circum-oesophageal commissure and different numbers of ganglia where one is dorsal another is situated ventrally and two or many more are lateral in their positions. Lateral and ventral nerves are responsible for sensory functions while and motor activities are controlled by dorsal and ventral counterparts (Chappell 1979).

Nervous system of platyhelminth parasites is formed of a nerve network consisting of unmyelinated neuron and most of that are involved with motor activity. Nervous system in acanthocephalan parasites is quite under-developed consisting of a cerebral ganglion from which single and paired neurons are liberated and the male acanthocephalan has an added ganglion related to their reproduction.

Data on nervous system of the nematodes are gathered specially from the study on free living forms like *Caenorhabditis elegans*. Two ganglionated, ventral nerves stretch towards posterior parts and joins to a nerve of dorsal part through a bunch of commissures. The organization of the nervous system shows a tendency to remain in tune throughout the entire phylum.

8.5.2 Sense Organs

The activity of the sense organs of helminthes is preliminarily determined from the morphological and ethological studies. The chief areas of sensory biology which have been studied are photoreception and chemoreception which are associated with the identification of host, mate, etc. along with the response to gravitational pull and temperature.

A number of larval helminthes react against the ambient light stimulus and use this response they come to the close proximity of suitable host in order to invade. Some larval forms of helminth possess eye spots. Without this sense organ it is problematic to distinguish between one stimulus from other in an experimental condition. It is doubtless that photosensitivity is crucial to orient invasive larvae to find appropriate host and often to initiate hatching of eggs.

Ability to respond to thermal stimulus is also important to infect a homoeothermic host (e.g. Hookworm larvae in land and schistosomes in aquatic medium). It is still not clear that which sensory structure is associated in the above mentioned responses. The sensory organs involved in the reaction against gravity and chemical gradients are still to be pointed out. Surface receptors involve ciliary sensillae (Monogenea), papillae (Digenea), ciliated pits in the Nematoda, etc. Cestodes lack sense organs. Sedentary life in a sheltered habitat (like endoparasites) might be correlated with the lacking of developed sensory apparatus.

8.5.3 Neurotransmission and Neurosecretion

Parasitic helminth produces many putative neurotransmitter agents like adrenalin, acetylcholine, Dopamine, GABA, etc. Synapses releasing acetylcholine are reported in all the helminthes and acetylcholinesterase has been identified in several species also. Nematode's neuromuscular junctions are cholinergic also. GABA is the chief

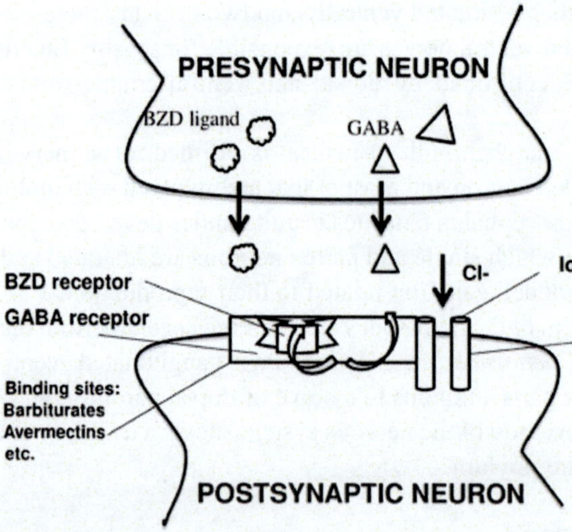

Fig. 8.2 Interaction between nematode synapse and drug ivermectin

inhibitory neurotransmitter in case of nematode parasites (Fig. 8.2) and Ivermectin acts as an antihelminthic agent which acts on such synapse.

For the purpose of chemotherapy the chemical messengers like neuropeptides, amino acids, acetylcholine, etc. are of great attention as they might differ in hosts and parasites. Some recent observations indicate the complexity of this phenomenon. As, for example, in tapeworms, aminergic, cholinergic neurons are found. At least –29 neurotransmitters are recognized that includes pancreatic polypeptide of cattle, growth hormone releasing factors, gastrin, gastrin releasing peptide, oxytocin, etc. No function of this complex of neurotransmitters is revealed in detail till now in cestodes but it will create new avenue for the generation of the novel drug that will act at neurophysiological level.

A number of parasites are cause endocrinological disorders in the body hosts through the liberation of components that mimic the hormones of host and act as xenohormones. As, for example, it is observed in *Trichobilharzia* sp., a parasitic flat worm of birds that induces host gonadal degeneracy by a factor named 'schistosomin' which is produced by the snail host itself under parasitic influence. A number of digenean larvae bring about huge growth and sexual alterations in the snail host mostly mediated endocrinologically. Plerocercoids of tapeworm, *Spirometra mansonoides*, secrete a platelet growth factor (PGF) that can mimic growth hormones of mammal remarkably. Plerocercoid stage causes sparganosis in a vast range of host animals along with man as this stage has minimal host specificity. The closeness between human growth hormone and PGF indicates that human gene for growth hormone is obtained by such parasite. Viral transduction might be a probable way of this assumed genetic interchange.

AChE and nAChRs are significant for different cell signalling reactions of parasitic helminths and therefore recently used most antihelminthic drugs target these. AChE is an essential enzyme in schistosomes that is involved in the process of glucose utilization from body fluid of host organism. Metrifonate has been employed as an effective chemotherapeutic agent preliminary for treating urinary schistosomiasis as this drug targets AChE. Subsequent researches during the period of 1980/90s on AChE in *S. mansoni,* etc. disclosed the neuromuscular impacts of AChE through experiments that obstruct the interaction between ACh and the nAChR components controlling gated positively charged ion channels in the nervous system of such parasitic creature. Several observations emphasized the crucial function of nAChRs in creating inhibitory post synaptic potential in schistosomes and their properties for being chief targets for medicines those mimic the functions of acetylcholine (You et al. 2017).

8.5.4 The Neuromuscular Junction in Helminth

Observations on *Ascaris* let out that the muscle cells of nematode are unusual as they possess both nervous and contractile elements: muscle arm makes synaptic junctions with motor neurons of the nerve tissue. Contractile part possesses normal arrangements of thin and thick myofilaments, feature of striated muscle, where A and I bands are prominent while no Z band is detectable. Neuromuscular junctions of *Ascaris lumbricoides* have been studied under light and electron microscopes. The electron microscopic observation reveals that the neuromuscular junctions consist of numerous muscle cell processes which are situated very close to single axon. Intersynaptic cleft is almost 350–500 Å in width. The sarcolemma and axolemma both are triple layered membranes of 75–80 Å thickness. The efficacy of the unique cyclo-depsipeptide antihelminthic drug on the muscle of body wall of *Ascaris* has also been revealed. The efficacy of this emodepside has not been found in a denervated muscle layer and this observation suggests that this drug may employ its action through the neuron and not on muscle directly. It has also been suggested that the response to emodepside depends on extracellular potassium ions (J. Willson et al. 2003). The neuromuscular junction of other helminth parasites requires further studies.

References

Cézilly F, Thomas F, Médoc V et al (2010) Host-manipulation by parasites with complex life cycles: adaptive or not? Trends Parasitol 26:311–317

Chappell LH (1979) Nervous systems, sense organs and behavioural coordination. In: Physiology of parasites. Tertiary level biology. Springer, Boston. https://doi.org/10.1007/978-1-4684-7808-2_9

Haas W (2003) Parasitic worms: strategies of host finding, recognition and invasion. Zool 106 (4):349–364

Mclaren DJ (1976) Nematode sense organs. Adv Parasitol 14:195–265

Willson J, Amliwala K, Harder A, Holden-Dye L, Walker R (2003) The effect of the anthelmintic emodepside at the neuromuscular junction of the parasitic nematode *Ascaris suum*. Parasitology 126:79–86

You H, Liu C, Du X, McManus DP (2017) Acetylcholinesterase and nicotinic acetylcholine receptors in schistosomes and other parasitic helminths. Molecules 22(9):1550

Immunology

9

9.1 Types of Immunity

Immunity is the potentiality of protecting ourselves from different pathogens. The immune system can discriminate between self and non-self therefore provides us the ability to tolerate the self ones and reject the foreign particles. Basically it is of two types-

9.1.1 Innate Immunity

Innate immunity generally comprises four different protective walls: anatomical barrier (integument along with mucous layer), physiological barrier (body heat, pH and chemical components), ability of phagocytosis and inflammation. Cells and techniques which are important for potent innate or non-specific immunity to foreign particles that dodge the anatomic protection are extensively documented. Pattern recognition receptors (PRRs) are critical for this innate immunity where these receptors permit a restricted numbers of cells to recognize and react quickly to a broad spectrum of pathogens which share similar structural pattern, designated as pathogen associated molecular patterns. As, for example, bacterial cell wall substances like lipopolysaccharides and double-stranded RNA are generated when virus attacks.

Innate immunity promptly employs different immune cells to the places of inflammation through the generation of chemokines as well as cytokines. A number of immune-processes throughout the body are mobilized by the generation of cytokines while localized cellular reactions against the infection or physical trauma are also observed. Chief inflammatory cytokines produced in course of the primary response against bacteria are: tumour necrosis factor, interleukin 1 and interleukin 6. These above mentioned cytokines are important for stimulating cell engagement

P. K. Bandyopadhyay et al., *Biochemical, Immunological and Epidemiological Analysis of Parasitic Diseases*, https://doi.org/10.1007/978-981-16-4384-2_9

and the inflammatory reaction at infected site that is crucial for the removal of many pathogens which in addition are responsible for occurrence of pyrexia.

Biochemical machinery that acts for identification and opsonization (coat) of bacteria and/or other pathogens is considered as complement system. Then phagocytic cells internalize microbes and clear cellular scrap as well as kill some parasites and infected and/or injured cells the moment. The phagocytic activity of the non-specific immune response assists removal of dead cells or antibody complexes and eliminates alien components present in different organs, tissues or body fluid. It is able to trigger the acquired immunity by transporting and stimulating antigen presenting cells (APCs) also.

Numerous cells like macrophages, neutrophils, dendritic cells, mast cells, natural killer (NK) cells, etc. participate in this non-specific immunity. Phagocytic cells are chiefly of two types: macrophages and neutrophils, both of which share alike role that is internalization and destruction of microbes through a number of immunological mechanisms.

Phagocytosis is also performed by dendritic cells which act as APCs and initiate the specific immunity by playing as crucial signal in between non-specific and specific immune response. Mast cells and basophils are very effective in the commencement of acute inflammatory reactions which are observed in allergic manifestations and/or asthma. Again eosinophils are granulocytes which acquire phagocytic ability and act to smash pathogens that are generally quite sizable for being engulfed. Eosinophils also take part in the processes related with allergic manifestations. A paramount role in the rejection of tumours and the killing of infected cells are played by natural killer cells. NK cells release perforins and granzymes that are responsible for the destruction of infected cells by the induction of apoptosis. NK cells also produce one more cytokine, interferon-gamma (IFN-γ) that assists in the migration of APCs and generates immunity against virus. Innate lymphoid cells (ILCs) participate in a more regulatory function. Different types of innate lymphoid cells (like ILC-1, ILC-2, etc.) also develop cytokines selectively.

9.1.2 Acquired Immunity

The development of acquired immunity is supported by the non-specific or innate immune action and plays crucial role when such non-specific immune reaction fails to eliminate pathogens. The chief roles of the acquired immunity are as follows: identification of specific foreign antigens, differentiating those from native proteins with production of foreign antigen-specific effectors cascade that remove distinct immunogens or pathogen-affected cells; and then generation of an immunogenic impressions that is able to eliminate a particular pathogen quickly when same infections take place in future. Acquired immune system includes the following cells: pathogen-specific T lymphocytes that proliferate under the stimulation of APCs, and B lymphocytes that produce plasma cells by differentiation and ultimately develops immunoglobulins.

9.1.2.1 T Lymphocytes and APCs

T lymphocytes are developed in bone marrow and then migrate to thymus for maturation. These lymphocytes exhibit a number of receptors on their plasmalemma to bind specific antigens. These are recognized as the T cell receptors (TCRs). T cells have the potential to differentiate and proliferate after receiving pertinent signal. It is already mentioned that T cells need the assistance of antigen presenting cells to perceive a specific antigen.

Major histocompatibility complex (MHC) is a group of proteins that are exhibited on the outer side of antigen presenting cells. MHC are classified as either class I that are shown on all cells possessing nucleus, or class II which are shown only on some specific cells like macrophages, B lymphocytes, etc. Class I MHC molecules exhibit internal peptides whereas class II molecules show extrinsic ones. The MHC protein expresses parts of antigens (peptides) when a cell is infected with a virus or has engulfed non-self-organisms or proteins.

T lymphocytes possess a broad domain of specific TCRs that can recognize distinct particles which are non-self. T cells which recognized self-antigens are mostly removed when the immune system is developed. This phenomenon is also considered as central tolerance. T lymphocytes are stimulated when they encounter an APC with phagocytosed antigen and which is expressing the proper fragments of antigen attached to its MHC components. The TCRs are stimulated by complex of antigen-MHC and then the T cell produces cytokines that regulate the immune reaction later. This antigen displaying activates T cells to develop primarily into either cytotoxic T cells (CD8+ cells) or T-helper cells (CD4+ cells). Cytotoxic T cells are fundamentally associated in killing of infected cells as well as the killing of tumour cells displaying suitable peptides. Proliferation of cytotoxic T cells develops effector cells to stimulate programmed cell death at target site.

Helper T lymphocytes are crucial in establishing and enhancing immunity. These cells lack the ability to kill infected cells or to remove foreign peptides directly as these have no cytotoxic and phagocytic function. However, they are active components of immune system as they stimulate by directing other cells to execute these roles. Helper T (Th) cells are stimulated when TCR identifies antigen–MHC complex. Being stimulated, helper T cells produce cytokines that shape the functions of variety of immune cells. The response of Th1 is marked by the synthesis of IFN-γ that stimulates the antibacterial functions of macrophages and magnifies the immune response against virus and other pathogens present within the cell also. Again Th1 participates in the process to differentiate the B cells to produce immunoglobulins that amplify the activity of phagocytic cells. Therefore, autoimmune disorders might be caused by the improper signalling of Th1. The Th2 activity is recognized by the production of interleukins like IL-4, etc. that are associated with the production of IgE forming B lymphocytes along with the synthesis and utilization of eosinophils as well as mast cells which take crucial role in the fruitful reaction towards a number of parasitic agents. Most of the helper T cells do expire when infection is over and just a few of these lymphocytes stay as memory T cells.

A group of the CD4+ T lymphocyte, recognized as the regulatory T lymphocytes or T reg, also takes part in the immunity. These T reg lymphocytes suppress immune responses; hence these cells are important to develop 'immune tolerance' against some exotic or non-native proteins, which may enter through edibles.

9.1.2.2 B Cells

Bone marrow helps in the development of B lymphocytes from its haematopoietic stem cells. Matured lymphocytes depart from the bone marrow exhibiting specific receptor on their plasmalemma for the antigens. B lymphocytes have the ability to identify the foreign antigens directly where the help of APCs are not required as specific antibodies are manifested on the outer surface of these B cells. The chief role of B lymphocytes is the generation of immunoglobulins for non-self-antigens. In some situations, B lymphocytes perform the role of antigen presenting cells.

B lymphocytes proliferate and generate immunoglobulin-producing plasma cells or memory B lymphocytes by differentiation when they are stimulated by the exotic proteins for which they possess specific receptors. These memory cells have a good lifespan and are able to exhibit those specific receptors for subsequent infections. Therefore, B lymphocytes take an important part in the humoral immunity whereas cell-mediated immunity is performed chiefly by T lymphocytes (Fig. 9.1) (Marshall et al. 2018).

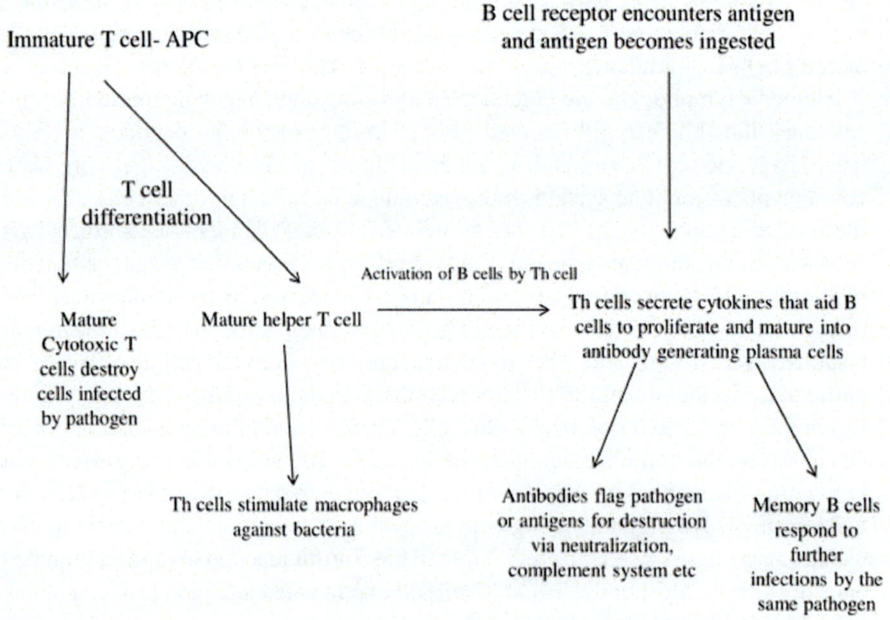

Fig. 9.1 Adaptive immunity: T-cell and B-cell activation and function (*APC* antigen-presenting cell, *TCR* T-cell receptor, *MHC* major histocompatibility complex)

9.2 Immunity to Microorganisms

9.2.1 Immunity to Virus and Bacteria

A number of different factors are involved in the immunity to viruses. Virus is obligate intracellular pathogen and the main defence mechanism against this is the development of interferon-alpha (IFN-alpha) that prevents the arrival of virus non-specifically within its host cells. Still when virus enters the host cells, the viral pathogens expressed on the surface of these host cells and hence are marked and killed by cytotoxic T cells. Specific antibodies like Ig M and Ig G (in serum) as well as Ig A (as in milk) take a key role in defence against recurrence of viral infection.

As bacteria are usually larger in size than that of viruses, they have an inclination to remain outside the cells instead of being the intracellular pathogen. The primary lines of defence are innate blockade like skin, gastric juices or lysozyme present in tears and saliva. Still, when the bacteria can cross these barriers and enter the body they are immediately recognized and phagocytosed non-specifically by neutrophils or macrophages. Within the phagolysosome, macrophages and neutrophils digest bacteria by means of a combination of and oxygen-dependent as independent mechanisms and the O_2 dependent process is amplified by IFN-y produced with the help of T cells. Mandatorily intracellular bacteria like the causative agents of leprosy and tuberculosis are destroyed also by IFN-y activated macrophages. Complement also takes important role in the immune responses against bacterial infections.

9.2.2 Immunity to Protozoa

9.2.2.1 Intestinal Protozoa

Most of the amoebae and flagellated parasites residing in the lumen of the gut follow commensalism and evoke no defence mechanism. *Entamoeba histolytica* shows the same thing until it enters the intestinal layer and later the liver and/or other parts of host's body. After its invasion immunoglobulins are stimulated by complement system and become able to lyse the trophozoites. In hepatic amoebiasis, recurrence is infrequent. Again, serum immunoglobulins in individuals with amoebic liver abscess develop in 1 week and sustain up to 10 years. Though a mucosal IgA reaction against *E histolytica* is observed in the course of invasive amoebiasis but no affirmation marks that invasive amoebiasis is amplified in patients with deficiency of IgA. Experiments on several animals reveal spontaneous resistance which is also correlated with the activity of T lymphocytes. This cell-mediated immunity is crucial in limiting the reinfection. Lymphokines, including interferon-delta stimulate the destruction of the trophozoites of *E. histolytica* with the help of the macrophages (Fig. 9.2).

The mucus layer covers IEC layer in the large intestinal lumen and this layer consists of mucin and Immunoglobulin A of host itself as well as gut-microbiota. But the proteases and glycosidases released from the *Entamoeba* are able to degrade

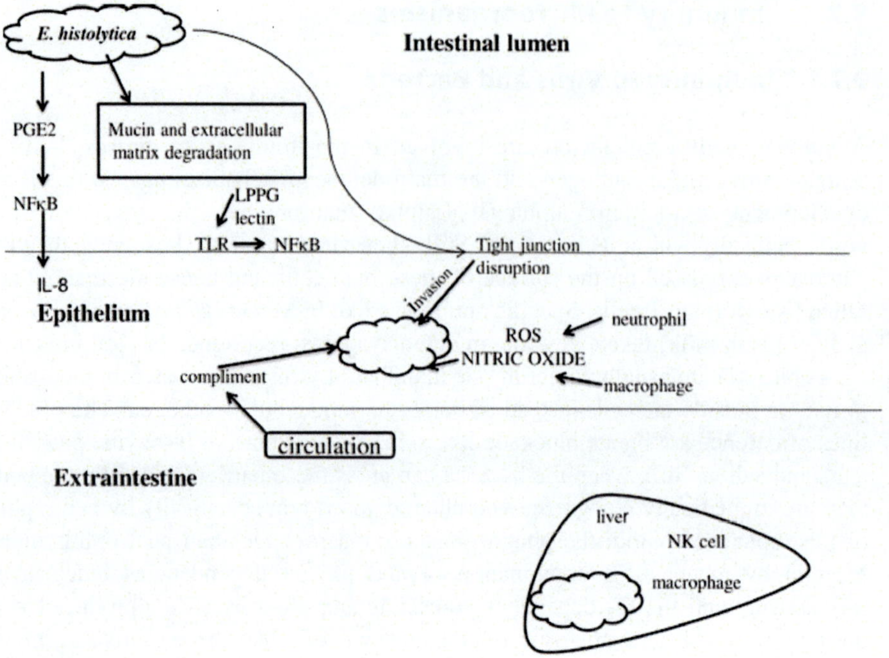

Fig. 9.2 Mechanisms of colonization and invasion by *E. histolytica* trophozoites and host immune responses to suppress and control amebic infection

ECM (extra cellular matrix) and mucin. The EhCP-A5 becomes attached to stimulate integrin as well as intensifies the inflammasome generation that leads an inflammataory reaction. Again, Prostaglandin E2 generated from the *Entamoeba* results in the over-secretion and depletion of mucin. In addition, PGE2 promotes a chain of reaction that causes stimulation of NFκB and secretion of IL-8. The lectin and lipopeptidophosphoglycan on the surface of ameba attaches to TLR2 and causes stimulation of NFκB and cytokine secretion as pro-inflammatory responses. Invading trophozoites are striked by complement as well as by ROS and NO. Neutrophils, macrophages etc take part in this response. Inducing different kinds of T lymphocytes the lectin and lipopeptidophosphoglycan increase protective cell mediated immune reaction. CD4 T cells are responsible to generate IFN-γ, Interleukin 4 etc while CD8 T lymphocytes generate Interleukin-17. When dispersed to the hepatic tissues, the amebae are removed by the IFN-γ produced by Natural killer T lymphocytes. (Tsukui and Nozaki 2016).

Human giardiasis occurs in different shapes without invading tissues; some persons conquer the infection speedily, while others suffer from a persistent problem. It indicates that immune reaction plays an important role against this giardiasis and this immune response is regulated genetically and this is reinforced by documentation from *Giardia muris*. Immunoglobulins like IgA perform as mucosal Ig A when human suffers from infections and at that time antibody dependent cell-mediated cytotoxicity has also been expressed.

9.2.2.2 Leishmaniasis

Leishmaniasis is considered as a complex of diseases from the angle of immunology and it falls into three key types: visceral, mucocutaneous as well as cutaneous. Cutaneous types, as represented by L. *major* and L. *tropica* in the Old World and New World L. *mexicana*, indicate slow recovery rate and this is characterized by little antibody production. Final recovery occurs in mucocutaneous pattern, caused by L. *braziliensis,* but recovery may be quite slow-going in a few patients. *L. donovani* causes visceral type where recovery is infrequent but this infection is distinguished by the synthesis of considerable amount of antibody. These studies suggest that chief role is taken by T cells whereas requirement of antibody is not the case. All forms of leishmaniasis specifically by *L. donovani* go hand in hand with prolonged and intense immunosuppression.

Immunity is mainly governed by cell-mediated immunity where macrophages are stimulated by interferon-y. Though the parasite or pathogen can live in resting macrophage but are unable to reside in activated macrophages as they are destroyed by ROS and NO. In affected hosts, this immune reaction can be counteracted by cytokines like IL-4 that are associated in inducing B cells for the generation of immunoglobulins. Documentation reveals that homogeneous processes are operated in leishmaniasis of man.

9.2.2.3 American Trypanosomiasis (Chagas Disease)

American trypanosomiasis, created by *Trypanosoma cruzi,* shows in a sequence of distinct phases, an incubation stage of 14–21 days then an acute stage staying almost a month and a persistent phase that continues for unlimited period. Both the non-specific and specific immune reactions of the host interact with *Trypanosoma cruzi* in a complex fashion. The fact of host–parasite co-evolution has provided the power of resisting and escaping of the host's immunity and hence it becomes able to develop a chronic infection. In spite of enormous research for many decades in this arena, the infection is not curable wholly, and the components that govern chronic Chagas disease to a clinically detected stage from the asymptomatic conditions are quite fuzzy still now. Many of the disease preventing mechanisms are quite well-documented, but to know all the components as a whole, further research is indispensable.

After entering through the faecal matter of the vector, the parasitic molecule throw itself into the macrophages, where they stay by running away from the phagosome prior to its amalgamation with lysosome, and some added cells like neural cells. Inside such cells, amastigote stages are produced by the multiplication of parasite and then trypomastigotes are produced that enter the blood. Some of these pathogenic stages survive because they are proficient to split immunoglobulin molecules along with hindering the breakdown of complement. *Trypanosoma cruzi* can impede the production of IL-2 and slow down the manifestation of IL-2 receptors that cause generalized immunocompromisation accompanying the infection. Nevertheless, parasitic immunogens on the external surface of infected cells can be traced by immunoglobulins and then they are killed by lytic actions mediated by complement system.

9.2.2.4 Malaria

Generally malarial infection is prolonged and chronic; lasting up to 4 years in the infection caused of *P. vivax, P. ovale* and *P. falciparum* and might be long lasting when one becomes infected by *P. malariae* but this disease might be quite deadly and acute, specifically in young children and immigrants to different endemic regions. Studies indicate that immune responses of man are able to control the effect of this disease but it is unable to remove the pathogens completely. Observations like recovery of individuals, their resistance to reinfection with the homologous strain and passage of protective antibodies from mother to their offspring reveal the role of acquired or specific immunity in this regards. *Plasmodium* in general shows host specificity and a number of studies have been performed on P. *falciparum* culturing in squirrel as well as owl monkeys and on *P. knowlesi* maintaining in monkeys and *P. berghei,* etc. that can be maintained in the animals of the order rodentia. These observations suggest that there are great variations among the immune reactions even against non-identical parasitic strains in the same host. Current findings mainly bother what is literally taking place in human hosts and such study becomes quite smooth as *P. falciparum* can easily be cultured in laboratory.

The sporozoite contains a circumsporozoite (CS) protein that is a dominant coat protein which shows tandem repeats of as little as four amino acids and causes strong immune response because it elicits immunoglobulin reaction vigorously that is responsible to inhibit the entry of pathogen within hepatic cells. Still, only one pathogen in the hepatic cells can instigate the infection. The processes of immune reaction against the hepatic phases are ill-defined but it is thought that T cell's participation is associated in this response. The erythrocytic phases contain huge antigens recognizable from the antigens of the sporozoite or hepatic stages; few are features of the early ring stages, few remain attached to the surface of the infected RBCs and others pass into the blood plasma by rupturing the RBCs. The lion's share of antigens elicits specific immunoglobulins that prohibit entry to uninfected RBCs or assist intracellular killing. However, none is apt to stimulate complete immune reaction. A substitute probability is that the products of induced macrophages like reactive oxygen species or tumour necrosis factors are able to destroy the pathogen within RBCs. The gametocytes of *Plasmodium* show their specific immunogens possessing tandem repeats.

9.3 Immunity to Macroorganisms

9.3.1 Filariasis

Within the human host the filarial worms grow slowly from the contagious larval phases into adults that fix in the lymph vessels or subintegumentary tissues in due course. Microfilarial worm can migrate through the body fluid or settle in the cutaneous layer for a number of years. Epidemiological researches reveal that gradual aggregation of helminthic parasite occurs and reaching a highland suggests that the worms may escape the immune response or they do not generate this reaction at high level but in both cases a minimum level of immune reactions develops certainly.

In vitro experiments reveal that the microfilariae can bind immunoglobulins and can participate in ADCC reactions requiring eosinophils. Antibodies contribute notably in the defence against several pathogens, and this humoral immunity against filarial worm is also well-documented. Still, disease protective response as suggested against filarial worm has been confronted by some different observations. As, for example, in *O. volvulus* infection, B cells or immunoglobulins have not been found to affect intensely against microfilarial worm. These contradictory findings possibly indicated that activity of antibodies may significantly be related with the infected area. In lymphatic filariasis, microfilareae predominate in peripheral circulation and lymphatic channels while in Onchocerciasis, microfilareae are located in integumentary site. These differential transmission areas could take an important part in initiating immunity. Overexpression of serological IgG4 isotype in affected individuals is thought to be steered by T reg lymphocytes. In unaffected persons, the IgG variety consists of 5% of the complete amount of immunoglobulins in circulation whereas it increases markedly to about 95% in filarial infection. Experiments suggest the role of IgG4 as an indicator for immunoregulation in this infection. IgG4 immunoglobulins are upregulated when different cytokines like IL-10 are present. Field studies indicate a relationship between the levels of IgG4 and MF. IgG4 is involved in different obstructing activities that interfere with the immune reactions required to remove microfilariae from the body of host. This kind of deviation in immune response is controlled by the adult female helminth. A powerful correlation between IgG4 and regulatory T cells in a co-maintenance assay has been demonstrated in an experiment almost 10 years ago. This observation also reveals that Tr-1 clones drive the generation of IgG4 vigorously. Excessive amount of interleukin-10 by modified monocytes that is macrophages has now been revealed to diminish immune reactions in murine model. Therefore adult female worms are suggested to assist the interplay of T–B cells to generate IL-10 and then IgG4 to develop a favourable atmosphere needed for the staying of microfilariae.

IgE is among the major observed immunoglobulins in case of filariasis. In presence of IL-4 production of IgE is magnified. It has been observed impressively that IgG4 choke the immune role of IgE by competing with it for its binding sites. Other scientists have suggested that IgA and IgE show important role in the immune response against filarial infection. Nonetheless, a prudent regulation of IgE is pivotal as it aids pathology in helminth contamination and the regulators of the class-switching process in between these antibodies are still to be demonstrated.

Interleukin-6 shows both the anti- and pro-inflammatory roles. In an observation it is investigated that IL-6 takes a key part in immune reaction to the *Litomosoides sigmodontis*. Hence, IL-6$^{-/-}$ mice show a markedly escalated worm load in comparison to wild-type. Escalated vascular permeability might help in the movement of microfilareae, but obstruction in histamine receptors did not show any impact on worm load and permeability of vessels remained alike in between IL-6$^{-/-}$ and wild-type. In comparison it has been observed that clogging of mast cell degranulation decreases the worm load in IL-6$^{-/-}$ mice in part which indicates the enhancement in larval movement by the mast cell-derived mediators. Study of the cytological configuration by means of flow cytometry and polymerase chain reaction in the skin after exposure to filarial extract or third larval stage reveal that slow employment of phagocytic cells to the area of primary infection is observed when there is no IL-6.

While different antibody isotypes have specific biological role, an isotype is able to influence the sharp specificity of an immunoglobulin. IgG1 and IgG3 show remarkable role in the neutralization of parasites at the entrance to the body of host. IgG1 and IgG3 are involved in pathogen removal by means of opsonization, stimulation of natural killer cells as well as induction of the complement pathway. IgG2 is activated against polysaccharides, while pathogens of protein in nature activate IgG1 and IgG3 reactions in general. IgG3 has been suggested to be associated in protective immune reactions for the filarial pathogens as it is able to fix complement strongly. These observations clearly indicate a doable combinatorial modulation of antibodies that may significantly shape the virulence of filarial infection (Kwarteng et al. 2017).

9.3.2 Schistosomiasis

Beside man different mammals like rat, monkey, etc. are infected by *Schistosoma* and in all of these cases, at least a minimum level of immune response is observed that develops rapidly. In most of our populations, the occurrence of schistosomiasis might be associated with age. First 15–20 years of life is more susceptible to the infection and then parasitic load declines by the acquisition of some degree of immunity to recurrence. Both the in vivo and in vitro observations indicate that two possible modes of immune reactions are there: (1) ADCC that involves antibodies, neutrophils, eosinophils, macrophages, etc. which may act alone or as a whole (Fig. 9.3) and/or (2) immunoglobulin-independent reaction utilizing induced macrophages.

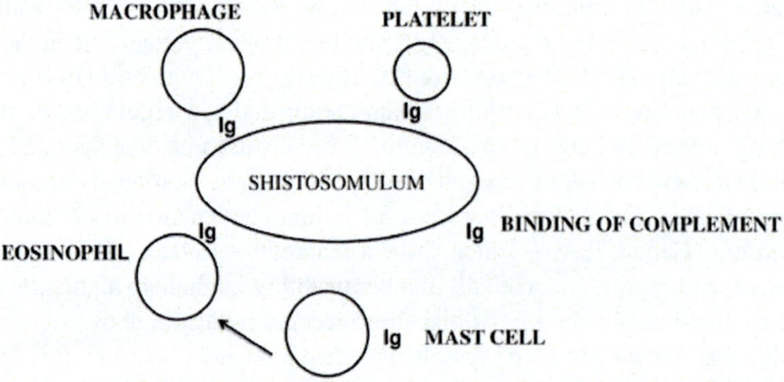

Fig. 9.3 Possible interactions of immune cells and immunoglobulins against schistosomula

Children might be affected and pathogen can be recognized at the first 5 years of their life but ten more years are needed to develop effective immunity against this parasite as the protective immunity develops very slowly in man. Protection against schistosomiasis involves IgE and eosinophils but in infancy or early childhood the IgE activities are inhibited by other antibodies. During the infection by *S. haematobium*, IgE and IgG4 are protective and blocking immunoglobulins, respectively. Nevertheless, schistosomiasis is a complicated disorder and, when the total immune reaction is realized finally, it is not straightforward at all.

9.4 Evasion of the Immune Response

Parasites show complex life cycles where they have evolved numerous pathways to evade the immune response of its hosts. These mechanisms are so powerful that most of the pathogens can sustain throughout the complete lifespan of the host organism. African trypanosomes, able to show antigenic variation and the schistosomes, able to camouflage themselves by picking the antigens of host individual, are some well-known examples. However, some additional instances are there and practically all parasitic organisms have developed some way of escaping the immune reactions of their hosts.

- *By suppressing or not stimulating immune cells*: The easiest strategy to escape the immune reaction is just not to induce it and most of the parasitic individuals can achieve this by throwing themselves into the host cells which are immunologically non-reactive. Ultimately, pathogens become able to escape from the host's immunity by obstructing the bodily response collectively, for example, suppression of immune cells.
- *By invading immune cells and blocking the action of phagolysosome*: By means of the invasion within the cells of the immune system a number of parasitic organisms can be escaped from the immune reaction of the host. As, for example, *Leishmania* sp., *Toxoplasma gondii*, etc. stay within different immune cells possessing phagocytic ability and therefore these pathogens inhibit the phagocytosis of these cells in different manners. In general, these cells grasp pathogens by phagocytosis and then lysosomes merge to these phagosomes to generate a phagolysosome where assimilation takes place. Even within such phagolysosome *Leishmania* sp. can reside. *T. cruzi* causes lysis of the phagosome prior to its fusion with the digestive bag and can stay in the cellular cytosol; however, *T. gondii* prevents the union of phagosome and lysosome. *Theileria* sp. are the unique parasitic individuals which affect lymphocytes, occupy the lymphoblasts and convert them into perennially dividing cells where all the newly formed daughter cells gain some of the newly developed parasites, thus the host's immune cells are accustomed with the sustained boost of the parasite.
- *By inappropriate signalling*: Regulation of the immune reaction can also be hampered by the parasites by stimulating helper T lymphocytes to generate inappropriate cues or by repressing synthesis of specific cytokines and/or their

receptors molecules. The stimulation of improper cues is distinctly observed in case of leishmaniasis where immune response is driven by the help of T_H1 cells and by macrophages which are stimulated by IFN-y but in the course of the parasitic infection generation of IFN-y is subdued by interleukin 4 and 10 synthesized by T_H2 lymphocytes. Disturbances in the equilibrium between T_H1 and T_H2 cells is also seen in the infections in mice by *S. mansoni, Trichuris muris*, etc. Generation of IL-1 and 2 is obstructed by *Trypanosoma cruzi*, again infections caused by *Plasmodium falciparum* go along with diminished amount of the IL-2. Improper induction of T_H2 lymphocytes also aids to synthesize non-functional immunoglobulins due to their wrong isotypes or as they do possess no specificity. In many parasitic infections antibody producing cells might be exhausted probably due to this misdirected control mechanisms.

- *By antigenic variation or shedding surface antigen*: The best studied example of escaping from an effective response of immunoglobulins is the antigenic variation that is observed in case of African trypanosomiasis. Due to this variation antibody cannot recognize antigens and the parasites evade our immune response. Humoral immunity can also be dodged by shedding surface antigens which is documented in several parasites like *Entamoeba histolytica, Ancylostoma caninum*, etc. and some additional species together with the African trypanosomes. In malarial infection, circumsporozoite shedding of antigens may happen and in addition to that there is strong affirmation that some of the antigens that are repeated tandemly may also be removed in host's RBC stage. Shedding of the surface antigens helps in the generation of antigen antibody complexes which inhibits the interaction between the immune cells and the actual pathogen or parasite.
- *By cleaving antibody*: Protease can inhibit the activity of immunoglobulins as these specific proteases cleave the antibody molecule and hence these prevent the activation of ADCC or complement system. As, for instance, Schistosomula synthesizes a schistosome derived inhibitory factor (SDIF) that inhibits IgG and again antibody molecules can be inactivated by *T. cruzi*. Other parasite like *E. histolytica* can defend themselves from the complementation response of host because of some anticomplementary factors present within these parasites.

9.5 Immunopathology

Mainly three kinds of damage are observed due to parasitic infections: (I) damage due to parasites directly; (2) damage as a consequence of the immune reaction and (3) damage caused by an amalgamation of the two above mentioned damages. The chief factor accountable for pathological effect is the host's own immunity as the alloy of inhabiting antigens and operative immune reaction causes immunopathological damages which might be more harmful than the infection itself. Complex interaction between the host's immunity and parasitic individuals, in which parasitic antigens can be recognized but the parasites are not eradicated, causes persistent infections escorted by malfunction in overall regulation. Therefore, unlike normal scenario, the immune reaction is not switched off even after the removal of this stimulant. Rather, circulating antigens blanket immune cells or present in vessels or tissues that leads to the development of hypersensitivity response of all the different types; immediate type I (anaphylaxis), type II (stimulation of complement and lysis of cell), type III (formation of immune complex) and type IV (delayed hypersensitivity reaction).

In case of malarial infection immunopathological effects are certain as irrelevant antigens, correlated with the schizogonic phases are released that bind to uninfected erythrocytes, causing anaemia, or due to the formation of antigen–antibody complexes that stay in blood vessels leading to obstruction and the complement-mediated impairment also. IL-1 is also responsible for the development of fevers associated with this disease partly. African trypanosomiasis is characterized by consecutive parasitaemia as parasite of a particular antigenic type is replaced by parasite with different coat proteins after its own destruction. Antigens liberated from the disintegrated trypanosomes cause anaemia, develop antigen–antibody complexes and lead to complement-mediated damage as well as commence numerous inflammatory responses like coagulation reaction. Chagas disease may develop long-term autoimmune disorders as the antigens of *T. cruzi* cross-react with proteins of the muscle and neuron of host body and if the heart is involved, cardiac failure may take place. Inflammatory reactions are initiated after the death of filarial worm and microfilariae cause type I hypersensitivity reactions (immediate) and eosinophilia. In case of cutaneous leishmaniasis, interferon-γ helps in the immune response but IL-4 and 10 inhibit interferon-y synthesis, hence they promote the disease effectively. Though tumour necrosis factor causes immunity against parasites they cause damage to adjacent cells again interleukin-l which is crucial in the stimulation of macrophage, also develops fever.

9.6 Immunization to Combat with Parasitic Infections

Most of researches have been devoted to the development of vaccines against these parasitic infections. Vaccination is a crucial part of all integrated programme to control parasitic infection and it shows numerous superiorities over medications or in the avoidance of transmission. Vaccination against a lot of viral and bacterial infections has shown outstanding results like the elimination of smallpox and eradication of poliomyelitis in several nations. Vaccines might be classified into five types: (1) live normal varieties with little virulence; (2) weakened parasitic organism; (3) disabled or destroyed parasites; (4) subunits, together with extracts or metabolic derivatives and (5) antigens that are manufactured and recombinant. Nonetheless, the best immunity could be generated by the use of pathogens that are alive and vaccines established on such parasites are unlikely to be allowable to utilize for man. In spite of different difficulties, significant progress is observed particularly in the arena of veterinary studies. The vaccine produced to combat with lungworm of cattle, *Dictyocaulus viviparus*, has been considered as the earliest vaccine against parasite. This above mentioned vaccine that contains irradiated lungworm larvae which stays for adequate time to induce an immune reaction but not for sufficient span to develop a disease has victoriously been used for more than 30 years in European countries and the vaccine against D. *filariae* of sheep is being utilized in India also. There are also some vaccines that act against protozoan parasites successfully and are used in veterinary science. The vaccines comprising of whole blood containing pathogens with truncated lethality are usually utilized against *Babesia bovis*, etc. that affect cattle. A defence against *Theileria annulata* might be developed by using affected lymphoid cells and in the India and entire Middle East such vaccine is used very successfully. Again immunization of some poultry birds against coccidia might be done by introducing small numbers of oocysts.

Molecular biology has enhanced the probability of the development of vaccines against some major parasites of man where no alternatives are there. Great progress is observed in the generation of vaccine against the disease malaria and in this case human trials have been done using circumsporozoite antigen. The observations revealed that a number of volunteers show great immunity, while others do not. However, the manufactured peptide vaccine was widely utilized in South American countries where around 27,000 individuals have been immunized with a substantial level of efficacy. A different strategy is the production of vaccines which are able to obstruct the sexual stage in the vector mosquito though this approach is in experimental phase till now. But these would have to be used as part of a cocktail vaccine. To eradicate malaria, the modified vaccine technology planning to 2030 (Moorthy et al. 2013) presently demands a next-generation immunization to attain around 75% success against *P. falciparum* and/or *P. vivax* within a couple of year (in the age of revived worldwide target to eradicate malaria) along with its primary 2015 'landmark' target of a first generation vaccine with a success rate of >50%. Along with the victory of ongoing pre-erythrocytic component and total sporozoite-based vaccines different approaches are taken which include blood-phase or transference-blocking immune reaction (Draper et al. 2018).

For last couple of decades a number of vaccines against dengue are processed but have not been proved as a triumphant till now. Generating a dengue vaccine is not a straightforward assignment. The main reason is that there are 4 dengue virus serotypes (DENV-1 to DENV-4) that migrate, among which any one shows prevalence in a specific season of dengue. Chemotherapy to attenuate virus is still quite difficult. This hindrance could be overcome by S. Pasteur, which has generated the first authorized vaccine (Dengvaxia®) against dengue globally. Dengvaxia® is also suggested by the WHO for the persons aged between 9 and 45 years living in dengue endemic regions. Dengvaxia® is developed by genetic engineering and has also been assessed in a 3-dose scheme (at 0, 6 and 12 months) in the third phase of medical trials. Indian endeavours to produce dengue vaccine are also in progress. In global scale two dengue vaccines are presently underway. One is TetraVax-DV, a live but attenuated one, which has been generated by National Institutes of Health, USA. This has already been permitted to Indian agencies Panacea Biotec, Serum Institute of India, etc. Till date in India, Panacea Biotec has launched phase I/II clinical trials solely. Two hundred volunteers have participated in this trial (Bharati and Jain 2019).

The production of vaccines against two types of leishmaniasis (cutaneous and visceral) is one of the prime concerns of WHO but experimentation is still at the preliminary level. At present the antigens are being scrutinized as they induce response of T and B cells. Development of vaccines to combat with Chagas disease is quite problematic because of the autoimmune factors as well as antigenic variation. More than 880 million persons in almost 52 countries are presently under threat of getting this filariasis (LF) throughout the globe. Several researches reveal that human show immune response against lymphatic filariasis. Hence, generation of prophylactic vaccine to combat with this filariasis is feasible for us. A number of effective candidate vaccines have been marked and studied to prove their functional ability for this filariasis. A potent prophylactic vaccine along with MDA has outstanding efficacy (Khatri et al. 2018). Not only the immunological factors but also the integrated control strategy, practicability in delivery as well as interplay with the chemotherapy should be considered in the development of effective vaccine.

References

Bharati K, Jain H (2019) Dengue vaccines: current status and future prospects. J Clin Diagn 13(6): AB01–AB02

Draper SJ, Sack BK et al (2018) Malaria vaccines: recent advances and new horizons. Cell Host Microbe 24(1):43–56

Khatri V, Chauhan N, Vishnoi K et al (2018) Prospects of developing a prophylactic vaccine against human lymphatic filariasis - evaluation of protection in non-human primates. Int J Parasitol 48:773–783

Kwarteng A, Ahuno S (2017) Immunity in filarial infections: lessons from animal models and human studies. Scand J Immunol 85(4). https://doi.org/10.1111/sji.12533

Marshall JS, Warrington R, Watson W et al (2018) An introduction to immunology and immunopathology. Allergy Asthma Clin Immunol 14(2):49

Moorthy VS, Newman RD, Okwo-Bele JM (2013) Malaria vaccine technology roadmap. Lancet 382:1700–1701

Tsukui KN, Nozaki T (2016) Immune response of Amebiasis and immune evasion by *Entamoeba histolytica*. Front Immunol 12(7):175. https://doi.org/10.3389/fimmu.2016.00175

Ecology of Parasites 10

The word ecology was first coined by German zoologist Ernst Haeckel. It is composed of two words 'Oikos' (residence) and 'logos' (study). Scientist Tansley first introduced the word ecology. Ecology is the study of relationships among the organisms and their biotic and abiotic environment.

The ecology of parasites is the habitat of a parasite where the parasite interacts and spends a part or whole of its life. The host is the parasites environment. The study of ecology of parasites, therefore, is the study of the infective forms of the parasite from the point of taxonomy, transmission, population dynamics and evolutionary history.

The local environment influences the parasitic transmission directly or through vector. The transmission is the infection capability of the parasite which depends upon its environment and interaction with the different hosts which leads to ecological relationship.

Parasitism is an age old relation. Host–parasite conflict depends upon the habitat of the parasite like ectoparasite, staying on the surface of the body of the host permanently or visits occasionally for food only, while the endoparasites reside throughout their life within the body of the host in any organ, they may be termed coeloparasite, histoparasite, intestinal parasite, blood or tissue parasite, intestinal-tissue parasite, intracellular parasite, etc., these terms are given according to their address of the residence.

Though the environment of a parasite is the host, different stages of transmission like spores, eggs, cysts, larva must survive outside in the abiotic condition during their life outside the hosts.

For a parasite primary need is the food or nutrients which parasite easily finds in the body fluids of the host having dissolved proteins, amino acids, carbohydrates and nucleic acid precursors. In the intestine of a host digested food material is at hand. Parasites exploit such living environment.

P. K. Bandyopadhyay et al., *Biochemical, Immunological and Epidemiological Analysis of Parasitic Diseases*, https://doi.org/10.1007/978-981-16-4384-2_10

We find that with the evolutionary changes in hosts parallel changes have occurred in their parasites also. The transmission stages also changed to resistant cysts for the survival in the outside abiotic environment.

Besides there are some freak of nature, in birds, females of certain social parasites lay their eggs in the nest of host species, thereby parasitizing the parental care of the host. Certain Dipteran and Hymenopteran insects like females of some flies and wasps deposit their eggs upon the larvae of other insects. The eggs hatch and larvae feed on the host, pupate and then come out as adults. This relationship is termed parasitoidism and they regulate or control the population of the hosts in nature.

The ecological niche of parasite is the nutrients provided by the living body of another species and outside environment confronted by the transmission stages like eggs, cysts, spores, larvae. So, most of the parasites have to face a wide range of micro- and macro-environmental conditions during their life cycles.

A journey through the alimentary canal of host may be described in terms of different symbionts confronted along the way starting from *Entamoeba gingivalis* in the mouth, fourth stage larvae of *Ascaris lumbricoides* in the stomach to Taenia and other helminthes in the small intestine to *Dientamoeba fragilis, Entamoeba coli, Endolimax nana* and *Trichuris trichiura* in the large intestine and ultimately to pinworms (*Enterobius vermicularis*) crawling around the anus. Sometimes the site specific parasites may be found not in the main route but into the lungs, in the bile ducts, etc.

Intensive research gives us the idea of distribution of different parasitic amoebae, parasitic nematodes, parasitic helminthes within the entire length of the intestine. The distribution is motivated by the diet of host, physiological condition and the presence of other parasites. Furthermore fine difference occurs in oxygen, carbon dioxide tension, pH and other chemical and physical factors from the intestinal wall to the centre of the lumen of intestine. Because of such differences different habitats are present for the parasites to colonize.

The population biology of the parasites indicates the phenomenon of 'r'-selection and 'k'-selection strategies. 'r'-selection occurs when selective forces upon organisms are unstable and environmental conditions are variable. Conversely, 'k'--selection prevails when environmental condition is more stable over a period of time. 'r'-strategies create high fecundity, high density-independent mortality, short lifespan, effective dispersal mechanism and size of the population vary over time and always remain below the level of carrying capacity of the habitat. 'k'-strategies usually have low fecundity, density dependent mortality, longer lifespan and population size is more stable. Digenetic trematodes are considered 'r'-strategists. Their biotic potentials for increase in population and high mortality rate put selective pressure on the organisms.

Parasites may exert strong control over the population of their hosts over a period of time. That is why the parasitic pathogens are used to control the population of pests.

Besides, the geographical distribution of parasites is controlled by the presence or absence of certain biological factors, chemicals, weather, physical factors, availability of hosts, vector and stress created by the manmade eradication programme.

Epidemiology is the study of factors responsible for the transmission and distribution of the disease causing organisms. The distribution of hosts, vectors, host specificity and density of hosts directly influences the epidemiology of the pathogenic parasites.

The population dynamics of the disease causing parasites affect the population dynamics of hosts as the parasites depend on the hosts for their life and sustenance.

When one species is dependent on the other for food and shelter, the relationship may be controlled by making the host resistant or by destroying the parasites and for that their population, immunization is necessary to check the reproductive rate of infection. Two pronged attacks: one by immunizing unaffected individual by administrating vaccine and other by destroying parasites by chemotherapy in the affected person will control the diseases and parasites with different reproductive rates of infection.

A complete understanding of host–parasite relationship needs a careful consideration of the ecological context of the relationship. This will ultimately provide us an idea of host–parasite interactions in nature which will lead to a better management of parasite infection in human population.

10.1 Evolution of Parasitism

Evolution of parasitism is the pattern of association among parasites, hosts and the ecological and geographical distribution. The main factors are descent and colonization. Descent is the association of a parasite with a host for a long evolutionary period. Colonization is the association of host and parasite and has undergone evolutionary change together and parasite has colonized the host like the people make colony in a new country or inland far from his own settled place. The host–parasite association may be the product of descent, colonization physical separation of population and extinction. Any of the four factors or all of them may induce parasitism and parasites may be specific to their hosts due to evolution. The high degree of host specificity suggests association by descent.

Evolution of virulence of any parasite depends on pathogenesis and transmission dynamics. Parasites are transmitted horizontally and vertically. Vertical transmission means transmission between generations and horizontal transmission means between the members of same generation. The research work suggests that less virulent strains transmit vertically, while more virulent strains transmit horizontally with high transmission rate.

It can be said that parasitism is a secondarily adapted relation. It may be that some free living organisms accidentally or casually became associated with another organism. This temporary association or accidental association due to pre-adaption found a place suitable to reside as the environment of their old home became hostile. Then one of the associates developed increasing dependency on the other slowly and gradually.

The pre-adaption, in this context, may be said that potential capability of a free living organism for adaptation in a parasitic lifestyle.

This accidental association at initial period remains free living and just an association at one point of time. This association became very necessary for one to survive for some reason. Now this association with a potential host became very important for their survival because of hostile environment.

It is believed that parasites of intestine probably got entrance having been swallowed accidentally by the host. If they were pre-adapted they became gradually more and more dependent upon the new associate and environment. If not might have been migrated to find a more suitable site such as lungs or liver.

It is quite evident that multihost parasites developed their life cycle by trial and error method and this was possible due to their high fecundity and high rate of reproduction.

The blood parasites at first instance entered into the alimentary canal of insects and during the feeding of vertebrates and adapted to that environment secondarily. The present intermediate host may be definitive host but due to increasing number, adaptability reaches the highest point and so elimination started. Elimination of certain hosts proves the natural selection and a more successful life cycle enhances the chance of reaching definitive host.

An example may be cited for evolution of parasites in numerous animal groups. In nematodes there are two such related groups free living *Caenorhabditis elegans* and parasitic *Pristionchus pacificus*. In the free living nematode, Caenorhabditis, certain larvae are produced which are called diapauses (dauer) larvae. This can be explained as arrested developmental stages of a free living nematode which is the result of hostile environmental conditions. Pheromones signal these adverse changes in the environment and compelled the worm to stop development and produce dauer larvae. These larvae attach themselves with other animals and are dispersed to a new environment. This existence of such dauer larvae may be an instance of parasitic nematode life cycle evolved from free living ones.

It is seen that L3 larvae of many parasite nematodes which are infective may have many similarities with dauer larvae of free living nematode. Both are in arrested stage of development, both are in dispersive stage and both will emerge from their arrested condition under congenial environmental condition. There is another example, *Strongyloides stercoralis* having both free living and parasitic life cycle.

The virulence of parasitic infections is the result of evolutionary changes. The parasitologists are trying to have answer why some parasites are so much virulent and others not.

The answer is parasites over a long period evolve into less virulent forms because the death of a host due to virulence of the parasite will ultimately have a negative effect on them for their survival. So a median path is chosen by the parasite being optimum virulent.

Several studies have been made on the evolution of virulence of the parasites. These studies support the idea that genetic diversity of both host and parasite, difference of time period of transmission and individual host-parasite interactions influence the evolution of virulence.

Control

11

11.1 Components of Control

Control of the diseases is dependent on the knowledge of the causative organism, epidemiology and vector. The method of transmission should also be considered.

Components of control: The components of control are:

1. Diagnosis of the disease
2. Recognition of disease causing organism
3. Identification of vector
4. Chemotherapy
5. Immunoprophylaxis—vaccines
6. Vector control
7. Environmental management
8. Health education

Diagnosis of the disease: The very important part of the control of parasitic diseases is the diagnosis of the disease itself. There are several methods to diagnose the disease.

Recognition of disease causing organism: If the disease causing organism is identified, then it is very easy to diagnose the disease itself. There are several methods for identification of the parasites in diagnostic laboratory. The latest method is serodiagnosis. The recognition of parasites by observing through microscope and/or electron microscope after proper fixation and staining the sample where the parasites may reside in like urine, stool, blood, body fluid, etc.

Identification of the vector: If the disease is vector borne, then at first the disease causing organism is to be identified. After identification of vector it is to be controlled as part of an integrated control strategy. This is important if the strategy is aimed at a community basis, at whatever level community is defined—local, regional or national rather than at the level of individual.

P. K. Bandyopadhyay et al., *Biochemical, Immunological and Epidemiological Analysis of Parasitic Diseases*, https://doi.org/10.1007/978-981-16-4384-2_11

Integrated control strategies based on long-term planning, resource and knowledge derived from different scientific disciplines.

Chemotherapy: To kill the parasites by administrating chemicals used as antiparasitic drugs. The drug should be selected on the basis of cost, side effects due to toxicity and method of delivery. Dosages of drugs should be judged by experiment and at the same time monitoring is required for development of resistance.

Immunoprophylaxis: That is use of vaccines. Vaccination is an artificial method by which the lymphocytes of a healthy person are excited to form antibodies by injecting specific antigens either living or attenuated or toxin of the particular parasite.

Vector control: To control the vector ecology of vector is to be studied, then most of the times insecticides are applied or sprayed to kill the insect vector. Sometimes genetic and biological control are also applied to control parasitic diseases in cost-effective way.

The objective is to raise the level of health in the broadest sense by a political commitment to the poorest and most deprived population.

WHO recognized that the inadequate resources are trained man power, facilities, equipment and above all finance. The limited resources have to be used appropriately to give the most relevant and cost-effective intervention to reduce the burden of morbidity and mortality and must be targeted on those diseases which are most important in each community.

Environmental management: Environmental management is the change of environment in order to prevent or minimize vector propagation and human contact with the vector pathogen by destroying, altering, removing or recycling nonessential containers that provide larval habitat.

The environmental management is of three types: environmental modification, environmental manipulation and changes to human habitation.

Role of water resource management is important to prevent spreading a number of parasitic diseases which are transmitted by invertebrate organisms where life cycle either partly or wholly is associated with the aquatic environment. They may be flying insects, aquatic snails, etc. known as intermediate hosts.

The role of engineers is very important for the design and construction of dams and other hydrolic projects.

Insecticides, larvicides, rodenticides, lethal ovitraps and repellents are used to control vectors. Vector control is any method to limit or eradicate the mammals, birds, insects or other arthropods which transmit disease pathogens.

Health education: Parasitic disease control should be integrated with primary health care system if it is technically possible. The health care system must be capable of responding to and controlling parasitic infections, preferably integrating control into the health care delivery system by providing education concerning the tools to control parasitic diseases in cost-effective way.

The objective is to raise the level of health in the broadest sense by a political commitment to the poorest and most deprived population.

WHO recognized that the inadequate resources are trained man power, facilities, equipment and above all finance. The limited resources have to be used appropriately to give the most relevant and cost-effective intervention to reduce the burden of morbidity and mortality and must be targeted on those diseases which are most important in each community.

11.2 Vector Control

11.2.1 Control of Lice

Louse is an ectoparasite. It is under the genus Pediculus. The head louse of human beings is *Pediculus humanus capitis*. There is another type of louse called body louse residing on the body of human beings and is known as body louse, *Pediculus humanus linnaeus*. There is another louse called crab louse (*Pthirus pubis*) living in the pubic hairs of both the sexes of human beings.

To control lice a number of preparations are in the market which contain insecticides.

In hair care products now insecticides like permethrin are added to control louse. These types of delousing products are proved to be effective at the same time safe for human use. In some countries different antilouse shampoo are used which have the ability to kill both adults and eggs of lice.

Nowadays naturally occurring phytochemicals are used to kill a number of arthropod vectors. It is also discovered that hot hair from hair dryers has the ability to kill the head lice.

To control body louse personal hygiene and washing of garments frequently including dry cleaning will control the body lice. Blowing of insecticide dust on the clothing and garments in large scale will control the lice effectively.

The body louse spreads from clothing and linen. The crab louse utilizes public toilets and is commonly spread by sexual intercourse because the partners become so close.

The head louse spread by using combs, brush, hats, caps or any type of headgear used by so many persons.

The head louse can be eliminated by an alternative method by using 0.5% malathion lotions on the body for 24 h and then washing. The body lice can be killed by spraying insecticide like 1% malathion powder. Another method of control of body lice is dipping of cloths in 5% DDT solution.

11.2.2 Control of Kissing Bugs

Usually BHC or Dieldrin spraying within the houses is very effective in controlling the house infesting bugs. It is seen that spraying of 5% Dieldrin emulsion or kerosene can control the infestation of kissing bugs up to a year. Repeated use of insecticides can eradicate Chagas' disease in many towns and villages of Brazil.

The population of kissing bugs (Triatomines) in a house depends upon the number of people residing there. But the population of kissing bugs Triatomines can be minimized by reducing the number of hiding places of bugs in the house. Improved construction and repeated repairing of cracks and crevices in the wall and roof can control the population of kissing bugs. In a survey, it was seen that removal of stacked firewood from the nearby place of a house replacing the floor of the house by concrete eliminated *Triatoma dimidiata* infestation.

In some cases the replacement of thatched roofs with asbestos or metal sheet helped much in control of the kissing bug populations.

It is seen that these bugs use the signalling of faeces of the other bugs to colonize or to aggregate in a particular place.

Spraying of insecticides in their hiding places is very much useful in control or painting with paint mixed with insecticide (Chlorpyrifos) used successfully in controlling the kissing bug infestations.

A phytochemical called Precocene II which is an extraction of *Ageratum sp.* named plant used successfully as a fumigant against Triatoma.

In a research it is found this phytochemical is cytotoxic to the neurosecretory organ, Corpora allata, of the insect which damages the organ preventing the production of juvenile hormone (JH). This Precocene II prevents oogenesis in adult female of the insects and also causes immature to become sterile adults.

11.2.3 Control of Fleas

The control is necessary around our home or in our pets. They act as vector of the diseases like plague, murine typhus, etc.

In the human inhabitations one should be very careful about the dust under carpet, in the cracks and crevices of floor and in the bedding of pets. Insecticides should be applied indoor frequently. These insecticides act on the eggs and larvae of fleas. The insecticides like diflubenzuron kill the fleas with eggs and larvae under the carpets for a year. The eggs of the fleas can be made infertile when hosts are treated with certain pyriproxyfen or methoprene based formulations. Fleas can be controlled by placing light trap with yellow green filter. In the market different types of flea control powder are available for use on dogs and cats, but repeated application is necessary.

11.2.4 Control Measure of Housefly

The housefly is a menace to public health. They can be controlled in a number of ways:

1. Destruction of adult flies: The adhesive paper, hanging cards with resin and castor oil are used to trap the adult and then they may be destroyed.

2. Chemical control: Excreta, vomit, etc. of diseased person which contains germs of the disease must be treated chemically like phenyl, Lysol, dilute Dettol, etc. so that flies cannot sit on them.
   The cow dung, human faeces, etc. should be treated with lime, copper sulphate, formaldehyde, like chemicals so flies cannot sit on them and to prevent breeding.
3. Destruction of breeding grounds: The decaying organic matters on which flies lay eggs should be properly treated with chemicals and be removed by municipality or corporation.
   In villages and in farms the manure should be kept in covered container. The container should be sprayed with cyanide solution and debris should be treated with calcium borate to destroy eggs and larvae.
4. Personal hygiene: Cleanliness is godliness. Every household should be cleaned twice in a day. Leftover food, skin of the vegetables and fruits, etc. should be removed quickly or to be disposed in a covered waste box kept in the corner of the kitchen and the container should be removed and washed every day. To be careful not to eat food or drink liquid contaminated by housefly.

11.2.5 Control of Tsetse Fly

The species of *G. palpalis* is easy to control than the other species of Glossina because they prefer to remain close to the source of water.

The bank of rivers and lakes must be cleared by eliminating ruthlessly the low branches of trees for mile after mile and to note that reinfestation does not occur in wet season. In this way the species is eliminated from rivers, streams and lakesides. Cleaning range should be 500 to 800 yards from the bank of rivers or water source.

The adult flies can be destroyed by fumigation and spraying of insecticides such as pyrethrum like substances DDT and BHC.

The reservoir animals should be destroyed.

11.2.6 Control of Sandfly

The sandflies bite during night and at day time they remain hidden in cracks and crevices of walls, holes in the trees and in cattle stables.

First repairing and filling of cracks and crevices of the wall is very much necessary. Chemical control by spraying insecticides must be done in the vicinity of breeding places of sandfly. DDT should be sprayed to the houses, cattle sheds, poultry sheds, etc. The spraying of DDT should be 50 m from the breeding places of sandfly as they usually remain confined to 50 m from their breeding place.

11.2.7 Control of Head Louse

Head louse can be eradicated or controlled by application of 0.5% malathion lotion. This lotion should be applied to the hairs and left for 12 to 24 h followed by washing of hair.

Body louse can be eradicated by using 1% malathion powder mixing with talcum powder. It is observed that DDT and HCH have been of no use now as the louse becomes resistant to these chemicals. Personal hygiene washing of cloths is very essential to keep body lice away.

Daily combing for at least 3–4 times and use of shampoo every alternate day are essential to keep lice free hairs.

Crab lice most of the time remain in public toilets and is commonly spread by sexual intercourse because the partners become so close. The crab lice spread through comb as people with crab lice comb their beards. The brush, hats, caps or any head gear used by several persons spread the lice. These should not be used by many persons. People should not share these articles.

The clothing is disinfected by using disinfectant solutions and dipping them in 5% DDT solution.

11.2.8 Control of Deer Fly

Cattle can be protected by spraying emulsions of pyrethrins with a synergist chemical. Aerial sprayings of insecticide in oil have proved successful in killing the larvae of deer fly. Repellent like diethyltoluamide is also very useful against the adult flies.

11.2.9 Control of Ticks and Mites

Ticks and mites can be controlled by several methods:

A. *General control*:

1. *Insecticidal or chemical control*: The ticks and mites can be controlled by dusting or spraying of insecticides like malathion, lindane, chlordane at the rate of 0.5–1 kg per acre floors and walls up to height of two feet.
2. *Environmental protection*: The normal habitat of ticks and mites like cracks and crevices in the walls and floors, in the field particularly near human inhabitation should be filled up. As far as possible animal hosts must be kept away.
3. *Protection of workers*: Workers should be encouraged to wear protective clothing impregnated with an insect repellent like benzyl benzoate, indalone or diethyltoluamide, etc.

B. *Specific control*:

1. Chigger mite infestations in gardens, lawns and premises can be destroyed by applying or by dusting in the infested areas.
2. Rodents should be avoided or controlled.

C. *Personal control*:

1. Avoid infected person.
2. Avoid sleeping in the same bed with infected person.
3. Use of Dettol soap during bathing.

11.2.10 Control of Mosquito

Mosquito not only causes nuisance by its biting habit but also a more serious consequence is the transmission of serious diseases such as malaria, dengue, Zika, filarial, etc. Mosquitoes not only affect man but they also can transmit a number of diseases and parasites to the dogs and horses as, for example, dog heartworms, eastern equine encephalitis, etc.

Insecticide Resistance Since 2000, specifically in sub-Saharan Africa mosquito control is one of the main focuses to control malaria. But the development of insecticide resistance in mosquito could lead to a significant increase in malaria occurrence as well as death of human host. The WHO *Global report 2010–2016* indicated that resistance to the 4 regularly used insecticide classes—pyrethroids, carbamates, organochlorines and organophosphates—is widespread in all important malaria vectors across various parts of Africa, the Americas, Southeast Asia, the Eastern Mediterranean and the Western Pacific (WHO 2018).

It is suggested that leucine to phenylalanine substitution mutation at residue L1014 is the most common *kdr* mutation that is found in several insects including *Anopheles* (Singh et al. 2015). As the development of insecticide resistance is the dominant challenge to control dissemination of mosquito-borne diseases (Chatterjee et al. 2018), hence alternative strategies must be adopted. *Ae. aegypti* population in different areas has been observed to be susceptible to type II pyrethroids and malathion but shows high level of resistance against DDT. Collected mosquitoes reveal a high rate of polymorphisms in the *VGSC* gene. Hence, use of type II pyrethroids for the management of dengue vectors replacing DDT is a pertinent choice that has been taken by the national programme. Mutations in the *VGSC* gene might be a marker of development of resistance against pyrethroid insecticides and it should be watched at a regular gap. Again, *Culex* mosquitoes of northeastern India as, for example, have been observed as resistant to DDT and it has led 11.9–41.2% mortality, whereas the assay of knockdown resistance (kdr) indicates complete susceptibility to the deltamethrin in most of the studied areas. The result of AS-PCR confirmed the occurrence of three genotypes: susceptible (SS), resistant

(RR) and heterozygous (SR) in the studied population. This observation has an important role in the control of elephantiasis in India (Sarkar et al. 2009).

Control by Phytochemicals In general the active toxic components of plant extracts are secondary metabolites that are developed to protect themselves from primary consumers. The insects consume these phytochemicals potentially encountering toxic substances with relatively non-specific impacts on a broad spectrum of molecular targets. These targets span from proteins (enzymes, receptors, signalling molecules, ion channels and structural proteins), nucleic acids, biomembranes and other histological substances (for example, mitotic poisoning by azadirachtin, etc.).

Control by Bacteria A number of microorganisms inhabit in the digestive tracts and other parts of the body of almost all insects. The midgut microbe is responsible for numerous interactions as, for example, pathogenesis to the symbiotic relationship. These gut-bacterial populations can inhibit the growth of *Plasmodium* directly by the production of free radicals and indirectly through the genesis of the non-specific immunity of the mosquito. In Europe, mosquito control programmes chiefly depend on the use of larvicides containing the toxins of *Bacillus thuringiensis* subsp. *israelensis* (Bti) and/or *Lysinibacillus sphaericus*. The utilization of microbial toxins can effectively bring down the population densities of the target species like floodwater vectors *Aedes vexans* and container breeders *Culex pipiens* or *Aedes albopictus* to a justifiable level. Since 2011, researchers have been injecting *Wolbachia* into the eggs of *Aedes aegypti* vectors and releasing the hatched insects, which spread this immunoprotection to the next generation. In several release sites, experiments conducted by the nonprofit World Mosquito Program (WMP) reveal about 76% reduction in the occurrence of dengue (Servick 2019). Therefore, the alternative strategies show rays of hope for the control of vector mosquitoes.

References

Chatterjee M, Ballav S, Maji AK et al (2018) Polymorphisms in voltage-gated sodium channel gene and susceptibility of *Aedes albopictus* to insecticides in three districts of northern West Bengal, India. PLoS Negl Trop Dis 12(1):e0006192

Sarkar M, Bbhattacharyya IK, Bborkotoki A et al (2009) Insecticide resistance and detoxifying enzyme activity in the principal bancroftian filariasis vector, Culex quinquefasciatus, in north-eastern India. Med Vet Entomol 23:122–131

Servick K (2019) Bacteria-armed mosquitoes make dent in dengue. Science 366(6469):1056. https://doi.org/10.1126/science.366.6469.1056

Singh OP, Dykes CL, Sharma G, Das MK (2015) L1014F-kdr Mutation in Indian *Anopheles subpictus* (Diptera: culicidae) arising from two alternative transversions in the voltage-gated sodium channel and a single pira-PCR for their detection. J Med Entomol 52(1):24–27

World Health Organization (2018) Global report on insecticide resistance in malaria vectors: 2010–2016. Geneva